T0340231

Quality Management in Construction Projects

Second Edition

Industrial Innovation Series

Series Editor
Adedeji B. Badiru
Air Force Institute of Technology (AFIT)—Dayton, Ohio

Published Titles

Additive Manufacturing Handbook: Product Development for the Defense Industry, *Adedeji B. Badiru, Vhance V. Valencia, and David Liu*

Carbon Footprint Analysis: Concepts, Methods, Implementation, and Case Studies, *Matthew John Franchetti and Defne Apul*

Cellular Manufacturing: Mitigating Risk and Uncertainty, *John X. Wang*

Communication for Continuous Improvement Projects, *Tina Agustiady*

Computational Economic Analysis for Engineering and Industry, *Adedeji B. Badiru and Olufemi A. Omitaomu*

Conveyors: Applications, Selection, and Integration, *Patrick M. McGuire*

Culture and Trust in Technology-Driven Organizations, *Frances Alston*

Design for Profitability: Guidelines to Cost Effectively Management the Development Process of Complex Products, *Salah Ahmed Mohamed Elmoselhy*

Global Engineering: Design, Decision Making, and Communication, *Carlos Acosta, V. Jorge Leon, Charles Conrad, and Cesar O. Malave*

Global Manufacturing Technology Transfer: Africa–USA Strategies, Adaptations, and Management, *Adedeji B. Badiru*

Guide to Environment Safety and Health Management: Developing, Implementing, and Maintaining a Continuous Improvement Program, *Frances Alston and Emily J. Millikin*

Handbook of Construction Management: Scope, Schedule, and Cost Control, *Abdul Razzak Rumane*

Handbook of Emergency Response: A Human Factors and Systems Engineering Approach, *Adedeji B. Badiru and LeeAnn Racz*

Handbook of Industrial Engineering Equations, Formulas, and Calculations, *Adedeji B. Badiru and Olufemi A. Omitaomu*

Handbook of Industrial and Systems Engineering, Second Edition, *Adedeji B. Badiru*

Handbook of Military Industrial Engineering, *Adedeji B. Badiru and Marlin U. Thomas*

Industrial Control Systems: Mathematical and Statistical Models and Techniques, *Adedeji B. Badiru, Oye Ibidapo-Obe, and Babatunde J. Ayeni*

Industrial Project Management: Concepts, Tools, and Techniques, *Adedeji B. Badiru, Abidemi Badiru, and Adetokunboh Badiru*

Inventory Management: Non-Classical Views, *Mohamad Y. Jaber*

Kansei Engineering—2-volume set

• Innovations of Kansei Engineering, *Mitsuo Nagamachi and Anitawati Mohd Lokman*
• Kansei/Affective Engineering, *Mitsuo Nagamachi*

Kansei Innovation: Practical Design Applications for Product and Service Development, *Mitsuo Nagamachi and Anitawati Mohd Lokman*

Knowledge Discovery from Sensor Data, *Auroop R. Ganguly, João Gama, Olufemi A. Omitaomu, Mohamed Medhat Gaber, and Ranga Raju Vatsavai*

Learning Curves: Theory, Models, and Applications, *Mohamad Y. Jaber*

Managing Projects as Investments: Earned Value to Business Value, *Stephen A. Devaux*

Modern Construction: Lean Project Delivery and Integrated Practices, *Lincoln Harding Forbes and Syed M. Ahmed*

Moving from Project Management to Project Leadership: A Practical Guide to Leading Groups, *R. Camper Bull*

Project Feasibility: Tools for Uncovering Points of Vulnerability, *Olivier Mesly*

Project Management: Systems, Principles, and Applications, *Adedeji B. Badiru*

Project Management for the Oil and Gas Industry: A World System Approach, *Adedeji B. Badiru and Samuel O. Osisanya*

Project Management for Research: A Guide for Graduate Students, *Adedeji B. Badiru, Christina Rusnock, and Vhance V. Valencia*

Project Management Simplified: A Step-by-Step Process, *Barbara Karten*

Quality Management in Construction Projects, Second Edition *Abdul Razzak Rumane*

Quality Tools for Managing Construction Projects, *Abdul Razzak Rumane*

A Six Sigma Approach to Sustainability: Continual Improvement for Social Responsibility, *Holly A. Duckworth and Andrea Hoffmeier*

Social Responsibility: Failure Mode Effects and Analysis, *Holly Alison Duckworth and Rosemond Ann Moore*

Statistical Techniques for Project Control, *Adedeji B. Badiru and Tina Agustiady*

STEP Project Management: Guide for Science, Technology, and Engineering Projects, *Adedeji B. Badiru*

Sustainability: Utilizing Lean Six Sigma Techniques, *Tina Agustiady and Adedeji B. Badiru*

Systems Thinking: Coping with 21st Century Problems, *John Turner Boardman and Brian J. Sauser*

Techonomics: The Theory of Industrial Evolution, *H. Lee Martin*

Total Productive Maintenance: Strategies and Implementation Guide, *Tina Agustiady and Elizabeth A. Cudney*

Total Project Control: A Practitioner's Guide to Managing Projects as Investments, Second Edition, *Stephen A. Devaux*

Triple C Model of Project Management: Communication, Cooperation, Coordination, *Adedeji B. Badiru*

Quality Management in Construction Projects

Second Edition

Abdul Razzak Rumane

CRC Press
Taylor & Francis Group
Boca Raton London New York

CRC Press is an imprint of the
Taylor & Francis Group, an **informa** business

CRC Press
Taylor & Francis Group
6000 Broken Sound Parkway NW, Suite 300
Boca Raton, FL 33487-2742

First issued in paperback 2019

ISBN-13: 978-1-4987-8167-1 (hbk)
ISBN-13: 978-0-367-89003-2 (pbk)

Library of Congress Cataloging-in-Publication Data

Names: Rumane, Abdul Razzak, author.
Title: Quality management in construction projects / Abdul Razzak Rumane.
Description: Second edition. | Boca Raton : Taylor & Francis, CRC Press, 2017. | Series: Industrial innovation series | Includes bibliographical references and index.
Identifiers: LCCN 2017019041|ISBN 9781498781671 (hardback: alk. paper) | ISBN 9781498781688 (ebook)
Subjects: LCSH: Building--Quality control.
Classification: LCC TH438.2 .R86 2017 | DDC 690.068/5--dc23
LC record available at https://lccn.loc.gov/2017019041

Visit the Taylor & Francis Web site at
http://www.taylorandfrancis.com

and the CRC Press Web site at
http://www.crcpress.com

To

My Parents

For their prayers and love

My prayers are always for my father who encouraged me all the times.

I wish he would have been here to see this book and give me blessings.

My prayers and love for my mother who is always inspiring me.

Contents

List of Figures xvii
List of Tables xxiii
Foreword xxvii
Acknowledgments xxix
Author... xxxi
Synonyms... xxxiii
Abbreviations xxxv
Introduction xxxvii

1. Overview of Quality... 1
   1.1   Quality History... 1
   1.2   Quality Definition 6
   1.3   Quality Inspection... 10
   1.4   Quality Control... 11
   1.5   Quality Assurance... 20
   1.6   Quality Engineering.. 22
   1.7   Quality Management.. 25
   1.8   Quality Gurus and Their Philosophies 29
          1.8.1   Philip B. Crosby.. 30
          1.8.2   W. Edwards Deming... 30
                   1.8.2.1   PDCA Cycle.................................... 31
                   1.8.2.2   Statistical Process Control 32
                   1.8.2.3   14 Principles for Transformation...... 34
                   1.8.2.4   The Seven-Point Action Plan 34
          1.8.3   Armand V. Feigenbaum 34
          1.8.4   Kaoru Ishikawa 34
          1.8.5   Joseph M. Juran.. 42
          1.8.6   John S. Oakland.. 45
          1.8.7   Shigeo Shingo 45
          1.8.8   Genichi Taguchi.. 46
          1.8.9   Summary of Philosophies................................ 47
   1.9   Total Quality Management 49
          1.9.1   Changing Views of Quality 49
          1.9.2   Principles of Total Quality Management.......... 51
   1.10   Quality Information System 57
          1.10.1   Quality in Use of CAD Software 59
          1.10.2   Building Information Modeling....................... 62

1.11   Customer Relationship..62
        1.11.1 Customer Satisfaction..62
        1.11.2 Customer Relationship in Construction Projects 63
1.12   Supply Chain Management...64
        1.12.1 Supply Chain Management Process 64
        1.12.2 Supply Chain Stakeholder.......................................66
        1.12.3 Supply Chain Management Processes.........................67
                 1.12.3.1 Product Specification 68
                 1.12.3.2 Supplier Selection...................................68
                 1.12.3.3 Purchasing..68
                 1.12.3.4 Product Quality 68
                 1.12.3.5 Product Handling...................................69
                 1.12.3.6 Improvement..70
        1.12.4 Supply Chain in Construction Projects.......................70
1.13   Risk Management 72
        1.13.1 Risk Management Process 72
                 1.13.1.1 Identify Risk..73
                 1.13.1.2 Analyze Risk 74
                 1.13.1.3 Prioritization 77
                 1.13.1.4 Plan Risk Response 77
                 1.13.1.5 Reduce Risk..80
                 1.13.1.6 Monitor and Control Risk 80
        1.13.2 Risks in Construction Projects.................................80
1.14   Six Sigma 87
        1.14.1 Introduction 87
        1.14.2 Six Sigma Methodology 88
                 1.14.2.1 Leadership Principles 90
                 1.14.2.2 Six Sigma Team.....................................91
        1.14.3 Analytic Tool Sets..93
                 1.14.3.1 The DMAIC Process..............................96
        1.14.4 Six Sigma in Construction Projects...........................99
                 1.14.4.1 The DMADV Process.............................99
1.15   TRIZ 105
        1.15.1 TRIZ Methodology 105
        1.15.2 Application of TRIZ...107
        1.15.3 TRIZ Process..107
1.16   Value Methodology..108
        1.16.1 Value 108
        1.16.2 Value Engineering...108
                 1.16.2.1 Objective of Value Engineering 109
                 1.16.2.2 Value Engineering Team.......................109
                 1.16.2.3 Approach 110
                 1.16.2.4 Timings of Value Engineering 112
                 1.16.2.5 Stages of Value Engineering Study.......115

1.16.2.6 Function Analysis System Technique (FAST)... 117
1.16.2.7 The Benefits of Value Engineering 117

2. Quality Management System 121
2.1   Introduction 121
2.2   Quality Standards 121
2.3   Standards Organizations 123
2.4   International Organization for Standardization 126
2.5   ISO 9000 Quality Management System 131
       2.5.1   Quality System Documentation 140
2.6   ISO Certification 146
2.7   ISO 14000 Environmental Management System 150
       2.7.1   Benefits of ISO 14000 150
2.8   Occupational Health and Safety Assessment Series 18000 151
       2.8.1   Benefits of OHSAS Management System 151
2.9   Quality Audit 151
       2.9.1   Introduction 151
       2.9.2   Categories of Auditing 152
       2.9.3   Quality Auditing Process 152
       2.9.4   Quality Auditing in Construction Projects 152
               2.9.4.1   Quality Audit by Designer 153
               2.9.4.2   Quality Audit by Contractor 154
               2.9.4.3   External Quality System 154
       2.9.5   Auditing versus Inspection 155
       2.9.6   Risk in Auditing 156
2.10   Quality Cost 156
       2.10.1 Introduction 156
       2.10.2 Categories of Costs 157
       2.10.3 Reasons for Poor Quality 159
       2.10.4 Quality Cost in Construction 162
               2.10.4.1   Prevention Costs 163
               2.10.4.2   Appraisal Costs 163
               2.10.4.3   Internal Failure Costs 164
               2.10.4.4   External Failure Costs 164
       2.10.5 Quality Performance Management System 165
2.11   Integrated Quality Management 166

3. Construction Projects 169
3.1   Project Definition 169
3.2   Construction Projects 170
3.3   Construction and Manufacturing 174
3.4   Tools for Construction Projects 177
       3.4.1   Introduction 177

3.4.2   Design Tools...178
         3.4.2.1   Quality Function Deployment.........................178
         3.4.2.2   Six Sigma DMADV Tool in Design of
                    Construction Projects...181
         3.4.2.3   Design of Experiments 189
3.4.3   Project Control..190
3.4.4   Lean Tools 191
         3.4.4.1   Concurrent Engineering......................................191
         3.4.4.2   Value Stream Mapping.......................................192
         3.4.4.3   5S for Construction Projects.............................192
3.4.5   Improvement Tools...197
         3.4.5.1   PDCA..197
         3.4.5.2   Kaizan 198
         3.4.5.3   Mistake Proofing 200
         3.4.5.4   Corrective Action..200
         3.4.5.5   Preventive Action 202
3.5   Systems Engineering 202
       3.5.1   Introduction..202
       3.5.2   System Definition..203
       3.5.3   Systems Engineering 204
3.6   Construction Project Life Cycle..206

4. Quality in Construction Projects 217
   4.1   Quality in Construction Projects....................................217
          4.1.1   Design–Bid–Build 229
          4.1.2   Design–Build 230
          4.1.3   Build–Own–Operate–Transfer 230
          4.1.4   The Turnkey Contract..231
          4.1.5   Project Manager 231
          4.1.6   Construction Manager 232
          4.1.7   Integrated Project Delivery...............................233
   4.2   Conceptual Design...237
          4.2.1   Identification of Need..237
          4.2.2   Feasibility 240
          4.2.3   Identification of Alternatives...........................242
          4.2.4   Identification of Project Team...........................243
          4.2.5   Development of Concept Design......................245
          4.2.6   Estimate Time Schedule 252
          4.2.7   Estimate Project Cost..252
          4.2.8   Manage Quality...253
          4.2.9   Estimate Resources 253
          4.2.10   Identify/Manage Risk.......................................253
          4.2.11   Finalize Concept Design 254

4.3   Preliminary Design 254
    4.3.1   Identification of Preliminary Design Requirements 256
    4.3.2   Identification of Project Team.............................256
    4.3.3   Develop Preliminary Design 256
    4.3.4   Regulatory Approval....................................257
    4.3.5   Contract Terms and Conditions...........................257
    4.3.6   Schedule 258
    4.3.7   Budget...258
    4.3.8   Manage Quality......................................258
    4.3.9   Estimate Resources 259
    4.3.10   Identify/Manage Risk..................................259
    4.3.11   Perform Value Engineering Study.......................260
    4.3.12   Finalize Preliminary Design 261
4.4   Design Development 261
    4.4.1   Identification of Design Development
         Requirements 261
    4.4.2   Identification of Project Team...........................263
         4.4.2.1   Identify Design Team Members 263
    4.4.3   Develop Detail Design of the Works 263
         4.4.3.1   Architectural Design.......................266
         4.4.3.2   Concrete Structure........................267
         4.4.3.3   Elevator Works...........................269
         4.4.3.4   Fire Protection System 269
         4.4.3.5   Plumbing Works..........................270
         4.4.3.6   Drainage System.........................270
         4.4.3.7   HVAC Works.............................271
         4.4.3.8   Electrical System.........................272
         4.4.3.9   Fire Alarm System 273
         4.4.3.10   Telephone/Communication System................274
         4.4.3.11   Public Address System 274
         4.4.3.12   Audiovisual System 274
         4.4.3.13   Security System/CCTV 275
         4.4.3.14   Security System Access Control 275
         4.4.3.15   Landscape.............................276
         4.4.3.16   External Works (Infrastructure and Road) 277
         4.4.3.17   Bridges 277
         4.4.3.18   Highways 279
         4.4.3.19   Furnishings/Furniture (Loose) 279
    4.4.4   Regulatory/Authorities' Approval.......................287
    4.4.5   Prepare Contract Documents and Specifications 287
    4.4.6   Detail Plan..293
    4.4.7   Budget...293
         4.4.7.1   Cash Flow 295

4.4.8 Manage Quality..295
       4.4.8.1 Plan Quality 295
       4.4.8.2 Quality Assurance...................................297
       4.4.8.3 Control Quality.......................................297
4.4.9 Estimate Resources 297
4.4.10 Identify/Manage Risk..298
4.4.11 Finalize Detail Design 300
4.5 Construction Documents 300
    4.5.1 Identification of Construction Documents
        Requirements...300
    4.5.2 Identification of Project Team...300
    4.5.3 Development of Construction Documents 301
    4.5.4 Prepare Project Schedule...301
    4.5.5 Estimate Cost/Budget..301
    4.5.6 Quality Management...301
       4.5.6.1 Plan Quality 301
       4.5.6.2 Quality Assurance...................................302
       4.5.6.3 Control Quality.......................................302
    4.5.7 Estimate Resources 303
    4.5.8 Risk Management 303
    4.5.9 Finalize Construction Documents 304
4.6 Bidding and Tendering..304
    4.6.1 Organize Tender Documents...309
    4.6.2 Identification of Project Team...309
    4.6.3 Identification of Bidder 309
    4.6.4 Manage Tender Documents...309
    4.6.5 Identify/Manage Risk..311
    4.6.6 Select Contractor 312
    4.6.7 Award Contract 312
4.7 Construction 313
    4.7.1 Mobilization...315
    4.7.2 Identification of Project Team...317
    4.7.3 Planning and Scheduling..319
    4.7.4 Execution of Works...331
    4.7.5 Management of Resources/Procurement 335
    4.7.6 Monitoring and Control...340
       4.7.6.1 Subcontractors Submittal and
           Approval Log 346
       4.7.6.2 Shop Drawings and Materials Logs—E1 347
       4.7.6.3 Procurement Log—E2...............................347
       4.7.6.4 Equipment and Manpower Logs 350
       4.7.6.5 Project Payment/Progress Curve (S-Curve) 350
       4.7.6.6 Time Control 353
       4.7.6.7 Cash and Time Control....................................353

4.7.6.8 Progress Reports and Meetings..........................353
4.7.6.9 Variation Orders 361
4.7.7 Quality Management..370
4.7.7.1 Shop Drawings Approval 374
4.7.7.2 Materials Approval 382
4.7.7.3 Contractor's Quality Control Plan...................399
4.7.7.4 Inspection of Works 415
4.7.8 Risk Management 415
4.7.9 Contract Management..416
4.7.10 Site Safety...419
4.8 Testing, Commissioning, and Handover 425
4.8.1 Develop Testing and Commissioning Plan.....................425
4.8.2 Identification of Project Team..............................426
4.8.3 Testing and Commissioning Quality Process 426
4.8.4 Testing and Commissioning..................................426
4.8.4.1 Testing...429
4.8.4.2 Commissioning...431
4.8.5 As-Built Drawings 433
4.8.6 Technical Manuals and Documents 433
4.8.7 Training of User's Personnel 434
4.8.8 Regulatory/Authorities' Approval.....................434
4.8.9 Move-In Plan 434
4.8.10 Handover of Facility to Owner/End User.......................435
4.8.11 Substantial Completion..435

5. Operation and Maintenance 441
5.1 Operation and Maintenance...441
5.2 Categories of Maintenance 441
5.2.1 Preventive Maintenance..442
5.2.2 Scheduled Maintenance..442
5.2.3 Breakdown Maintenance 442
5.2.4 Routine Maintenance...442
5.2.5 Replacement of Obsolete Items............................442
5.2.6 Predictive Testing and Inspection.......................443
5.3 O&M Program 443

6. Facility Management...449
6.1 Facility Management 449
6.2 Facility Management Outsourcing................................453
6.3 Computer-Aided Facility Management........................454
6.4 Benefits of Facility Management...................................454
6.5 Quality Requirements of Facility Management............457

7. Assessment of Quality 459
   7.1   Introduction 459
   7.2   Assessment Categories.. 459
          7.2.1   Cost of Poor Quality.. 459
          7.2.2   Organization's Standing in the Marketplace 460
          7.2.3   Quality Cultures in the Organization 464
          7.2.4   Operation of Quality Management System 464
   7.3   Self-Assessment.. 465

Appendix A: Design Review Checklists 467

**Appendix B: Major Activities during the Construction Process
            in Building Construction Project** 485

Appendix C: Content of Contractor's Quality Control Plan.................... 495

Bibliography 521

Index 523

List of Figures

Figure 1.1 Birth of total quality. 6

Figure 1.2 Construction project trilogy. 9

Figure 1.3 The feedback loop. 13

Figure 1.4 Cause and effect diagram for bad concrete. 15

Figure 1.5 Control chart for A/V system. 16

Figure 1.6 Flow chart for concrete casting. 17

Figure 1.7 Histogram for manpower. 17

Figure 1.8 Pareto analysis for variation cost. 18

Figure 1.9 Pie chart for content of construction project cost 18

Figure 1.10 Run chart for manpower. 19

Figure 1.11 Scatter diagram. 19

Figure 1.12 Quality engineering triangle. 23

Figure 1.13 Total quality system 26

Figure 1.14 Development of PDCA cycle. 33

Figure 1.15 PDCA cycle for construction projects (design phases). 33

Figure 1.16 Categories of control charts. 35

Figure 1.17 The technology triangle. 39

Figure 1.18 The Ishikawa "fishbone" diagram. 41

Figure 1.19 TQM model: John S. Oakland. 46

Figure 1.20 Basic components of TQM 54

Figure 1.21 Phases of the TQM journey. 55

Figure 1.22 Elements of TQM 56

Figure 1.23 QIS process. 58

Figure 1.24 Customer relationship management process. 63

Figure 1.25 Supply chain management process. 66

Figure 1.26 Supply chain process in construction project. 69

Figure 1.27   Purchasing process...70

Figure 1.28   Risk management cycle. 73

Figure 1.29   Six Sigma roadmap. 89

Figure 1.30   Ishikawa diagram for CCS data. 104

Figure 1.31   Effectiveness of VE. 112

Figure 1.32   Schematic for target costing..116

Figure 1.33   VE summary process. 117

Figure 1.34   VE study process activities. 118

Figure 1.35   FAST model "classic." 119

Figure 2.1   QMS pyramid. 140

Figure 2.2   ISO certification process flow diagram. 148

Figure 2.3   Quality management system certification schedule. 149

Figure 2.4   Quality auditing process. 153

Figure 2.5   Categories of cost of quality..158

Figure 2.6   Primary reasons for poor quality. 160

Figure 2.7   Implementation of quality management (Roberts, 1991). 162

Figure 2.8   Logic flow diagram for development of IQMSIs....................167

Figure 3.1   Traditional Construction Project Organization......................172

Figure 3.2   The house of quality. 179

Figure 3.3   House of quality for smart building system............................180

Figure 3.4   Project design organization chart...183

Figure 3.5   Project monitoring and controlling process cycle..................191

Figure 3.6   Concurrent engineering for construction life cycle...............192

Figure 3.7   Value stream mapping for emergency power system............193

Figure 3.8   PDCA cycle for preparation of shop drawing.........................197

Figure 3.9   Corrective action process. 201

Figure 3.10   Black Box. 204

Figure 3.11   The product life cycle. 205

Figure 3.12   Application areas for systems engineering. 206

Figure 3.13   The system life cycle process. 207

Figure 3.14 Project life cycle. 208

Figure 3.15 Representative construction project life cycle; per Morris. 209

Figure 4.1 Divisional values of building construction project. 218

Figure 4.2 Juran's triple role concept applied to construction. 221

Figure 4.3 Primary reasons for quality. 222

Figure 4.4 Juran's triple role-functional relationship. 224

Figure 4.5 Design-bid-build (traditional contracting system). 229

Figure 4.6 Design–build contracting system. 230

Figure 4.7 Project manager contracting system. 232

Figure 4.8 Agency construction management contractual relationship. 233

Figure 4.9 Construction manager at risk contractual relationship
(CM-at-Risk). 234

Figure 4.10 Integrated project delivery system. 234

Figure 4.11 Logic flow diagram for construction projects-design-bid-
build system. 235

Figure 4.12 Logic flow diagram for construction projects-design-
build system. 236

Figure 4.13 Logic flow in the conceptual design phase. 238

Figure 4.14 Steps in project identification. 239

Figure 4.15 Project feasibility stage (modified). 241

Figure 4.16 Alternative study and impact analysis process. 243

Figure 4.17 Division of responsibility between promoter, engineer,
and contractor. 246

Figure 4.18 House of quality for office building project. 249

Figure 4.19 Typical time schedule of construction project. 252

Figure 4.20 Major activities in the detailed design phase. – – – –
Functional relationship. 262

Figure 4.21 Design management team. 264

Figure 4.22 Room layout. 280

Figure 4.23 Furniture index. 281

Figure 4.24 Finish index. 282

Figure 4.25 Specification for desk units. 283

Figure 4.26 Specification for desk lamp. 284

Figure 4.27 Specification for sofa. 285

Figure 4.28 Design data review cycle. 288

Figure 4.29 Preliminary work program. 294

Figure 4.30 Project S-curve budgeted. 296

Figure 4.31 A simple contract schema for ex ante incentivization. 306

Figure 4.32 Bid clarification. 310

Figure 4.33 Contract award process. 312

Figure 4.34 Major activities during construction phase. 316

Figure 4.35 Job site instruction. 320

Figure 4.36 Project planning steps. 324

Figure 4.37 Logic flow diagram for firefighting works. 327

Figure 4.38 Logic flow diagram for plumbing works. 328

Figure 4.39 Logic flow diagram for HVAC works. 329

Figure 4.40 Logic flow diagram for electrical works. 330

Figure 4.41 Summary construction program. 332

Figure 4.42 Request for staff approval. 337

Figure 4.43 Manpower plan. 338

Figure 4.44 Equipment schedule. 339

Figure 4.45 Material approval and procurement procedure. 341

Figure 4.46 Request for subcontractor approval. 348

Figure 4.47 Contractor's submittal status log. 349

Figure 4.48 Contractor's shop drawing submittal log. 349

Figure 4.49 Contractor's procurement log. 350

Figure 4.50 Equipment status. 351

Figure 4.51 Manpower comparison histogram. 352

Figure 4.52 Planned S-curve. 354

Figure 4.53 Project progress status. 355

Figure 4.54 Planned versus actual. 356

Figure 4.55 (a–c) Daily progress report. 357

Figure 4.56 Work in progress..360

Figure 4.57 Daily checklist status. 361

Figure 4.58 Request for information. 363

Figure 4.59 Flow diagram for processing RFI. 364

Figure 4.60 Site work instruction...365

Figure 4.61 Request for variation..366

Figure 4.62 Request for modification. 367

Figure 4.63 Variation order proposal. 368

Figure 4.64 Flow diagram for processing variation order
(owner initiated)..369

Figure 4.65 (a) Variation order. (b) VO attachment....................................371

Figure 4.66 Site transmittal for work shop drawings. 383

Figure 4.67 Logic flow diagram for shop drawings/composite
drawings. 384

Figure 4.68 Site transmittal for material. 385

Figure 4.69 Specification comparison statement. 386

Figure 4.70 (a, b) Request for substitution. 387

Figure 4.71 Material inspection report..389

Figure 4.72 Checklist for form work. 391

Figure 4.73 Notice for daily concrete casting...392

Figure 4.74 Checklist for concrete casting. 393

Figure 4.75 Quality control of concreting..394

Figure 4.76 Report on concrete casting. 395

Figure 4.77 Notice for testing at lab. 396

Figure 4.78 Concrete quality control form. 397

Figure 4.79 Checklist. 398

Figure 4.80 Remedial note. 399

Figure 4.81 Nonconformance report. 400

Figure 4.82 Sequence of execution of works...401

Figure 4.83 PDCA cycle (Deming wheel) for execution of works. 413

Figure 4.84 Safety violation notice. 421

Figure 4.85   Accident report. 424

Figure 4.86   Logic flow process for testing, commissioning, and
handover phase. 427

Figure 4.87   Development of inspection and test plan. 428

Figure 4.88   Electrical power supply connection procedure. 430

Figure 4.89   Checklist for testing of electromechanical works. 432

Figure 4.90   Handing over certificate. 436

Figure 4.91   Handing over of spare parts. 437

Figure 4.92   Project substantial completion procedure. 438

Figure 6.1   Areas under facility management. 450

Figure 6.2   Distinction between maintenance management and
facility management. 452

Figure 6.3   Elements of computer-aided facility management. 455

Figure C.1   Site quality control organization. 498

Figure C.2   Method of sequence for concrete structure work. 512

Figure C.3   Method of sequence for block masonry work. 513

Figure C.4   Method of sequence for false ceiling work. 513

Figure C.5   Method of sequence for mechanical work (fire protection)....514

Figure C.6   Method of sequence for mechanical work (public health). 514

Figure C.7   Method of sequence for HVAC work. 515

Figure C.8   Method of sequence for electrical work. 516

Figure C.9   Method of sequence for external works. 517

List of Tables

Table 1.1 A Sample of Ancient Chinese Writings on Quality Management 2

Table 1.2 Check Sheet.. 15

Table 1.3 Work Elements...24

Table 1.4 Fourteen-Step Quality Program: Philip B. Crosby 31

Table 1.5 Comparison of Some Control Charts 36

Table 1.6 Fourteen Principles of Transformation: W. Edwards Deming 37

Table 1.7 Seven Points Action Plan: Deming 38

Table 1.8 Four Points for Formation of Quality Circles: Kaoru Ishikawa 40

Table 1.9 Seven Tools of Quality Control by Kaoru Ishikawa..................40

Table 1.10 The Quality Control Steps...43

Table 1.11 Steps to Continuous Quality Improvements.............................44

Table 1.12 The Quality Planning Steps 44

Table 1.13 Ten Points for Senior Management: John S. Oakland...............45

Table 1.14 The Quality Gurus Compared 48

Table 1.15 Periodical Changes in Quality System.....................................49

Table 1.16 Cultural Changes Required to Meet Total Quality Management 51

Table 1.17 Questions and Major Response for Development of QIS 58

Table 1.18 Software Quality Factors..61

Table 1.19 Customer Satisfaction Survey in Construction..........................65

Table 1.20 Stakeholder's Responsibilities in Supply Chain Management 67

Table 1.21 Risk Register 75

Table 1.22 Risk Probability Levels...76

Table 1.23 Risk Probability and Impact Matrix 78

Table 1.24 Risk Response Strategy 79

Table 1.25 Typical Categories of Risks in Construction Projects.................82

Table 1.26 Potential Risks on Scope, Schedule, and Cost, during Construction Phase and Its Effects and Mitigation Action 84

Table 1.27 Fundamental Objectives of Six Sigma Tools 95

Table 1.28 Level of Inventiveness...106

Table 1.29 Area of VE Area of Value Study by Design Stage 114

Table 2.1 Most Common Standards Used in Building Construction Projects..124

Table 2.2 Correlation between IS0 9000:1994 and ISO 9000:2000............129

Table 2.3 Correlation between ISO 9001:2015 and ISO 9001:2008 132

Table 2.4 List of Quality Manual Documents-Consultant (Design and Supervision) 142

Table 2.5 List of Quality Manual Documents-Contractor.......................144

Table 2.6 Check List for Design Drawings..155

Table 2.7 Audit versus Inspection 156

Table 2.8 Cost of Quality..161

Table 2.9 Cost of Quality during Design Stage 161

Table 3.1 Types of Construction Projects...175

Table 3.2 5S Concept...194

Table 3.3 5S for Construction Projects...195

Table 3.4 Mistake Proofing for Eliminating Design Errors.....................200

Table 3.5 Construction Project Life-Cycle (Design-Bid-Build) Phases...213

Table 4.1 Key Quality Assurance Activities on a Typical Construction Project 225

Table 4.2 Principles of Quality in Construction Projects..........................227

Table 4.3 Categories of Project Delivery Systems 228

Table 4.4 Conceptual Alternatives...242

Table 4.5 Typical Requirements of Project Team Members 244

Table 4.6 Typical Responsibilities of Project Team Members 245

Table 4.7 Contribution of Various Participants 247

Table 4.8 Major Causes of Rework, by Phase 286

Table 4.9 MasterFormat® 2016 Division Numbers and Titles.................289

Table 4.10 Checklist for Design Drawings....................................298

Table 4.11 Mistake Proofing for Eliminating Design Errors......................299

Table 4.12 Contract Documents...304

Table 4.13 Contract Forms and Ex Ante Incentivization.........................307

Table 4.14 Contract Forms and Flexible, Farsighted, Ex Post Governance 308

Table 4.15 Major Risk Factors Affecting Contractor 311

Table 4.16 Responsibilities of Supervision Consultant 314

Table 4.17 Matrix for Site Administration and Communication 318

Table 4.18 Benefits of Project Planning and Scheduling 321

Table 4.19 Key Principles for Planning and Scheduling 325

Table 4.20 Contents of Progress Report...342

Table 4.21 Monthly Progress Report..343

Table 4.22 Quality Assurance and Quality Control Typical Responsibilities...373

Table 4.23 Reasons for Rejection of Executed Works..............................402

Table 4.24 Responsibility for Site Quality Control................................411

Table 4.25 Major Risks during Construction Phase and Mitigation Action..417

Table 4.26 Project Closeout Checklist 439

Table 5.1 Preventive Maintenance Program 445

Table 6.1 FM Market General Characteristics—Europe versus Middle East...451

Table 6.2 Typical Requirements of Facility Management Professional 452

Table 6.3 Quality Control Requirements in Facility Management 456

Table 7.1 Customer Satisfaction Questionnaire.................................461

Foreword

Several years ago, it was my honor to write the foreword to the first edition of this book. At that time, I mentioned how impressed I had been with Abdul Razzak Rumane, even before actually meeting him. He has once again asked me to contribute some thoughts to his second edition; it is my distinct pleasure to do so.

Dr. Rumane struck me early on as a hardworking, determined, passionate, compassionate individual. Over the years, as an expert in quality assurance processes, particularly those applicable to the construction industry, he has provided valuable information and insights from his extensive background and education, thus providing a manageable framework for the practitioners and gatekeepers in the industry. He has put to practical use the tenet so expertly articulated by esteemed educational philosopher and healthcare administrator Patrick M. Vance, who once stated (relative to working with students):

> They're doing what they *have* to do to get by.
> What *we* have to do is make it so they *can't* get by without doing some worthwhile things.

The wisdom shared by Abdul Razzak Rumane helps construction professionals do worthwhile things. Applying concepts described in his manuscripts, they influence quite literally the standards of quality of our societal infrastructure, which in turn influences the quality (and in some cases the actual quantity) of life for millions of citizens. What a relief it is to feel secure, to know that our buildings, roads, bridges, power plants, schools, and homes are constructed in mindful, deliberate ways that have the potential to enhance and perhaps even preserve our lives.

Managing risk, ensuring safety, and providing integrity and reliability are, for the gatekeepers who plan and supervise construction, priorities to be taken very seriously. To those who value and apply the level of expertise and care promoted by Dr. Rumane in this volume, our sincere thanks.

Ted Coleman, PhD, CHES, MS (Counseling)
Professor and Department Chair, Health Science and Human Ecology
California State University, San Bernardino

Acknowledgments

"Share the knowledge with others" is the motto of this book.

Many colleagues and friends extended help while preparing this book by arranging reference material—many thanks to all of them for their support.

I thank publishers and authors, whose writings are included in this book, for extending their support by allowing me to reprint their material.

I thank reviewers from various professional organizations for their valuable input to improve my writing. I thank members of ASQ Design and Construction Division, the Institution of Engineers (India), and the Kuwait Society of Engineers for their support to bring out this book.

I thank Dr. Adedeji B. Badiru, series editor; Cindy Renee Carelli, executive editor; Joette Lynch, project editor; Renee Nakaash, editorial assistant-engineering and other staff of CRC Press for their support and contribution to make this construction-related book a reality.

I thank Raymond R. Crawford of Parsons Brinckerhoff for his valuable and timely suggestions during my writing. I thank Dr. N. N. Murthy of Jagruti Kiran Consultants for his support.

I extend my thanks to Dr. Ted Coleman for his everlasting support and thoughtful foreword.

My special thanks to H.E. Sheikh Rakan Nayef Jaber Al Sabah for his support and good wishes.

I thank my well-wishers whose inspiration helped me complete this book.

Most of the data discussed in this book are based on my practical and professional experience and are accurate to the best of my knowledge. However, if any discrepancies are observed in the presentation, I would appreciate if they were communicated to me.

The contribution of my son Ataullah, my daughter Farzeen, and daughter-in-law Masum is worth mentioning here. They encouraged me and helped me in my preparatory work to achieve the final product. I thank my mother, brothers, sisters, and family members for their support, encouragement, and good wishes at all times.

Finally, my special thanks go to my wife, Noor Jehan, for her patience, as she had to endure a lot because of my busy schedule.

Abdul Razzak Rumane

Author

Abdul Razzak Rumane, PhD, is a registered senior consultant with the Chartered Quality Institute (UK) and a certified consultant engineer in electrical engineering. He obtained a bachelor of engineering (electrical) degree from Marathwada University (now Dr. Babasaheb Ambedkar Marathwada University), India, in 1972 and received his PhD from Kennedy Western University, USA (now Warren National University), in 2005. His dissertation topic was "Quality Engineering Applications in Construction Projects." Dr. Rumane's professional career exceeds 44 years, including 10 years in manufacturing industries and more than 34 years in construction projects. Presently, he is associated with SIJJEEL Co., Kuwait, as advisor and director for construction management.

Dr. Rumane is associated with a number of professional organizations. He is a chartered quality professional-fellow of the Chartered Quality Institute (UK) and fellow of the Institution of Engineers (India) and has an honorary fellowship of the Charter Management Association (Hong Kong). He is also senior member of the Institute of Electrical and Electronics Engineers (USA), senior member of the American Society for Quality, member of the Kuwait Society of Engineers, member of SAVE International (The Value Society), and member of the Project Management Institute. He is also associate member of the American Society of Civil Engineers, member of the Diplomatic Academy of London, member of the International Diplomatic Academy, and member of the Board of Governors of International Benevolent Research Forum.

As an accomplished engineer, Dr. Rumane has been awarded an honorary doctorate in engineering from the Yorker International University, USA (2007). The World Quality Congress honored him with the "Global Award for Excellence in Quality Management and Leadership." The Albert Schweitzer Leadership for Life Foundation honored him with gold medal for "Outstanding contribution in the field of construction quality management" and "Outstanding contribution in the field of electrical engineering/consultancy in construction projects in Kuwait." He was selected as one of the top 100 engineers of IBC (International Biographical Centre, Cambridge, UK) in 2009. The European Academy of Informatization honored him with "World Order of Science-Education-Culture" and a title of "Cavalier," and the Sovereign Order of the Knights of Justice, England, honored him the Meritorious Service Medal.

Dr. Rumane has attended many international conferences and has made technical presentations at various conferences. Apart from *Quality Management in Construction Projects*, Dr. Rumane is author of *Quality Tools for Managing Construction Projects* and editor of the *Handbook of Construction*

Management: Scope, Schedule, and Cost Control, published by CRC Press (a Taylor & Francis Group Company), Boca Raton, Florida.

He was honorary chairman of the Institution of Engineers (India), Kuwait chapter, for the years 2016–2017, 2013–2014 and 2005–2007. He is secretary of ASQ, LMC, Kuwait for the year 2017.

Synonyms

Owner	Client, Employer
Consultant	Architect/Engineer (A/E), Designer, Design Professionals, Consulting Engineers, Supervision Professional
Engineer	Resident Project Representative
Engineer's Representative	Resident Engineer
Project Manager	Construction Manager (Agency CM)
Contractor	Constructor, Builder
Quantity Surveyor	Cost Estimator, Contract Attorney, Cost Engineer, Cost and Works Superintendent
Main Contractor	General Contractor

Abbreviations

AAMA	American Architectural Manufacturers Association
ACI	American Concrete Institute
AISC	American Institute of Steel Construction
AMCA	American Composites Manufacturers Association
ANSI	American National Standards Institute
API	American Petroleum Institute
ARI	American Refrigeration Institute
ASCE	American Society of Civil Engineers
ASHRAE	American Society of Heating, Refrigerating, and Air-Conditioning Engineers
ASQ	American Society for Quality
ASTM	American Society of Testing and Materials
BMS	Building Management System
BREEAM	Building Research Establishment Environmental Assessment
BSI	British Standards Institution
CDM	Construction (Design and Management)
CEN	European Committee for Standardization
CIBSE	Chartered Institution of Building Services Engineers
CIE	International Commission on Illumination
CII	Construction Industry Institute
CMAA	Construction Management Association of America
CSC	Construction Specifications Canada
CSI	Construction Specifications Institute
CTI	Cooling Tower Industry
DIN	Deutsches Institute fur Normung
EIA	Electronic Industries Association
EJCDC	Engineering Joint Contract Documents Committee
EN	European Norms
FIDIC	Federation Internationale des Ingenieurs-Conseils
HQE	Higher Quality Environmental
ICE	Institution of Civil Engineers (UK)
IEC	International Electrotechnical Commission
IEEE	Institute of Electrical and Electronics Engineers
IP	Ingress Protection
ISO	International Organization for Standardization
LEED	Leadership in Energy and Environmental Design
NEC-USA	National Electrical Code
NEC-UK	New Engineering Contract
NEMA	National Electrical Manufacturers Association (USA)
NFPA	National Fire Protection Association

NWWDA	National Wood, Window, and Door Association
PMBOK	Project Management Body of Knowledge
PMI	Project Management Institute
QS	Quantity Surveyor
RFID	Radio Frequency Identification
SDI	Steel Door Institute
TIA	Telecommunications Industry Association
UL	Underwriters Laboratories

Introduction

It is almost 5 years since my book entitled *Quality Management in Construction Projects* was published. The response was encouraging and many people appreciated the book as it was totally devoted to quality management in construction projects. I am happy and feel honored that I could share my knowledge, which helped many readers in their day-to-day professional activities related to quality in construction projects.

Since then, ISO 9000:2008 has been revised and ISO 9000:2015 already published. The revised QMS focuses extensively on risk-based thinking that has to be considered from the beginning and throughout the life cycle of a project. There were certain quality-related topics such as customer relationship, supplier management, quality audits, tools for construction projects, and quality management during each phase of the project life cycle that were missing in the first edition, which I feel are important. Further, some of the figures and tables need to be updated to make for a comprehensive understanding of the subject.

The new edition is divided into seven chapters as before; however, additional sections have been added to update the information provided in the earlier edition.

Chapter 1 presents an overview of quality and is focused on the historical importance and background of quality that has moved in its different forms through distinct phases under a bewildering array of names such as inspection, statistical quality control, quality assurance, and strategic quality management, leading to the concept of total quality management (TQM). A brief description of various quality control tools with representative figures is included in this part to illustrate their applications in practical usage. Periodical changes in the quality system and cultural changes needed in various areas to meet TQM are also discussed. The chapter also includes principles and basic components of TQM. There are prominent researchers and practitioners, known as "quality gurus," whose works have dominated the quality movement. This part elaborates the contribution of their philosophies, methods, and approaches in addressing specific quality issues. It also includes quality information system and factors related to quality in the use of the CAD software. Customer relationship, supplier management, and risk management are added in this chapter, which were not included in the previous edition. Quality function deployment technique is discussed in order to know how this can be applied by construction project design professionals. An introduction to Six Sigma and a brief introduction to TRIZ are also included. Basic value-engineering-related topics such as the objectives of value engineering, timings of value engineering, and the benefits of value engineering are also discussed. Chapter 2 is concerned with quality management system. It gives a brief introduction to standards and standardization bodies, ISO certification process, and integrated quality management system. It also includes information on quality audits not included in the earlier edition. The chapter also covers cost of quality. In Chapter 3, project definition, types of construction projects, and comparisons between construction and manufacturing are discussed. Tools for construction projects are added in the chapter, which were not covered in the earlier edition. This includes design tools, project control tools, material control tools, and lean tools; 5S for construction projects is also discussed. This part defines systems engineering, its principles, and its applications and approach to construction projects. It includes a chapter on construction project life cycle that is developed based on systems engineering principles. It details how construction projects can be divided into seven most common phases (five common phases were discussed in the previous edition), such as conceptual design, preliminary design, detailed design, construction, construction documents; bidding and tendering, and testing, commissioning, and handover, further subdividing into various elements/activities/subsystems having functional relationship, to conveniently manage the construction projects. Chapter 4 details quality in construction projects and discusses various concepts of quality, principles, methods, tools, and processes that can be applied from the inception of the project until the issuance of substantial completion certificate. Brief information on different types of contract deliverable systems, contract documents, and contract types based on the forms of payment is also given in this part. This chapter discusses the owner's role while preparing project definition and terms of reference (TOR). It includes information regarding the designer's role and responsibilities to prepare design drawings and documents by properly specifying important parameters and features of the required services/systems. It elaborates different procedures a contractor has to follow during a construction process. The part also includes guidelines for contractors about preparation and submission of construction programs, logs, product data, and shop drawings. Various procedures and principles followed during the construction phase are discussed and implemented. These include mobilization, execution, planning, scheduling, monitoring, control, quality, and testing to achieve the desired quality results for the project. A summary of probable reasons for rejection of works executed by the contractor is presented in a tabular form. The importance of work breakdown structure is discussed and a list of major activities used in the preparation of contractor's construction program is included in Appendix A. The general requirements to prepare a contractor's quality control plan are described and an example contractor's quality control plan is included in Appendix B. Brief information on site safety is covered in this chapter, newly included in this edition. The chapter covers quality management and risk management during each phase of the project life cycle. Chapter 5 presents brief information on post-handover activities such as operation and maintenance. Chapter 6 discusses facility management and Chapter 7 relates to assessment of quality.

The book, I am certain, will meet the requirements of construction professionals, students, and academics and satisfy their needs.

1

Overview of Quality

1.1 Quality History

Quality issues have been of great concern throughout the recorded history of humans. During the New Stone Age, several civilizations emerged, and some 4000–5000 years ago, considerable skills in construction were acquired. The pyramids in Egypt were built approximately 2589–2566 BCE. Hammurabi, the king of Babylonia (1792–1750 BCE), codified the law, according to which, during the Mesopotamian era, builders were responsible for maintaining the quality of buildings and were given the death penalty if any of their construction collapsed and their occupants were killed. The extension of Greek settlements around the Mediterranean after 200 BCE left records showing that temples and theaters were built using marble. India had strict standards for working with gold in the fourth century BCE.

According to *A History of Managing for Quality* (J. M. Juran, editor-in-chief), China's recorded quality history can be traced back to earlier than 200 BCE. China had instituted quality control in its handicrafts during the Zhou dynasty between 1100 and 250 BCE. During this period, the handicraft industry was mainly engaged in producing ceremonial artifacts. Table 1.1 presents a sample of surviving writings that are related to quality management in China during ancient days. This industry survived the long succession of dynasties that followed up to 1911 CE.

Scandinavian shipbuilders were using quality improvement techniques over the entire first two millennia BCE and the first millennium CE. Examples of specification and inspection can be found in the Bible dating from at least 500 BCE, and at about the same time, the Greeks were using tight quality control methods in the building of their temples. The Romans standardized nearly everything they touched from 300 BCE to 300 CE. According to Sebestyen (1998), "Marcus Vitruvius Pollo, the first century BCE military and civil engineer, published his 10 books (i.e., a book with 10 chapters) in Rome. This was the world's first major publication on architecture and construction, and it dealt with building materials, the style and design of building types, the construction process, building physics, astronomy and building machines" (p. 2).

TABLE 1.1

A Sample of Ancient Chinese Writings on Quality Management

Kao Gong Ji (Records in Inspecting the Works)	403 BC	A recognition of quality as the combined result of "the time of heaven, the energy of earth, the beauty of material, and the skill of the workman"
Tang Lu Shu Yi Za Lu Men (Introduction to the Laws of the Tang Dynasty, Miscellaneous Categories)	Compiled in AD 635–640	A law stipulated that measuring tools were to be checked every August, and were to be used only after the seals were affixed.
Wu Jing Zong Yao (Compendium of the Most Important Military Techniques)	AD 650–950	Subject: weapons manufacture
Ying Zao Yi Xun (Architecture Rules and Methods)	Song dynasty (AD 960–1219)	Subject: architecture
Zi Re Yi Xun (Teachings of the Deceased)	Yuan dynasty (AD 1279–1368)	Subject: textiles
Long Jiang Chuan Chang Zhi (Records of the Long Jiang Shipyard)	Ming Dynasty (AD 1368–1644)	Subject: shipbuilding
Cong Cheng Zuo Fa Gui Ze (Regulations in Engineering Projects)	Qing dynasty (AD 1644–1911)	Subject: construction
Tian Gong Kai Wu (Technology and Manufacture)	AD 1637 by Sung Yingxing	Subject: manufacturing

Source: Juran, J.M. & Godfrey, A.B. (1999). *Juran's Quality Handbook*, Reprinted with permission of The McGraw-Hill Education.

During the Middle Ages, guilds took the responsibility for quality control upon themselves. Guilds and governments carried out quality control; consumers, of course, carried out informal quality inspection throughout history.

The guilds' involvement in quality was extensive. All craftsmen living in a particular area were required to join the corresponding guild and were responsible for controlling the quality of their own products. If any of the items was found defective, then the craftsman discarded the faulty items. The guilds also initiated punishments for members who turned out shoddy products. They maintained inspections and audits to ensure that artisans followed quality specifications. The guild hierarchy consisted of three categories of workers: apprentice, journeyman, and master. The guilds had established specifications for input materials, manufacturing processes, and finished products, as well as methods of inspection and testing. They were active in managing quality during the Middle Ages until the Industrial Revolution marginalized their influence.

The Industrial Revolution began in Europe in the mid-nineteenth century. It gave birth to factories, and the goals of the factories were to increase productivity and reduce costs. Prior to the Industrial Revolution, items were produced by individual craftsman for individual customers, and it was possible for workers to control the quality of their products. Working conditions then were more conducive to professional pride. Under the factory system, the tasks needed to produce a product were divided among several or many factory workers. Under this system, large groups of workmen performed similar types of work, and each group worked under the supervision of a foreman who also took on the responsibility of controlling the quality of the work performed. Quality in the factory system was ensured by means of skilled workers, and the quality audit was done by inspectors.

The broad economic result of the factory system was mass production at low costs. The Industrial Revolution changed the situation dramatically with the introduction of a new approach to manufacturing.

In the early nineteenth century, the approach to manufacturing in the United States tended to follow the craftsmanship model used in European countries. In the late nineteenth century, Frederick Taylor's system of "scientific management" was born. Taylor's goal was to increase production. He achieved this by assigning planning to specialized engineers, and execution of the job was left to supervisors and workers. Taylor's emphasis to increase production had a negative effect on quality. With this change in the production method, inspection of finished goods became the norm rather than inspection at every stage. To remedy the decline in quality, factory managers created inspection departments having their own functional bosses. These departments were known as quality control departments.

The beginning of the twentieth century marked the inclusion of process in quality practices. During World War I, the manufacturing process became more complex. Production quality was the responsibility of quality control departments. The introduction of mass production and piecework created quality problems as workmen were interested in earning more money by the production of extra products, which in turn led to bad workmanship. This situation made factories introduce full-time quality inspectors, which marked the real beginning of inspection quality control and thus the introduction of quality control departments headed by superintendents. Walter Shewhart introduced statistical quality control in the process. His concept was that quality is not relevant to the finished product but to the process that created the product. His approach to quality was based on continuous monitoring of process variation. The statistical quality control concept freed manufacturers from the time-consuming 100% quality control system because it accepted that variation is tolerable up to certain control limits. Thus, the quality control focus shifted from the end of line to the process.

The systematic approach to quality in industrial manufacturing began in the 1930s when some attention was given to the cost of scrap and rework. With the impact of mass production, which was required during World War II, it became necessary to introduce a more stringent form of quality control. This was instituted by manufacturing units and was identified as Statistical Quality Control (SQC). SQC made a significant contribution in that it provided a sampling rather than 100% product inspection. However, SQC was instrumental in exposing the underappreciation of the engineering of product quality.

In his books, Juran (Juran and Godfrey 1999) has detailed that, prior to World War II, Japanese research on and applications of modern quality control were limited. Japanese product quality was poor relative to international levels. These poor products were sold only at ridiculously low prices, and it was difficult to secure repeat sales. Among the exceptions were the high-technology products of some Japanese companies, primarily for military use, which were manufactured without the successful application of mass production techniques.

> The concepts and techniques of modern quality control were introduced in Japan from the United States immediately after World War II. The General Headquarters (GHQ) of the allied occupation forces in Japan was experiencing difficulties with the poor state of the country's communication systems and the defective quality and late delivery of communication equipment and components ordered from Japanese manufacturers. The GHQ's Civil Communication Section (CSS) was instructed to provide communication equipment manufacturers with business management guidance, including advice on quality control. Many Japanese manufacturing companies received help from the members of the section. (Juran 1999, p. 41.2)

Juran further states:

> In Europe quality was considered as a cultural issue first and a technical issue second. In western Europe as in other areas of the Western world, from the 1950's to the 1970's, quality was considered a priority issue mainly in defense aerospace, telecommunications, electro-nuclear energy and energy in general, chemicals and their high technology sectors. An indication of the level of interest in quality disciplines (specifically, in statistical quality control) was the formation of national quality associations in many European countries during the early 1950s. In the area of consumer durables and consumer goods in general, which were enjoying a period of high demand, a capacity for innovation and the ability to produce large volumes at low cost were the main priorities. Healthy market performance meant staying more or less in line with the typical (and often mediocre) quality standards of the relevant price/performance class, which, given the lack of specific competitive estimate, tended to remain stable. Certain niches, brands or even entire geographical areas stood out for high quality and reliability of their products, but as a rule, higher quality positioned a product in a higher price class compared with products offering similar performance. Countries like Germany, for instance, had a reputation for superior product quality, especially in niche markets (e.g., luxury automobiles, electric home appliances), but since this quality stemmed from the high professional skills and craftsmanship, product prices were usually higher. (p. 38.2)

Harold Kerzner (2001) has given the quality history of the past 100 years:

> "During the past 100 years the views of quality have changed dramatically. Prior to World War I, quality was viewed predominantly as inspection, sorting out the good items from the bad. Emphasis was on problem identification. Following World War I and up to the early 1950s, emphasis was still on sorting good items from bad. However, quality control principles were now emerging in the form of
>
> - Statistical and mathematical techniques
> - Sampling tables
> - Process control charts"

He further states that from the early 1950s to the late 1960s, quality control evolved into quality assurance, with its emphasis on problem avoidance rather than problem detection. Additional quality assurance principles emerged, such as

- The cost of quality
- Zero-defect programs
- Reliability engineering
- Total quality control

Kerzner (2001) has gone further, saying

> Today, emphasis is being placed on strategic quality management, including such topics as
>
> - Quality is defined by the customer.
> - Quality is linked with profitability on both the market and cost sides.
> - Quality has become a competitive weapon.
> - Quality is now an integral part of the strategic planning process.
> - Quality requires an organization wide commitment. (p. 1087)

Thomas Pyzdek (1999) has stated that "in the last century, quality has moved through four distinct "quality eras": inspection, statistical quality control, quality assurance, and strategic quality management. A fifth era is emerging—complete integration of quality into the overall business system. Managers in each era were responding to the problems they faced at the time (p. 12)."

From the foregoing writings and many others on the history of quality, it is evident that the quality system in its different forms has moved through distinct quality eras, such as

1. Quality inspection
2. Quality control
3. Quality assurance
4. Total quality

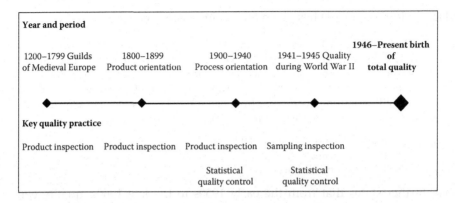

FIGURE 1.1
Birth of total quality. (*Source*: American Society for Quality (2001), Quality 101, Exhibit 17, Reprinted with permission of American Society for Quality © 2001, ASQ, www.asq.org.)

However, quality actually emerged as a dominant thinking only since World War II, becoming an integral part of the overall business system focused on customer satisfaction, and becoming known in recent times as "Total Quality Management" (TQM), with its three constitutive elements:

- Total: Organization-wide
- Quality: Customer satisfaction
- Management: Systems of managing

Figure 1.1 shows the birth of total quality.

1.2 Quality Definition

Quality has different meanings for different people. The American Society for Quality (ASQ) glossary defines quality as follows:

> A subjective term for which, each person has his or her own definition. In technical usage, quality can have two meanings:
>
> 1. The characteristics of a product or service that bear on its ability to satisfy stated or implied needs.
> 2. A product or service free of deficiencies.

It further states that it is

- Based on customers' perceptions of a product's design and how well the design matches the original specifications.
- The ability of a product and service to satisfy stated or implied needs.
- Achieved by conforming to established requirements within an organization.

The International Organization for Standardization (ISO 1994a) defines quality as "the totality of characteristics of an entity that bears on its ability to satisfy stated or implied needs."

Pyzdek (1999) views that there is no single generally accepted definition of quality. He has quoted five principal approaches to defining quality that have been described by Garvin (1988). These are as follows:

1. *Transcendent*—"Quality cannot be defined, you know what it is." (Persig 1974, p. 213)

2. *Product based*—"Differences in quality amount to differences in the quantity of some desired ingredient or attribute." (Abbott 1955, pp. 126–127)

   "Quality refers to the amounts of the unpriced attributes contained in each unit of the priced attribute." (Leflore 1982, p. 952)

3. *User based*—"Quality consists of the ability to satisfy wants." (Edwards 1968, p. 37)

   "In the final analysis of the market place, the quality of a product depends on how well it fits patterns of consumer preference." (Kuehn and Day 1954, p. 831)

   "Quality is fitness for use." (Juran 1974, p. 2-2)

4. *Manufacturing based*—"Quality (means) conformance to requirements." (Crosby 1979, p. 15)

   "Quality is the degree to which a specific product conforms to a design or specification." (Gilmore 1974, p. 16)

5. *Value based*—"Quality is the degree of excellence at an acceptable price and the control of variability at an acceptable cost." (Broh 1982, p. 3)

   "Quality means best 'for certain customer conditions.' These conditions are (a) the actual use, and (b) the selling price of the product." (Feigenbaum 1991, pp. 1, 25)

These definitions can further be summarized under the name of those contributors to the quality movement whose philosophies, methods, and tools have been proved useful in quality practices. They are called the "quality gurus." Their definitions of quality are as follows:

1. *Philip B. Crosby*—Conformance to requirements not as "goodness" or "elegance."
2. *W. Edwards Deming*—Quality should be designed into both product and the process.
3. *Armand V. Feigenbaum*—Best for customer use and selling price.
4. *Kaoru Ishikawa*—Quality of the product as well as after-sales services, quality of management, the company itself, and the human being.
5. *Joseph M. Juran*—Quality is fitness for use.
6. *John S. Oakland*—Quality is meeting customer's requirements.

Based on these definitions, it is possible to evolve a common definition of quality, which is mainly related to the manufacturing, processes, and service industries as follows:

- Meeting the customer's need
- Fitness for use
- Conforming to requirements

However, the definition of quality for construction projects is different from that of manufacturing or services industries as the product is not repetitive but a unique piece of work with specific requirements. Quality in construction projects is not only the quality of product and equipment used in the construction of a facility but the total management approach to complete the facility. The quality of construction depends mainly upon the control of construction, which is the primary responsibility of the contractor.

Quality in manufacturing is spread over a series of processes. Material and labor are input into these processes out of which a product is obtained. The output is monitored by inspection and testing at various stages of production. Any nonconforming product is identified as repaired, reworked, or scrapped, and proper steps are taken to eliminate problem causes. Statistical process control (SPC) methods are used to reduce the variability and increase the efficiency of the process. However, in construction projects, the scenario is not the same. If anything goes wrong, the nonconforming work is very difficult to rectify, and remedial action is sometimes not possible.

The authors of *Quality in the Constructed Project* (2000) by the American Society of Civil Engineers (ASCE) have defined quality as

> the fulfillment of project responsibilities in the delivery of products and services in a manner that meets or exceeds the stated requirements and expectations of the owner, design professional, and constructor.

Responsibilities refer to the tasks that a participant is expected to perform to accomplish the project activities as specified by contractual agreement and applicable laws and licensing requirements, codes, prevailing industry standards, and regulatory guidelines. Requirements are what a team member expects or needs to receive during and after his or her participation in a project. (p. xv)

Chung (1999) states: "Quality may mean different things to different people. Some take it to represent customer satisfaction, others interpret it as compliance with contractual requirements, yet others equate it to attainment of prescribed standards" (p. 3). As regards quality of construction, he further states: "Quality of construction is even more difficult to define. First of all, the product is usually not a repetitive unit but a unique piece of work with specific characteristics. Secondly, the needs to be satisfied include not only those of the client but also the expectations of the community into which the completed building will integrate. The construction cost and time of delivery are also important characteristics of quality" (p. 3).

Based on the foregoing, the quality of construction projects can be defined as follows: Construction project quality is the fulfillment of the owner's needs as per defined scope of works within a budget and specified schedule to satisfy the owner's/user's requirements. The phenomenon of these three components can be called the "construction project trilogy," and is illustrated in Figure 1.2.

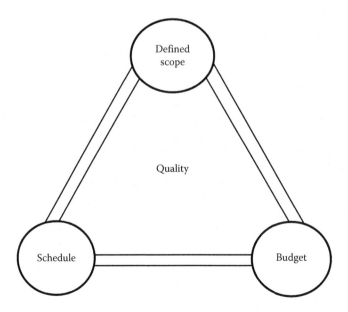

FIGURE 1.2
Construction project trilogy.

1.3 Quality Inspection

Prior to the Industrial Revolution, items were produced by an individual craftsman, who was responsible for material procurement, production, inspection, and sales. In case any quality problems arose, the customer would take up issues directly with the producer. The Industrial Revolution provided the climate for continuous quality improvement. In the late nineteenth century, Frederick Taylor's system of scientific management was born. It provided the backup for the early development of quality management through inspection. At the time when goods were produced individually by craftsmen, they inspected their own work at every stage of production and discarded faulty items. When production increased with the development of technology, scientific management was born out of a need for standardization rather than craftsmanship. This approach required each job to be broken down into its component tasks. Individual workers were trained to carry out these limited tasks, making craftsmen redundant in many areas of production. The craftsmen's tasks were divided among many workers. This also resulted in mass production at lower cost, and the concept of standardization started resulting in interchangeability of similar types of bits and pieces of product assemblies. One result of this was a power shift away from workers and toward management.

With this change in the method of production, inspection of the finished product became the norm rather than inspection at every stage. This resulted in wastage because defective goods were not detected early enough in the production process. Wastage added costs that were reflected either in the price paid by the consumer or in reduced profits. Due to the competitive nature of the market, there was pressure on manufacturers to reduce the price for consumers, which in turn required cheaper input prices and lower production costs. In many industries, emphasis was placed on automation to try to reduce the costly mistakes generated by workers. Automation led to greater standardization, with many designs incorporating interchanges of parts. The production of arms for the 1914–1918 war accelerated this process.

An inspection is a specific examination, testing, and formal evaluation exercise and overall appraisal of a process, product, or service to ascertain if it conforms to established requirements. It involves measurements, tests, and gauges applied to certain characteristics in regard to an object or an activity. The results are usually compared to specified requirements and standards for determining whether the item or activity is in line with the target. Inspections are usually nondestructive. Some of the nondestructive methods of inspection are

- Visual
- Liquid dye penetrant
- Magnetic particle

- Radiography
- Ultrasonic
- Eddy current
- Acoustic emission
- Thermography

The degree to which inspection can be successful is limited by the established requirements. Inspection accuracy depends on

1. Level of human error
2. Accuracy of the instruments
3. Completeness of the inspection planning

Human errors in inspection are mainly due to

- Technique errors
- Inadvertent errors
- Conscious errors
- Communication errors

Most construction projects specify that all the contracted works are subject to inspection by the owner/consultant/owner's representative.

1.4 Quality Control

The quality control era started at the beginning of the twentieth century. The Industrial Revolution had brought about the mechanism and marked the inclusion of process in quality practices. The ASQ termed the quality control era as process orientation that consists of product inspection and statistical quality control.

Thomas Pyzdek (1999) has described the start of the quality control era as follows:

> The Inspection-based approach to quality was challenged by Walter A. Shewhart. In 1931, Shewhart's landmark book *Economic Control of Quality of Manufacturing* introduced the modern era of quality management. In 1924, Shewhart was part of a group working at Western Electric's Inspection Engineering Department of Bell Laboratories. Other members of the group included Harold Dodge, Harry Romig, G.D. Edwards, and Joseph Juran, a veritable "who's who" of the modern quality movement. (p. 13)

Pyzdek further states:

> Quality continued to evolve after World War II. Initially, few commer-
> cial forms applied the new, statistical approach. However, those compa-
> nies that did, achieved spectacular results, and the results were widely
> reported in the popular and business press. Interest groups, such as the
> Society of Quality Engineers (1945), began to form around the country.
> In 1946, the Society of Quality Engineers joined with other groups to
> form the American Society for Quality (ASQ). In July 1944, the Buffalo
> Society of Quality Engineers published *Industrial Quality Control*, the
> first journal devoted to the subject of management discipline. (p. 15)

According to Feigenbaum (1991), the definition of control in industrial terminology is "a process for delegating responsibility and authority for a management activity while retaining the means of assuring satisfactory results."
He further states:

> The procedure for meeting the industrial goal is therefore termed quality
> "Control," just as the procedure for measuring production and cost goals
> are termed, respectively, production "Control" and cost "Control." There
> are normally four steps in such control:
>
> 1. *Setting Standards.* Determining the required cost–quality, perfor-
>   mance–quality, safety–quality, and reliability–quality standards
>   for the products.
> 2. *Appraising Conformance.* Comparing the conformance of the
>   manufactured product, or the offered service to the standards.
> 3. *Acting When Necessary.* Correcting problems and their causes
>   throughout the full range of those marketing, design, engineer-
>   ing, production, and maintenance factors that influence user
>   satisfaction.
> 4. *Planning of Improvements.* Developing a continuing effort to
>   improve the cost, performance, safety, and reliability standards.
>   (p. 10)

Kerzner (2001) describes that "quality Control is a collective term for
activities and techniques, within the process, that are intended to create
specific quality characteristics. Such activities include continually monitoring process, identifying and eliminating problem causes, use of SPC to
reduce the variability and to increase the efficiency of the process. Quality
control certifies that the organization's quality objectives are being met"
(p. 1099).

Gryna (2001) refers to quality control as the process employed to consistently meet standards. The control process involves observing actual per-
formances, comparing it with some standards, and then taking action if
observed performance is significantly different from the standard. The control process is in the nature of a feedback loop, as shown in Figure 1.3.

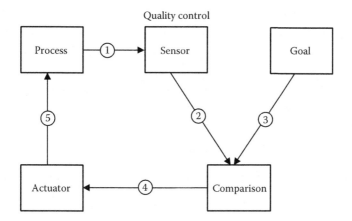

FIGURE 1.3
The feedback loop. (*Source*: Gryna, F.M. (2001), *Quality Planning & Analysis*, Reprinted with permission of The McGraw-Hill Education.)

Control involves a universal sequence of steps as follows:

1. Choose the control subject, that is, choose what we intend to regulate.
2. Establish measurement.
3. Establish standard of performance, product goal, and process goals.
4. Measure actual performance.
5. Compare actual measured performance against standards.
6. Take action on the difference.

Juran (1999, p. 4.2) defines quality control as a universal managerial process for conducting operation so as to provide stability to prevent adverse change and to "maintain the status quo." To maintain stability, the quality control process evaluates actual performance, compares actual performance to goals, and takes action on the difference.

Chung (1999, p. 4) defines quality control as referring to the activities that are carried out on the production line to prevent or eliminate causes of unsatisfactory performance. In the manufacturing industry, including production of ready-mixed concrete and fabrication of precast units, the major functions of quality control are control of incoming materials, monitoring of the production process, and testing of the finished product.

From the foregoing, quality control can be defined as a process of analyzing data collected through statistical techniques to compare with actual requirements and goals to ensure its compliance with some standards.

Quality control in construction projects is performed at every stage through the use of various control charts, diagrams, checklists, etc., and can be defined as

- Checking of executed/installed works to confirm that works have been performed/executed as specified, using specified/approved materials, installation methods, and specified references, codes, and standards to meet intended use
- Controlling budget
- Planning, monitoring and controlling project schedule

A control chart is a graphical representation of the mathematical model used to detect changes in a parameter of the process. Charting statistical data is a test of the null hypothesis that the process from which the sample came has not changed. A control chart is employed to distinguish between the existence of a stable pattern of variation and the occurrence of an unstable pattern. If an unstable pattern of variation is detected, action may be initiated to discover the cause of the instability. Removal of the assignable cause should permit the process to return to stable state.

There are a variety of methods, tools, and techniques that can be applied for quality control and the improvement process. These are used to create an idea, engender planning, analyze the cause, analyze the process, foster evaluation, and create a wide variety of situations for continuous quality improvement. These tools can also be used during various stages of a construction project. The following are the most commonly used quality control tools for a variety of applications to improve the quality process:

1. Cause-and-effect diagram
2. Check sheet
3. Control chart
4. Data collection
5. Flowchart
6. Histogram
7. Pareto analysis
8. Pie chart
9. Run chart
10. Scatter diagram

These tools can be applied at various stages of construction projects to analyze causes of rejection and take necessary preventive or remedial action; develop a system for the preparation and processing of design drawings and documents, execution/installation of work, processing of shop drawings; tabulation of data in the form of a checklist; preparation of construction schedule requirements; and many other applications.

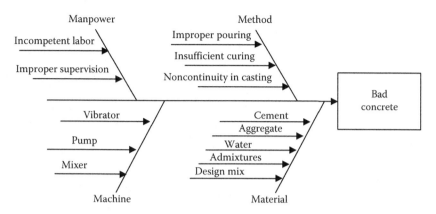

FIGURE 1.4
Cause and effect diagram for bad concrete.

A brief definition of these quality tools is as follows (values shown in the figures are indicative only):

1. The cause-and-effect diagram is also called an Ishikawa diagram or fishbone diagram. It is used to organize and graphically display multiple causes with a particular effect. Figure 1.4 illustrates an example of a cause-and-effect diagram for bad concrete (failure to comply with design concrete strength).

2. A check sheet is a structured list, prepared from the collected data, to indicate how often each item occurs. Table 1.2 illustrates a check sheet for approval record.

3. The control chart is the fundamental tool of SPC. It is a graph used to analyze variation in a process. By comparing current data to historically determined lines, one can arrive at a conclusion regarding whether the process is stable or is being affected by special cause of variation. There are many types of control charts. Each is designed for a specific kind of process or data. A control chart is employed to distinguish between the existence of a stable pattern of variation and

TABLE 1.2

Check Sheet

	Approval Record for a Particular Month			
	Approved	Not Approved	Total	% Not Approved
Shop drawing	〰〰///	///	15	20
Material	〰///	///	10	30
Checklists	〰〰〰 〰///	///	25	12

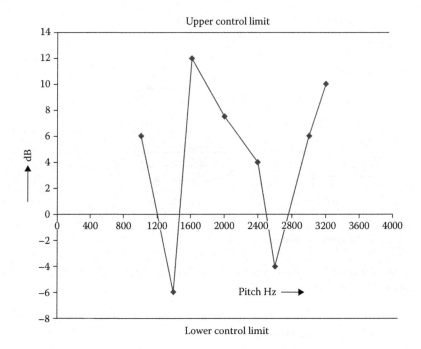

FIGURE 1.5
Control chart for A/V system.

the occurrence of an unstable pattern. Figure 1.5 illustrates sound level results of an audiovisual system.

4. Data collection objectives are to:
   - Identify the problem
   - Report the problem
   - Verify the problem
   - Analyze the problem
   - Correct the problem
5. A flowchart is a pictorial tool that is used for representing a process in sequential order. Flowcharts can be applied at all stages of the project life cycle. Figure 1.6 illustrates a flowchart for concrete casting.
6. The histogram is a pictorial representation of a set of data. It is created by grouping measurements into cells, and it displays how often the different values occur. Figure 1.7 illustrates a histogram for manpower for a period of 1 month.
7. Pareto analysis is a graphical representation of frequency of occurrence. Pareto charts are used to identify those factors that have the greatest

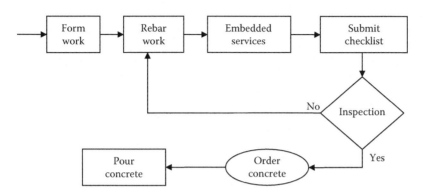

FIGURE 1.6
Flow chart for concrete casting.

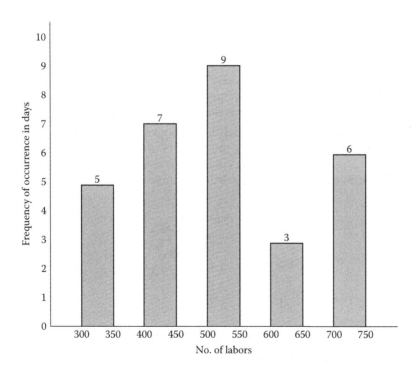

FIGURE 1.7
Histogram for manpower.

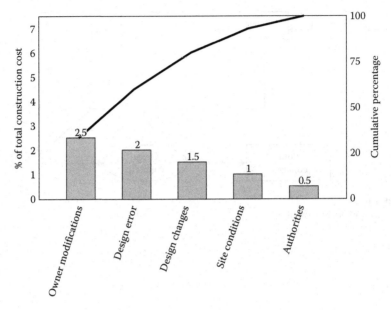

FIGURE 1.8
Pareto analysis for variation cost.

cumulative effect on the system, and, thus, less significant factors can be screened out from the process. Pareto analysis can be used at various stages in a quality improvement program to determine which step to take next. Figure 1.8 illustrates a Pareto chart for variation cost.

8. The pie chart is a circle divided into wedges to depict proportion of data or information in order to understand how they make up the whole. The portions of entire circle or pie sum up to 100%. Figure 1.9 illustrates the contents of construction project cost.

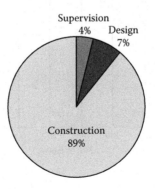

FIGURE 1.9
Pie chart for content of construction project cost.

9. The run chart is a graph plotted by showing measurement (data) against time. Run charts are used to know the trend or changes in the average and also to determine if the pattern can be attributed to common causes of variation, or if special causes of variation were present. Figure 1.10 illustrates a run chart for manpower at a site for a particular week. It is similar to the control chart but does not show control limits.

10. The scatter diagram is a plot of one variable versus another. It is used to identify potential root cause of problems and to evaluate cause-and-effect relationship. Figure 1.11 illustrates a scatter diagram for length versus height of rods.

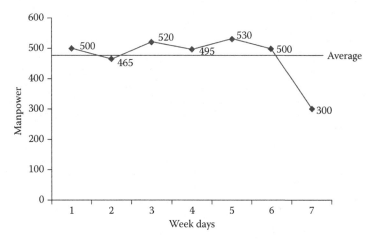

FIGURE 1.10
Run chart for manpower.

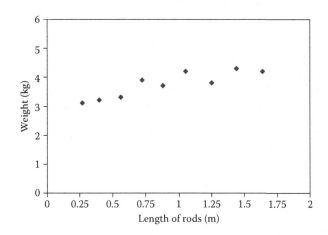

FIGURE 1.11
Scatter diagram.

1.5 Quality Assurance

Quality assurance is the third era in the quality management system.

The ASQ defines quality assurance as "all the planned and systematic activities implemented within the quality system that can be demonstrated to provide confidence a product or service will fulfill requirements for quality."

ASQ details this era:

> After entering World War II in December 1941, the United States enacted legislation to help gear the civilian economy to military production. At that time, military contracts were typically awarded to manufacturers who submitted the lowest competitive bid. Upon delivery, products were inspected to ensure conformance to requirements.
>
> During this period, quality became a means to safety. Unsafe military equipment was clearly unacceptable, and the armed forces inspected virtually every unit of product to ensure that it was safe for operation. This practice required huge inspection forces and caused problems in recruiting and retaining competent inspection personnel. To ease the problems without compromising product safety, the armed forces began to utilize sampling inspection to replace unit-by-unit inspection. With the aid of industry consultants, particularly the Bell Laboratories, they adapted sampling tables and published them in a military standard: Mil-Std-105. The tables were incorporated into the military contracts themselves. In addition to creating military standards, the armed forces helped their suppliers improve their quality by sponsoring training courses in Shewhart's statistical quality control (SQC) techniques. While the training led to quality improvements in some organizations, most companies had little motivation to truly integrate the techniques. As long as government contracts paid the bills, organizations' top priority remained meeting production deadlines. Most SQC programs were terminated once the government's contracts came to an end.

According to ISO 9000 (or BS 5750), quality assurance is "those planned and systematic actions necessary to provide adequate confidence that product or service will satisfy given requirements for quality." ISO 8402-1994 defines quality assurance as "all the planned and systematic activities implemented within the quality system, and demonstrated as needed, to provide adequate confidence that an entity will fulfill requirements for quality."

The third era of quality management saw the development of quality systems and their application principally to the manufacturing sector. This was due to the impact of the following external environment upon the development take-up of quality systems at this time:

- Growing, and more significantly, maturing populations
- Intensifying competition

These converging trends contributed greatly to the demand for more, cheaper, and better quality products and services. The result was the identification of quality assurance schemes as the only solution to meet this challenge.

Harold Kerzner (2001) has defined quality assurance as the collective term for the formal activities and managerial processes that are planned and undertaken in an attempt to ensure that products and services are delivered at the required quality level. Quality assurance also includes efforts external to these processes that provide information for improving the internal processes. It is the quality assurance function that attempts to ensure that the project scope, cost, and time function are fully integrated (p. 1098).

According to Frank M. Gryna (2001):

> Quality Assurance is the activity of providing evidence to establish confidence that quality requirements will be met. ... Quality Assurance is similar to the concept of the financial audit, which provides assurance of financial integrity by establishing, through "independent" audit, that the plan of accounting is (1) such that, if followed, it will correctly reflect the financial condition of the company, and (2) that it is actually being followed. Today, independent financial auditors (certified public accountants) have become an influential force in the field of finance. (p. 659)

Thorpe et al. (1996) have described quality assurance as the evolution of QA from techniques of final inspection in the 1930s, followed by quality control, mainly in the manufacturing industries, during the 1940s and 1950s, then a further extension of controls into the engineering/design phases of these industries during the 1960s.

They further state:

> The term "quality assurance" unfortunately tends to make people think of the finished product and services, whereas it is something far greater; in fact, today's quality system is not something imposed on top of other business systems; it is the system of the business. Our definition of quality assurance is a structural approach to business management and control, which enhances the ability to consistently provide products and services to specification, program and cost. (p. 9)

Quality assurance is the activity of providing evidence to establish confidence among all concerned that quality-related activities are being performed effectively. All these planned or systematic actions are necessary to provide adequate confidence that a product or service will satisfy given requirements for quality.

Quality assurance covers all activities from design, development, production/construction, installation, and servicing to documentation, and also includes regulations of the quality of raw materials, assemblies, products, and components; services related to production; and management, production, and inspection processes.

Quality assurance in construction projects covers all activities performed by the design team, contractor, and quality controller/auditor (supervision staff) to meet owners' objectives as specified and to ensure that the project/facility is fully functional to the satisfaction of the owners/end users.

1.6 Quality Engineering

Feigenbaum (1991) defines quality engineering technology as "the body of technical knowledge for formulating policy and for analyzing and planning product quality in order to implement and support that quality system which will yield full customer satisfaction at minimum cost" (p. 234).

Figure 1.12 illustrates the counterpart triangle for quality engineering.

Quality engineering becomes the field, or apex, of the triangle. The technical work area of the discipline—quality systems implementation—is shown in the first tier. The second tier then shows the principal techniques of quality engineering technology. Quality engineering relates the particular requirements of the plant and companies to the available quality technology—including both hardware equipment and planning and control actions—to put in place much of the ongoing operating detail of the quality systems framework for the firm.

Quality engineering technology thus provides the technical areas to deal with such questions as the following: What are the specific details of the control activities to take place during the development and production and service cycles? Will these quality activities best be accomplished through the use of quality information equipment or by the use of people guided by procedures? What information and material inputs will be needed? What type of information data is required? How should it be analyzed, and what sort of feedback should be used?

Depending upon the differences in the product-quality levels encountered, what criteria are there for alternative courses of corrective action?

Feigenbaum (1991) has further elaborated the entire range of techniques used in quality engineering technology by grouping them under three major headings:

1. *Formulating of quality policy.* Included here are techniques for identifying the quality objectives and quality policy of a particular company as a foundation for quality analysis and systems implementation.

2. *Product-quality analysis.* Techniques for analyzing include those for isolating and identifying the principal actors that relate to the quality of the product in its served market. These factors are then studied for their effects toward producing the desired quality result.

3. *Quality operations planning.* Techniques for implementing the quality system emphasize the development in advance of a proposed course of action and methods for accomplishing the desired quality result. These are the quality planning techniques underlying—and required by—the documentation of key activities of the quality system. (p. 237)

Table 1.3 lists the major work elements normally performed by quality specialists.

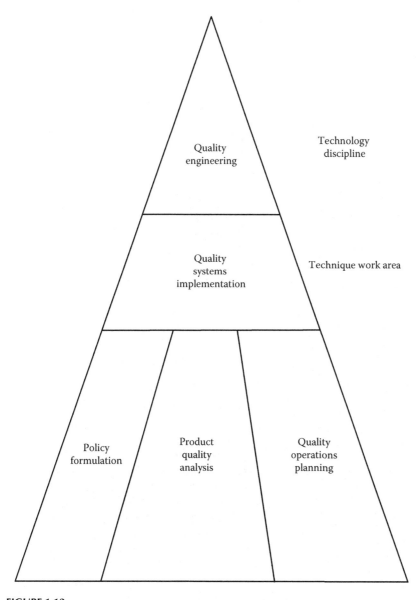

FIGURE 1.12
Quality engineering triangle. (*Source*: Feigenbaum, A.V. (1991), *Total Quality Control*, Reprinted with permission of The McGraw-Hill Education.)

TABLE 1.3

Work Elements

Reliability engineering	Establishing reliability goals
	Reliability apportionment
	Stress analysis
	Identification of critical parts
	Failure mode, effects, and criticality analysis (FMECA)
	Reliability prediction
	Design review
	Supplier selection
	Control of reliability during manufacturing
	Reliability testing
	Failure reporting and corrective action system
Quality engineering	Process capability analysis
	Quality planning
	Establishing quality standards
	Test equipment and gage design
	Quality troubleshooting
	Analysis of rejected or returned material
	Special studies (measurement error, etc.)
Quality assurance	Write quality procedures
	Maintain quality manual
	Perform quality audits
	Quality information systems
	Quality certification
	Training
	Quality cost systems
Inspection and test	In-process inspection and test
	Final product inspection and test
	Receiving inspection
	Maintenance of inspection records
	Gage calibration
Vendor quality	Pre-award vendor surveys
	Vendor quality information systems
	Vendor surveillance
	Source inspection

Source: Pyzdek, T. (1999), *Quality Engineering Handbook,* Reprinted with permission from Quality America Inc.

1.7 Quality Management

The ASQ glossary defines quality management as "the application of quality management system in managing a process to achieve maximum customer satisfaction at the lowest overall cost to the organization while continuing to improve the process."

Thomas Pyzdek (1999) has described the evolution of the quality management concept:

> The quality assurance perspective suffers from a number of serious shortcomings. Its focus is internal. Specifications are developed by the designers, often with only a vague idea of what customers really want. The scope of quality assurance is generally limited to those activities under the direct control of the organization; important activities such as transportation, storage, installation, and service are typically either ignored or given little attention. Quality assurance pays little or no attention to the competition's offerings. The result is that quality assurance may present a rosy picture, even while quality problems are putting the firm out of business. Such a situation existed in the United States in the latter 1970s.
>
> The approaches taken to achieve the quality edge vary widely among different firms. Some quality leaders pioneer and protect their positions with patents or copyrights. Others focus on relative image or service. Some do a better job of identifying and meeting the needs of special customer segments. And others focus on value-added operations and technologies.
>
> Once a firm obtains a quality advantage, it must continuously work to maintain it. As markets mature, competition erodes any advantage. Quality must be viewed from the customer's perspective, not as conformance to self-imposed requirements. Yet a quality advantage often cannot be obtained only by soliciting customer input, since customers usually are not aware of potential innovations. (p. 19)

Figure 1.13 depicts the Total Quality System (TQS) process. The TQS makes it clear that the quality specialist shares responsibility for quality with many others in the company.

According to Juran and Godfrey (1999, p. 41.7):

> At the end of World War II, the former Japanese military and political leaders were no longer in power, having been replaced in large part by relatively young industrialists who wanted Japan to advance as an industrialized country and not to fall back into the old agricultural economy of the type prevalent in some parts of developing countries. After this decision was made, however, they faced a difficult road: poor product quality was a principal obstacle; no one wanted to repeatedly buy such low quality goods. For a country so lacking in raw materials, the inability to sell finished goods for export also meant an inability to earn foreign

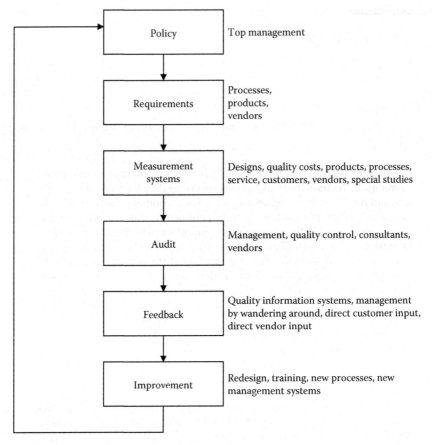

FIGURE 1.13
Total quality system. (*Source*: Pyzdek, T. (1999), *Quality Engineering Handbook*, Reprinted with permission from Quality America Inc.)

currency and hence an inability to buy the materials needed to create an upward spiral of industrial development. Thus, a revolution in product quality became essential. This quality revolution has been taking place in Japan since the early 1950s as the result of efforts to apply the concepts and techniques of statistical quality control on a company-wide scale.

Juran (Juran and Godfrey 1999, p. 41.11) further states:

Introduction and promotion of company-wide quality control led to a revolution in management philosophy, which required lengthy, preserving efforts in education and training. Thus, since the early 1950s, education and training in quality control have been continued for everyone from top management to first time workers in each and every department, including research and development designing, manufacturing, inspection, purchasing, marketing, sales and administration.

According to Feigenbaum (1991),

"Systems Engineering and Management is the foundation for Total Quality Control. ... Total-Quality-Control work requires effective ways to integrate the efforts of large numbers of people with large numbers of machines and huge quantities of information. Hence, it involves systems, questions of significant proportions, and a systems approach is inherent in total quality control.

Historically, the meaning of the word "systems" has varied over a complete spectrum—from a "paperwork" office procedure at one extreme through a "software" computer program to a "hardware" equipment system at the other extreme. In quality control, the term "systems" has meant anything from factory troubleshooting procedures to a shelf of operating "manuals" and "handbooks" covering all product inspection and test routines.

Experience has shown that these approaches have been too narrow. Effective quality control requires the strong coordination of all the relevant paperwork and software and hardware and handbook activities. It requires the integration of the quality actions of the people, the machines, and the information into strong total quality systems. This book refers to this comprehensive systems approach when it uses the phrase "quality system," or as a definition.

A quality system is the agreed-on, company-wide and plant-wide operating work structure, documented in effective, integrated technical and managerial procedures, for guiding the coordinated actions of the people, the machines, and the information of the company and plant in the best and most practical ways to assure customer quality satisfaction and economical costs of quality.

A clearly defined and thoroughly installed TQS is a powerful foundation for total quality control, organization-wide, and for total quality management. Without such systematic integration in a company, "quality management by anticipation" may remain a slogan and a conversation piece, but the actual condition can be quality management by crises and reaction to complaints. Quality can be a consequence rather than the result of carefully planned objectives and activities; it can be the end product of individual, sometimes unlinked, actions throughout the entire marketing–engineering–production–service–quality process. It can be based upon sincere intentions but not guided by firm, quantitative customer quality targets implemented by clear organization-wide programs.

In contrast to this, strong quality systems provide a management and engineering basis for effective prevention-oriented control that deals economically and soundly with the present levels of human, machine, and informational complexity that characterize today's company and plant operations.

The new technologies of systems engineering and systems management are important bases for the establishment and the continuing operation and administration of quality systems. That this is so has fundamental technical and managerial impacts upon the work of the quality-control function as follows:

> Systems engineering is likely to provide what might be thought of as the fundamental "design technology" of the modern quality engineer.
> Systems management is likely to become a fundamental management guide for the quality manager.
> Systems economics, particularly with respect to formalized total quality cost accounting, is likely to provide a major business guide-control point for the general manager."

Gryna (2001) states:

> Following World War II, two major forces emerged that have had a profound impact on quality. The first force was the Japanese revolution in quality. Prior to World War II, many Japanese products were perceived, throughout the world, to be poor in quality. To help sell their products in international markets, the Japanese took some revolutionary steps to improve quality:
>
> 1. Upper-level managers personally took charge of leading the revolution.
> 2. All levels and functions received training in the quality disciplines.
> 3. Quality improvement projects were undertaken on a continuing basis at a revolutionary pace.
>
> The Japanese success has been almost legendary. The second major force to affect quality was the prominence of product quality in the public mind. Several trends converged to highlight this prominence: product liability cases; concern about the environment; some major disasters and near disasters; pressure by consumer organizations; and the awareness of the role of quality in trade; weapons and other areas of international competition. (p. 2)

According to Gryna, "During the 20th century, a significant body of knowledge emerged on achieving superior quality. Many individuals contributed to this knowledge, and five names deserve particular attention: Juran, Deming, Feigenbaum, Crosby and Ishikawa." He further states, "The approaches of these matters have similarities as well as differences,—particularly in the relative emphasis as managerial, statistical, technological and behavioral elements" (p. 3).

According to J.L. Ashford (1989):

> After the second World War the economy of Japan was in ruins. To attain their military objectives, all available resources of capital and of technical manpower had been directed to armaments manufacture, while their civilian economy gained an unenviable reputation for producing poor

quality copies of products designed and developed elsewhere. Unless they were able to raise the quality of their products to a level which could compete, and win, in the international marketplace, they stood no chance of becoming a modern industrialized nation.

To learn how to regenerate their industries, they sent a team abroad to study the management practices of other countries and they invited foreign experts to provide advice. Among the latter were two Americans, J.M. Juran and W.E. Deming, who brought a new message, which can be summarized as follows.

1. The management of quality is crucial to company survival and merits the personal attention and commitment of top management.
2. The primary responsibility for quality must lie with those doing the work. Control by inspection is of limited value.
3. To enable a production department to accept responsibility for quality, management must establish systems for the control and verification of work, and must educate and indoctrinate the work force in their application.
4. The costs of education and training for quality, and other costs which might be incurred, will be repaid many times over by greater output, less waste, a better quality product and higher profits.

These are the basic principles of management concepts which have since become identified under the generic term of quality management. (p. 5)

Thus, the concept of quality management started after World War II, broadening into the development of initiatives that attempt to engage all employees in the systematic effort for quality. Quality management resulted from the work of the "quality gurus" and their theories. Extension of quality management concepts gave birth to TQM.

1.8 Quality Gurus and Their Philosophies

The TQM approach was developed immediately after World War II. There are prominent researchers and practitioners whose work has dominated this movement. Their ideas, concepts, and approaches in addressing specific quality issues have become part of the accepted wisdom in TQM, resulting in a major and lasting impact within the field. These persons have become known as "quality gurus." They all emphasize involvement of organizational management in the quality efforts. These philosophers are

1. Philip B. Crosby
2. W. Edwards Deming

3. Armand V. Feigenbaum
4. Kaoru Ishikawa
5. Joseph M. Juran
6. John S. Oakland
7. Shigeo Shingo
8. Genichi Taguchi

A brief summary of their philosophy and approaches is given next.

1.8.1 Philip B. Crosby

Crosby's philosophy is seen by many to be encapsulated in his five "Absolute Truths of Quality Management." These are

1. Quality is defined as conformance to requirement, not as "goodness" or "elegance."
2. There is no such thing as a quality problem.
3. It is always cheaper to do it right the first time.
4. The only performance measurement is the cost of quality.
5. The only performance standard is zero defects.

Crosby's perspective on quality has three essential beliefs:

1. A belief in qualification
2. Management leadership
3. Prevention rather than cure

Crosby's principal method is his 14-step program for quality management and is illustrated in Table 1.4. His main emphasis is the quantitative, that is, the performance standard of "zero defects."

1.8.2 W. Edwards Deming

Deming was perhaps the best-known figure associated with the quality field and is considered its founding father. His philosophy is based on four principal methods:

1. The plan–do–check–act (PDCA) cycle
2. Statistical process control
3. The 14 principles of transformation
4. The seven-point action plan

TABLE 1.4

Fourteen-Step Quality Program: Philip B. Crosby

Step	Description of Quality Program
Step 1	Establish management commitment
Step 2	Form quality improvement teams
Step 3	Establish quality measurements
Step 4	Evaluate the cost of quality
Step 5	Raise quality awareness
Step 6	Take action to correct problems
Step 7	Zero-defects planning
Step 8	Train supervisors and managers
Step 9	Hold a "Zero Defects" day to establish the attitude and expectation within the company
Step 10	Encourage the setting of goals for improvement
Step 11	Obstacle reporting
Step 12	Recognition for contributors
Step 13	Establish quality councils
Step 14	Do it all over again

Source: Crosby, P.B. (1979), *Quality is Free*, Excerpts from the book, Reprinted with permission of The McGraw-Hill Education.

1.8.2.1 PDCA Cycle

PDCA cycle is an iterative, four-step management method used for continuous improvement of business processes and products. Once it has been completed, it recommences without ceasing. The approach is seen as reemphasizing the responsibility of management to be actively involved in the organization's quality program. The PDCA cycle is also known as the plan–do–study–act (PDSA) cycle.

Nancy R. Tague (2005, p. 391) has elaborated on the PDCA cycle as follows:

Description
The PDCA or PDSA cycle consists of a four-step model for carrying out change. Just as a circle has no end, the PDCA cycle should be repeated again and again for continuous improvement. PDCA is a basic model that can be compared to the continuous improvement process, which can be applied on a small scale.

When to Use

- As a model for continuous improvement
- When starting a new improvement project
- When developing a new or improved design of process, product, or service
- When defining a repetitive work process

- When planning data collection and analysis in order to verify and prioritize problems or root causes
- When implementing any change

Procedure

1. *Plan*. Recognize an opportunity and plan the change.
2. *Do*. Test the change; carry out a small-scale study.
3. *Check*. Review the test, analyze the results, and identify learnings.
4. *Act*. Take action based on what you learned in the study step. If the change did not work, go through the cycle again with a different plan. If you were successful, incorporate the learning from the test into wider changes. Use what you learned to plan new improvements, beginning the cycle again.

PDCA is mainly used for continuous-process improvement. The PDCA cycle, when used as a process improvement tool for design improvement/design conformance in construction projects to meet owner's requirements, shall indicate the following actions:

Plan: Establish scope.

Do: Develop design.

Check: Review and compare.

Act: Implement comments, take corrective action, and/or release contract documents to construct/build the project/facility.

Figure 1.14 illustrates basic concepts to develop the PDCA cycle and Figure 1.15 illustrates the PDCA cycle model for conformance of construction projects designed to owner requirements/scope of work.

1.8.2.2 Statistical Process Control

SPC is a quantitative approach based on the measurement of process control. Deming believed in the use of SPC charts as the key method for identifying special and common causes and assisting diagnosis of quality problems. His aim was to remove "outliers"; that is, quality problems relating to the special causes of failure. This was achieved through training, improved machinery and equipment, and so on. SPC enabled the production process to be brought "under control."

The remaining quality problems were considered to be related to common causes; that is, they were inherent in the design of the production process. Eradication of special causes enabled a shift in focus to common causes to further improve quality.

Gryna (2001) describes SPC as the application of statistical methods to the measurement and analysis of variation in a process. This technique applies to both in process parameters and end process (product) parameters (p. 495).

● Corrective or preventive action.
● Implement approved changes.

● Identify goals and objectives.
● Establish requirements.

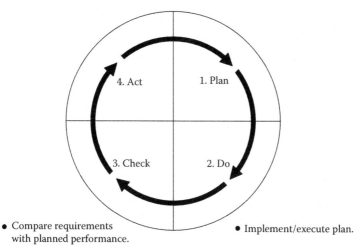

● Compare requirements
  with planned performance.
● Check for compliance.

● Implement/execute plan.

FIGURE 1.14
Development of PDCA cycle.

(Implement comments)
● Implement review comments (if any).
● Take corrective actions (if required).
        or
● Release documents
  for construction/bid.

(Establish scope)
● Establish owner's requirements.
● Determine/define scope of work.
● Establish standards and codes.
● Establish project schedule.
● Establish project budget.

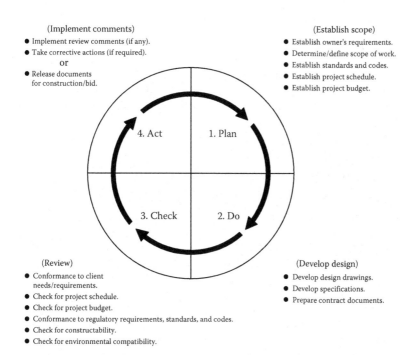

(Review)
● Conformance to client
  needs/requirements.
● Check for project schedule.
● Check for project budget.
● Conformance to regulatory requirements, standards, and codes.
● Check for constructability.
● Check for environmental compatibility.

(Develop design)
● Develop design drawings.
● Develop specifications.
● Prepare contract documents.

FIGURE 1.15
PDCA cycle for construction projects (design phases).

He further states that "a statistical control chart compares process performance data to computed 'statistical control limits,' drawn as limit lines on the chart" (p. 498).

There are two categories of control charts based on the type of data collected:

1. Variable control charts
2. Attributes control charts

Figure 1.16 illustrates types of control charts.
Table 1.5 compares three basic control charts.

1.8.2.3 14 Principles for Transformation

Deming's 14 principles for transformation are listed in Table 1.6.

1.8.2.4 The Seven-Point Action Plan

In order to implement the 14 principles, Deming proposed a seven-point action plan. These are listed in Table 1.7.

1.8.3 Armand V. Feigenbaum

Feigenbaum defines quality as "best for the customer use and selling price," and quality control as an effective method for co-coordinating the quality maintenance and quality improvement efforts at the various groups in an organization, so as to enable production at the most economical levels that allow for full customer satisfaction. Feigenbaum's philosophy of quality has a four-step approach:

Step 1. Set quality standards.
Step 2. Appraise conformance to standards.
Step 3. Act when standards are not met.
Step 4. Plan to make improvements.

Figure 1.17 illustrates Feigenbaum's technology triangle, relating the engineering technologies to the overall field of total quality control.

1.8.4 Kaoru Ishikawa

The founding philosophy of Ishikawa's approach is "company-wide quality control." He has identified 15 effects of company-wide quality control. Ishikawa's approach deals with organizational aspects and is supported by the "quality circles" technique and the "seven tools of quality control."

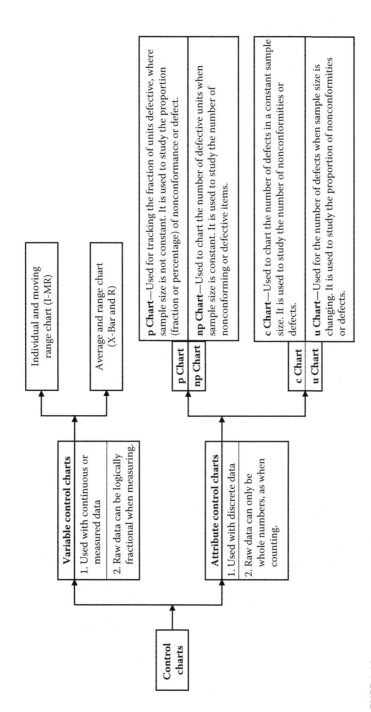

FIGURE 1.16
Categories of control charts.

TABLE 1.5

Comparison of Some Control Charts

Statistical Measure Plotted	Average X and Range R	Percentage Nonconforming (p)	Number of Nonconformities (c)
Type of data required	Variable data (measured values of a characteristics)	Attribute data (number of defective units of product)	Attribute data (number of defects per units of product)
General field of application	Control of individual characteristics	Control of overall fraction defective of a process	Control of overall number of defects per unit
Significant advantages	Provides maximum utilization of information available from data	Data required are often already available from inspection records.	Same advantages as p chart but also provides a measure of defectiveness
	Provides detailed information on process average and variation for control of individual dimensions	Easily understood by all personnel Provides an overall picture of quality	
Significant disadvantages	Not understood unless training is provided; can cause confusion between control limits and tolerance limits	Does not provide detailed information for control of individual characteristics	Does not provide detailed information for control of individual characteristics
	Cannot be used with go/no go type of data	Does not recognize different degrees of defectiveness in units of product	
Sample size	Usually four or five	Use given inspection results or samples of 25, 50, or 100	Any convenient unit of product such as 100 ft of wire or one television set

Source: Gryna, F. (2001), *Quality Planning & Analysis*, Reprinted with permission of The McGraw-Hill Education.

Quality circles are Ishikawa's principal method for achieving participation, composed of 5–15 workers from the same area of achieving, and led by a foreman or supervisor who acts as a group leader to liaison between the workers and the management. The function of quality circles is to identify local problems and recommend the solutions. The aim of quality circles is to

- Contribute to the improvement and development of the enterprise
- Respect human relations and build a happy workshop offering job satisfaction
- Deploy human capabilities fully and draw out infinite potential

TABLE 1.6

Fourteen Principles of Transformation: W. Edwards Deming

The Quality Gurus	
W. Edwards Deming	
Principle 1	Create constancy of purpose to improve product and service.
Principle 2	Adopt a new philosophy for the new economic age with management learning what their responsibilities are and by assuming leadership for change.
Principle 3	Cease dependence on mass inspection to achieve quality by building quality into the product.
Principle 4	End awarding business on price. Award business on total cost and move toward single suppliers.
Principle 5	Aim for continuous improvement of the system of production and service to improve productivity and quality and to decrease costs.
Principle 6	Institute training on the job.
Principle 7	Institute leadership with the aim of supervising people to help them to do a better job.
Principle 8	Drive out fear so that everyone can work effectively together for the organization.
Principle 9	Breakdown barriers between departments. Encourage research, design, sales, and production to work together to foresee difficulties in production and use.
Principle 10	Eliminate slogans, exhortations, and numerical targets for the workforce since they are divisive and, anyway, difficulties belong to the whole system.
Principle 11	Eliminate quotas or work standards and management by objectives or numerical goals; leadership should be substituted instead.
Principle 12	Remove barriers that rob people of their right to pride in their work.
Principle 13	Institute a vigorous education and self-improvement program.
Principle 14	Put everyone in the company to work to accomplish the transformation.

Source: Out of Crisis © 2000 W. Edwards Deming Institute. Reprinted with permission of MIT press.

Kerzner (2001) states:

> Quality circles are small groups of employees who meet frequently to help resolve company quality problems and provide recommendations to management. Quality circles were initially developed in Japan and only recently have achieved some degree of success in United States. The employees involved in quality circles meet frequently either at someone's home or at the plant before the shift begins. The group identifies problems, analyzes data, recommends solutions, and carries out management-approved changes. The success of quality circles is heavily based upon management's willingness to employ recommendations.

TABLE 1.7

Seven Points Action Plan: Deming

The Quality Gurus	
W. Edwards Deming	
Point 1	Management must agree on the meaning of the quality program, its implications, and the direction to take.
Point 2	Top management must accept and adopt the new philosophy.
Point 3	Top management must communicate the plan and the necessity for it to the people in the organization.
Point 4	Every activity must be recognized as a step in a process and the customers of that process identified. The customers are responsible for the next stage of the process.
Point 5	Each stage must adopt the "Deming" or "Shewhart" cycle—Plan, Do, Check, Action—as the basis of quality improvement.
Point 6	Teamworking must be engendered and encouraged to improve inputs and outputs. Everyone must be enabled to contribute to this process.
Point 7	Construct an organization for quality with the support of knowledgeable statisticians.

Source: Out of Crisis © 2000 W. Edwards Deming Institute. Reprinted with permission of MIT press.

He further states:
The key elements of quality circles include

- They are a team effort.
- They are completely voluntary.
- Employees are trained in group dynamics, motivation, communications, and problem solving.
- Members rely upon each other for help.
- Management support is achieved but as needed.
- Creativity is encouraged.
- Management listens to recommendations.

And the benefits of quality circles include

- Improved quality of products and services
- Better organizational communications
- Improved worker performance
- Improved morale (p. 1131)

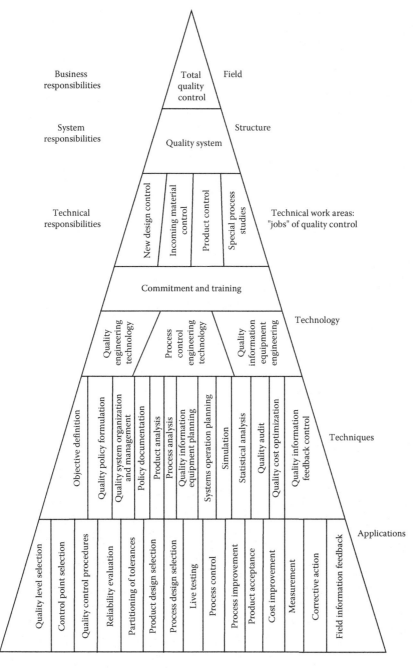

FIGURE 1.17
The technology triangle. (*Source*: Feigenbaum, A.V. (1991), *Total Quality Control*, Reprinted with permission of The McGraw-Hill Education.)

TABLE 1.8

Four Points for Formation of Quality Circles: Kaoru Ishikawa (Developed with Excerpts from Chapter II)

The Quality Gurus

Kaoru Ishikawa

1. Voluntarism: Circles are to be created on voluntary basis, and not by a command from above. Begin circle activities with people who wish to participate.
2. Self-development: Circle members must be willing to study.
3. Mutual development: Circle members must aspire to expand their horizons and cooperate with other circles.
4. Eventual total participation: Circles must establish their ultimate goal as full participation of all workers in the same workplace.

Source: Kaoru Ishikawa (Translated by David J. Lu), (1985), *What Is Total Quality Control? The Japanese Way*. Reprinted with permission from Pearson Education.

TABLE 1.9

Seven Tools of Quality Control by Kaoru Ishikawa

Serial Number	Name of Quality Tools	Usage
Tool 1	Pareto charts	Used to identify the most significant cause or problem.
Tool 2	Ishikawa/fishbone diagrams	Charts of cause and effect in processes.
Tool 3	Stratification	Layer charts that place each set of data successively on top of the previous one. How is the data made up?
Tool 4	Check sheets	To provide a record of quality. How often it occurs?
Tool 5	Histogram	Graphs used to display frequency of various ranges of values of a quantity.
Tool 6	Scatter diagram	To determine whether there is a correlation between two factors.
Tool 7	Control charts	A device in SPC to determine whether or not the process is stable.

Table 1.8 lists the points to be considered for the formation of the quality circle.

The quantitative techniques of the Ishikawa approach are referred to as "Ishikawa's Seven Tools of Quality Control," listed in Table 1.9. The approach includes both quantitative and qualitative aspects, which, taken together, focus on achieving company-wide quality.

These form a set of pictures of quality, representing in diagrammatic or chart form the quality status of the operation of process being reviewed. Ishikawa considered that all staff should be trained in these techniques as they have a useful role to play in managing quality.

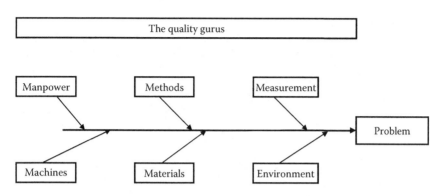

FIGURE 1.18
The Ishikawa "fishbone" diagram.

Ishikawa developed a technique of graphically displaying the causes of any quality problem. His method has many names, such as the Ishikawa diagram, fishbone diagram, and cause-and-effect diagram. The Ishikawa diagram is essentially an end- or goal-oriented picture of a problem situation. It is based around a set of "M" causes such as Manpower (personnel), Machine (plant and equipment), Material (raw material and parts), Method (techniques and technology), Measurement (sampling, instrumentation), and Mother Nature (environment). Figure 1.18 illustrates the Ishikawa diagram.

There are six steps that are used to perform a cause-and-effect analysis:

Step 1. Identify the problem to analyze its technical cause.

Step 2. Select an interdisciplinary brainstorm team.

Step 3. Draw a problem box and prime arrows.

Step 4. Specify major categories contributing to the problem.

Step 5. Identify a defect cause.

Step 6. Identify corrective action and perform the analysis in the same manner as for the cause and effect analysis.

Nancy R. Tague (2005, p. 248) has elaborated on the fishbone diagram as follows:

Description
The Fishbone Diagram identifies many possible causes for an effect or problem. It can be used to structure a brainstorming session. It immediately sorts ideas into useful categories.

When to use

- When identifying possible causes of a problem
- When the team's thinking tends to fall into ruts

Procedure

1. Agree on a problem statement (effect). Write it at the center right of the flipchart or whiteboard. Draw a box around it and draw a horizontal arrow running to it.
2. Brainstorm the major categories of causes of the problem. If there is difficulty here, use generic headings: method, machines (equipment), people (manpower), materials, measurement, and environment. Write the categories of causes as branches from the main arrow.
3. Brainstorm all the possible causes of the problem. Ask "Why does this happen?" As each idea is given, the facilitator writes it as a branch from the appropriate category. Causes can be written in several places if they relate to several categories or multiple relationships.
4. Ask again, "Why does this happen?" about each cause. Write subcauses branching off the causes. Continue to ask "Why" and generate deeper levels of causes. Layers of branches indicate causal relationships.
5. When the team runs out of ideas, focus attention on places on the fishbone where ideas are few.

Summarizing Ishikawa's approach, it can be seen to contain both quantitative and qualitative aspects, which, taken together, focus on achieving company-wide quality.

1.8.5 Joseph M. Juran

Juran's philosophy is perhaps best summed as "Quality does not happen by accident; it has to be planned."

The emphasis of Juran's work is on planning organizational issues, management's responsibility for quality, and the need to set goals and targets for improvement. Juran's definition of quality is "fitness for use or purpose." His thinking on quality is an operational framework of three quality processes:

1. Quality planning
2. Quality control
3. Quality improvement

These are best known as Juran's quality trilogy.

Shtub et al. (1994, p. 284) describe the Juran Trilogy as follows:

> *Quality planning.* In preparing to meet organizational goals, the end result should be a process that is capable of meeting those goals under operating conditions. Quality planning might include identifying internal and external customers, determining customer needs, developing a product or service that responds to

those needs, establishing goals that meet the needs of customers and suppliers at a minimum cost, and proving that the process is capable of meeting quality goals under operating conditions. A necessary step is for managers to engage cross-functional teams and openly supply data to team members so that they may work together with unity of purpose.

Quality control. At the heart of this process is the collection and analysis of data for the purpose of determining how best to meet project goals under normal operating conditions. One may have to decide on control subjects, units of measurement, standards of performance, and degrees of conformance. To measure the difference between the actual performance before and after the process or system has been modified, the data should be statistically significant and the process or system should be in statistical control. Task forces working on various problems need to establish baseline data so that they can determine if the implemented recommendations are responsible for the observed improvements.

Quality improvement. This process is concerned with breaking through to a new level of performance. The end result is that the particular process or system is obviously at a higher level of quality in delivering either a product or a service.

Juran's approach, like those of his colleagues, stresses the involvement of employees in all phases of a project. The philosophy and procedures require that managers listen to employees and help them rank the processes and systems that need improving.

Juran's quality trilogy of planning control and improvement offers the guidelines for his approach. Table 1.10 summarizes his steps on quality planning. Table 1.11 illustrates the quality control procedure, whereas Table 1.12 summarizes Juran's steps for continuous quality improvement.

TABLE 1.10

The Quality Control Steps

Step	Description
Step 1	Choose control subject.
Step 2	Establish standards/objectives.
Step 3	Monitor actual performance.
Step 4	Compare objectives with achievements.
Step 5	Take corrective action to reduce the differences.

Source: Juran, J.M. and Godfrey, A.B. (1999), *Juran's Quality Handbook*, Reprinted with permission of The McGraw-Hill Education.

TABLE 1.11

Steps to Continuous Quality Improvements

Step	Description
Step 1	Prove the need for quality improvement.
Step 2	Identify project.
Step 3	Set goals for continuous improvement.
Step 4	Build an organization team to achieve goals by establishing a quality council, identifying problems, selecting a project, appointing teams, and selecting facilitators.
Step 5	Train team members.
Step 6	Diagnose the causes.
Step 7	Prepare report.
Step 8	Formulate theories.
Step 9	Provide remedial action.
Step 10	Prove that the remedies are effective.
Step 11	Deal with resistance to change.
Step 12	Incorporate improvements into the company's regular systems and processes and control to hold the gain.

Source: Juran, J.M. and Godfrey, A.B. (1999), *Juran's Quality Handbook* (Excerpts from Section 5), Reprinted with permission of The McGraw-Hill Education.

TABLE 1.12

The Quality Planning Steps

Step	Description
Step 1	Establish the project.
Step 2	Identify the customers.
Step 3	Identify the needs of those customers.
Step 4	Analyze and prioritize customer needs.
Step 5	Develop a product that can respond to customer needs.
Step 6	Optimize the product features so as to meet the organization's product range as well as customer needs.
Step 7	Identify process and goals.
Step 8	Develop a process that is able to produce the product.
Step 9	Optimize the process features and goals.
Step 10	Prove that the process can produce the product under operating conditions.
Step 11	Identify control needs.
Step 12	Transfer the process to operations.

Source: Juran, J.M. and Godfrey, A.B. (1999), *Juran's Quality Handbook* (Excerpts from Section 3), Reprinted with permission of The McGraw-Hill Education.

1.8.6 John S. Oakland

Oakland's philosophy of quality is "We cannot avoid seeing how quality has developed into the most important competitive weapon, and many organizations have realized that TQM is the [sic] way of managing for the future (Oakland 1993, Preface)." He gives absolute importance to the pursuit of quality as the cornerstone of organizational success.

Oakland's view is that "quality starts at the top," with quality parameters inherent in every organizational decision. He offers his own overarching approach for TQM on the many well-established methods, tools, and techniques for achieving quality, and some new insight. The overarching method is his "Ten Points for Senior Management." Table 1.13 illustrates this.

Figure 1.19 illustrates Oakland's major features in his "Total Quality Management model."

1.8.7 Shigeo Shingo

Shingo's early philosophy embraced the "scientific management ideas" originated by Frederick Taylor in the early part of the twentieth century. Shingo believed that statistical methods detect error too late in the manufacturing process. He suggested that, instead of detecting errors, it was better to engage in preventative measures aimed at eliminating error sources. Shingo continues to believe in mechanizing the monitoring of error, considering that human assessment was "inconsistent" and prone to error and introduction

TABLE 1.13

Ten Points for Senior Management: John S. Oakland

The Quality Gurus	
John S. Oakland	
Point 1	Long-term commitment.
Point 2	Change the culture to "right first time."
Point 3	Train the people to understand the "customer–supplier relationship."
Point 4	Buy products and services on total cost (sic).
Point 5	Recognize that system improvement must be managed.
Point 6	Adopt modern methods of supervision and training and eliminate fear.
Point 7	Eliminate barriers, manage processes, improve communications and teamwork.
Point 8	Eliminate, arbitrary goals, standards based only on numbers, barriers to pride of workmanship, fiction (use the correct tools to establish facts).
Point 9	Constantly educate and retrain the in-house experts.
Point 10	Utilize a systematic approach to TQM implementation.

Source: Oakland, J.S. (2003), *TQM*, Reprinted with permission from Oakland Consulting plc.

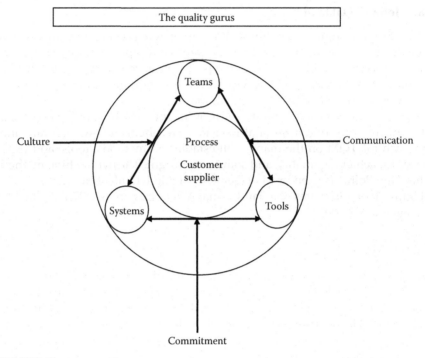

FIGURE 1.19
TQM model: John S. Oakland. (*Source*: Oakland, J.S. (2003), *TQM*, Reprinted with permission from Oakland Consulting plc.)

of controls within a process. He used people to identify underlying causes and produce preventative solutions. Shingo has a clear belief, like Crosby, in a "zero-defects" approach. His approach emphasizes zero defects through good engineering and process investigation and rectification.

Shingo is strongly associated with the "Just-in-Time" manufacturing philosophy. He was the inventor of the Single-Minute Exchange of Die (SMDE) system that drastically reduced the equipment setup time from hours to minutes. Just-in-Time is an integrated set of activities designed to achieve high-level volume production, with minimal inventories of parts that arrive at the workstation when they are needed.

Shingo is also associated with the Poka-Yoke system to achieve zero defects (fail-safe procedures). The Poka-Yoke system includes checklists that (1) prevent workers from making an error that leads to defects before starting, or (2) gives rapid feedback of abnormalities in the process to the worker in time to correct them.

1.8.8 Genichi Taguchi

Taguchi's two founding ideas of quality work are essentially quantitative. The first is a statistical method to identify and eradicate quality problems.

The second rests on designing products and processes to build in quality right from the outset.

Taguchi's prime concern is with customer satisfaction and with the potential for "loss of reputation and goodwill" associated with failure to meet customer expectation. Such a failure, he considered, would lead the customer to buy elsewhere in the future, damaging the prospects of the company, its employees, and society. He saw that loss not only occurred when a product was outside its specification but also when it varied from its target value. Taguchi recognized the organization as "open system," interacting with its environment. The principal tools and techniques espoused by Taguchi center on the concept of continuous improvement and eradicating, as far as possible, potential causes of "nonquality" at the outset. His concept of product development has three stages:

1. System design stage
2. Parameter design stage
3. Tolerance design stage

The first stage is concerned with system design reasoning involving both product and process. This framework is carried on to the second stage—parameter design. The third stage, tolerance design, enables the recognition of factors that may significantly affect the variability of the product.

1.8.9 Summary of Philosophies

Chase et al. (2001) have compared the philosophies of three of our gurus—Crosby, Deming, and Juran. This comparison is listed in Table 1.14.

Although there are differences in certain areas among these philosophers, all of them generally advocate the same steps. Their emphasis is on customers' satisfaction, management leadership, teamwork, continuous improvement, and minimizing defects.

Based on these, the common features of their philosophies can be summarized as follows:

1. Quality is conformance to the customer's defined needs.
2. Senior management is responsible for quality.
3. Institute continuous improvement of process, product, and services through the application of various tools and procedures to achieve higher level of quality.
4. Establish performance measurement standards to avoid defects.
5. Take a team approach by involving every member of the organization.
6. Provide training and education to everyone in the organization.
7. Establish leadership to help employees perform a better job.

Thus, their concept of quality forms the basic tenets of TQM.

TABLE 1.14

The Quality Gurus Compared

	Crosby	Deming	Juran
Definition of quality	Conformance to requirements	A predictable degree of uniformity and dependability at low cost and suited to the market	Fitness for use (satisfies customer's need)
Degree of senior management responsibility	Responsible for quality	Responsible for 94% of quality problems	Less than 20% of quality problems are due to workers.
Performance standard/ motivation	Zero defects	Quality has many "scales"; use statistics to measure performance in all areas; critical of zero defects.	Avoid campaigns to do perfect work.
General approach	Prevention, not inspection	Reduce variability by continuous improvement cease mass inspection.	General management approach to quality, especially human elements
Structure	14 steps to quality improvement	14 points for management	10 steps to quality improvement
Statistical process control (SPC)	Rejects statistically acceptable levels of quality (wants 100% perfect quality)	Statistical methods of quality control must be used.	Recommends SPC but warns that it can lead to tool-driven approach
Improvement basis	A process, not a program, improvement goals	Continuous to reduce variation; eliminate goals without methods	Project-by-project team approach; set goals
Teamwork	Quality improvement teams, quality councils	Employee participation in decision-making; break down barriers between departments	Team and quality circle approach
Costs of quality	Cost of nonconformance; quality is free	No optimum; continuous improvement	Quality is not free; there is no optimum.
Purchasing and goals received	State requirements; supplier is extension of business; most faults due to purchasers themselves	Inspection too late; sampling allows defects to enter system; statistical evidence and control charts required	Problems are complex; carry out formal surveys.
Vendor rating	Yes; quality audits useless	No; critical of most systems	Yes; but help supplier improve

1.9 Total Quality Management

The TQM concept was born following World War II. It was stimulated by the need to compete in the global market, where higher quality, lower cost, and more rapid development are essential to market leadership. Today, TQM is considered a fundamental requirement for any organization to compete, let alone lead, in its market. It is a way of planning, organizing, and understanding each activity of the process and removing all the unnecessary steps routinely followed in the organization. TQM is a philosophy that makes quality values the driving force behind leadership, design, planning, and improvement in activities. Table 1.15 summarizes periodical changes in the quality system.

1.9.1 Changing Views of Quality

Harold Kerzner (2001) states that, with the increased complexity of business, the cost of maintaining a meaningful level of quality had been steadily increasing. In order to reverse this trend, TQM is used to achieve a major competitive advantage. Kerzner has described this phenomenon: "During the past twenty years, there has been a revolution toward improved quality.

TABLE 1.15

Periodical Changes in Quality System

Period	System
• Middle Ages (1200–1799)	• Guilds-skilled craftsman were responsible to control their own products.
• Mid-eighteenth century (Industrial Revolution)	• Establishment of factories. Increase in productivity. Mass production. Assembly lines. Several workers were responsible to produce a product. Production by skilled workers and quality audit by inspectors.
• Early nineteenth century	• Craftsmanship model of production.
• Late nineteenth century (1880s)	• Frederick Taylor and "scientific management." Quality management through inspection.
• Beginning of twentieth century (1920s)	• Walter Shewhart introduced SPC. Introduction of full-time quality inspection and quality control department. Quality management.
• 1930s	• Introduction of sampling method.
• 1950s	• Introduction of statistical quality process in Japan.
• Late 1960s	• Introduction of quality assurance.
• 1970s	• Total quality control. • Quality management.
• 1980s	• TQM
• Beginning of twenty-first century	• Integrated quality management (IQM)

The improvements have occurred not only in product quality, but also in quality leadership and quality project management." He further states:

> The push for higher levels of quality appears to be customer driven. Customers are now demanding
>
> - Higher performance requirements
> - Faster product developments
> - Higher technology levels
> - Materials and processes pushed to the limit
> - Lower contractor profit managing
> - Fewer defects/rejects. (p. 1083)

Gryna (2001, p. 3) has described the changing business conditions the organization should understand to survive in competitive world market. He says:

> The prominence of product quality in the public mind has resulted in quality becoming a cardinal priority for most organizations. The identification of quality as a core concern has evolved through a number of changing business conditions. These include
>
> 1. Competition
> 2. The customer-focused organization
> 3. Higher levels of customer expectation
> 4. Performance improvement
> 5. Changes in organization forms
> 6. Changing workforce
> 7. Information revolution
> 8. Electronic commerce
> 9. Role of a quality department

The failure to address the culture of an organization is frequently the reason for management initiatives either having limited success or failing altogether. To understand the culture of the organization and using that knowledge to implement cultural change is an important element of TQM. The culture of good teamwork and cooperation at all levels in an organization is essential to the success of TQM. Table 1.16 describes cultural changes needed in an organization to meet TQM.

Juran (Juran and Godfrey 1999) described the introduction of TQM as follows:

> In the past two decades many organizations throughout the world have been under tremendous pressure. Some have been battered by international competition, others by new entrepreneurial companies that redefined business, and yet others were seriously challenged by new technologies which created formidable alternatives to their products and services. Some leading companies have changed rapidly. While some of the new companies have now become major players, other companies are still engaged in daily battles of survival, and many other companies have disappeared.
>
> Many companies have found that all of their restructuring, reengineering, downsizing and numerous quality programs may have helped them

TABLE 1.16

Cultural Changes Required to Meet Total Quality Management

From	To
• Inspection orientation	• Defect prevention
• Meet the specification	• Continuous improvement
• Get the product out	• Customer satisfaction
• Individual input	• Cooperative efforts
• Sequential engineering	• Team approach
• Quality control department	• Organizational involvement
• Departmental responsibility	• Management commitment
• Short-term objective	• Long-term vision
• People as cost burden	• Human resources as an asset
• Purchase of products or services on price alone basis	• Purchase on total cost minimization basis
• Minimum cost suppliers	• Mutual beneficial supplier relationship

survive, but they still do not have a distinctive quality advantage. Their future will be determined by key areas: alignment, linkage and replication. Combined with the fundamental concepts of quality management (continuous improvement, customer focus and the value of every member of the organization), their work in these three key areas is transforming the way they are managing the entire organization.

During these years, there has been an increasing global emphasis on quality management. In global competitive markets, quality has become the most important single factor for success. Quality management has become the competitive issue for many organizations.

Juran has gone so far as to state that "just as the twentieth century was the century for productivity, the twenty-first century will be the quality century" (p. 14.2).

1.9.2 Principles of Total Quality Management

Juran describes TQM in terms of the Juran Trilogy, which involves quality planning, quality control, and quality improvement.

In Japanese Union of Scientists and Engineers' view, as mentioned by Juran (Juran and Godfrey 1999, p. 14.3),

TQM is a management approach that strives for the following in any business environment:

- Under strong top management leadership established clear mid and long-term vision and strategies.
- Properly utilize the concepts, values, and scientific methods of TQM.
- Regard human resources and information as vital organizational infrastructures.

- Under an appropriate management system, effectively operate a quality assurance system and other cross-functional management systems such as cost, delivery, environment, and safety.
- Supported by fundamental organizational powers, such as core technology, speed, and vitality, ensure sound relationship with customers, employees, society suppliers, and stockholders.
- Continuously release corporate objectives in the form of achieving an organization's mission, building an organization with a respectable presence and continuously securing profits.

As per ASQ Quality Glossary:

Total Quality Management (TQM) is a term used to describe a management approach to quality improvement. Since then TQM has taken many meanings. Simply put, it is a management approach to long-term success through customer satisfaction. TQM is based on all members of organization in improving processes, products, services, and the culture in which they work. The methods for implementing this approach are found in the teachings of such quality leaders as Philip B. Crosby, W. Edwards Deming, Armand V. Feigenbaum, Kaoru Ishikawa, and Joseph M. Juran.

ASQ described the history of TQM as follows:

The history of Total Quality Management (TQM) began initially as a term coined by the Naval Air Systems Command to describe its Japanese-style management approach to quality improvement. An umbrella methodology for continually improving the quality of all processes, it draws on a knowledge of the principles and practices of

- The behavioral sciences
- The analysis of quantitative and nonquantitative data
- Economics theories
- Process analysis

It has further described the evolution of TQM as follows:

- 1920s
  - Some of the seeds of quality management were planted as the principles of scientific management swept through U.S. industry.
  - Businesses clearly separated the processes of planning and carrying out the plan and union opposition arose as workers were deprived of a voice in the conditions and functions of their work.
  - The Hawthorne experiments in the late 1920s showed how worker productivity could be impacted by participation.
- 1930s
  - Walter Shewhart developed the methods for statistical analysis and control of quality.

- 1950s
  - W. Edwards Deming taught methods for statistical analysis and control of quality to Japanese engineers and executives. This can be considered the origin of TQM.
  - Joseph M. Juran taught the concepts of controlling quality and managerial breakthrough.
  - Armand V. Feigenbaum's book Total Quality Control, a forerunner for the present understanding of TQM, was published.
  - Philip B. Crosby's promotion of zero defects paved the way for quality improvement in many companies.
- 1968
  - The Japanese named their approach to total quality company-wide quality control. It is around this time that the term quality management systems arose.
  - Kaoru Ishikawa's synthesis of the philosophy contributed to Japan's ascendency as a quality leader.
- Today
  - TQM is the name for the philosophy of a broad and systemic approach to managing organizational quality.
  - Quality standards such as the ISO 9000 series and quality award programs such as the Deming Prize and the Malcolm Baldrige National Quality Award specify principles and processes that comprise TQM.

TQM focuses on participative management and strong operational accountability at the individual contributor level. Total quality involves not just managers but everyone in the organization in a complete transformation of the prevailing culture. It is a change to the way people do things, and relies on trust between managers and staff. TQM is applicable to all kinds of organizations, both in the public and private sectors. It is also applicable to those providing services as well as those involved in producing goods or manufacturing activities. Figure 1.20 shows the basic components of TQM.

Jiju Antony and David Preece (2001) have linked TQM and performance as a strategic perspective toward competitive advantage. They have quoted Frangon et al. (1999) that "quality as a means of creating and sustaining a competitive advantage has been widely adopted by both public and private sector organization" (p. 4).

They further state that "to improve quality, businesses have applied 'Total Quality Management' (TQM) to their organizations to help them plan their efforts. The promise of superior performance through continuous quality improvement has attracted a wide spectrum of businesses to TQM, with applications reported in domains such as finance (Wilkinson et al. 1996), utilities (Candlin and Day 1993), federal agencies,

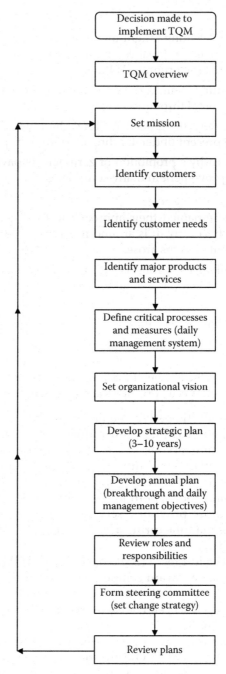

FIGURE 1.20
Basic components of TQM. (*Source*: Shtub, A., Bard, J.F., and Globerson, S. (1994), *Project Management*, Reprinted with permission from Pearson Education.)

healthcare, education and research, environment and manufacturing (Lakhe and Mohanty 1994)" (p. 4).

According to Construction Industry Institute (CII) source document 74: "Total Quality Management is often termed a journey, not a destination." This is because of its nature as a collection of improvement-centered processes and techniques that are performed in a transformed management environment. The concept of "continuous improvement" holds that this environment must prevail for the life of the enterprise and that the methods will routinely be used on a regular, recurring basis. The improvement process never ends; therefore, no true destination is ever reached. Figure 1.21 describes the phases of the TQM journey.

The document further states (p. 1) that from the viewpoint of the individual company, the strategic implications of TQM include

- Survival in an increasingly competitive world
- Better service to the customer
- Enhancement of the organization's "shareholder value"
- Improvement of the overall quality and safety of our facilities
- Reduced project duration and costs
- Better utilization of talents of the people

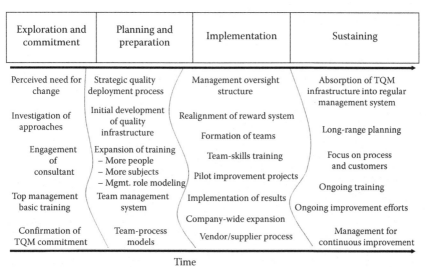

Exploration and commitment	Planning and preparation	Implementation	Sustaining
Perceived need for change	Strategic quality deployment process	Management oversight structure	Absorption of TQM infrastructure into regular management system
Investigation of approaches	Initial development of quality infrastructure	Realignment of reward system	
		Formation of teams	Long-range planning
Engagement of consultant	Expansion of training – More people – More subjects – Mgmt. role modeling	Team-skills training	Focus on process and customers
		Pilot improvement projects	
			Ongoing training
Top management basic training	Team management system	Implementation of results	Ongoing improvement efforts
		Company-wide expansion	
Confirmation of TQM commitment	Team-process models	Vendor/supplier process	Management for continuous improvement

Time

Note:
Phase boundaries are not sharply defined in time.
Read top to bottom, then left to right.

FIGURE 1.21
Phases of the TQM journey. (*Source*: CII Source document 74, Reprinted with permission of CII, University of Texas.)

Chase et al. (2001) have defined TQM as

> Managing the entire organization so that it excels on all dimensions of products and services that are important to the customer. This definition is more applicable than another commonly used one—"Conformance to specifications." Though valid for goods production, the second definition is problematic for many services. Precise specifications for service quality are hard to define and measure. It is possible, however to find out what's important to the customer and then create the kind of organizational culture that motivates and enables the worker to deliver a quality service.

They further state:

> The philosophical elements of TQM stress the operation of the firm using quality as the integrating element. The generic tools consist of (1) various statistical process control (SPC) methods that are used for problem solving and continuous improvement by quality terms, and (2) quality function deployment, which is typically used by managers to drive the voice of the customer into the organization. (p. 260)

Figure 1.22 illustrates the elements of TQM.

An ISO document has listed eight quality management principles on which the quality management system standards of the revised ISO 9000:2000 series are based. These are as follows:

Principle 1—Customer focus

Principle 2—Leadership

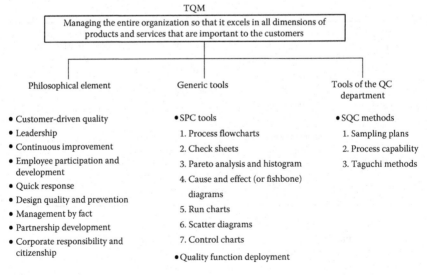

FIGURE 1.22

Elements of TQM. (*Source*: Chase, R., Aquilano, N., and Jacobs, F. (2001), *Operations Management*, Reprinted with permission of The McGraw-Hill Education.)

Principle 3—Involvement of people

Principle 4—Process approach

Principle 5—System approach to management

Principle 6—Continual improvement

Principle 7—Factual approach to design making

Principle 8—Mutual beneficial supplier relationship

Based on those principles, it can be summarized that TQM is a management philosophy that evolved in Japan after World War II. It places quality as a strategic objective and focuses on continuous improvement of products, processes, services, and cost to compete in the global market by minimizing rework, and maximizing profitability to achieve market leadership and customer satisfaction. It is a way of managing people and business processes to meet customer satisfaction. TQM involves everyone in the organization in the effort to increase customer satisfaction and achieve superior performance of the products or services through continuous quality improvement. TQM helps in

- Achieving customer satisfaction
- Continuous improvement
- Developing teamwork
- Establishing vision for the employees
- Setting standards and goals for the employees
- Building motivation within the organization
- Developing corporate culture

TQM is widely accepted as the basis for achieving sustainable competitive advantage. It is required to achieve customer satisfaction and ensure survival in the competitive world because of the global nature of competition, which never rests, and there is no end to product or service improvement.

1.10 Quality Information System

Quality information system (QIS) is an organized method of collecting, analyzing, storing, and reporting the quality-related information for appropriate decision-making.

Figure 1.23 illustrates the QIS process.

QIS is developed based on the organizational requirements to align with goals and objectives to achieve project quality. Table 1.17 illustrated summary of questions and major response evolved by the researchers about QIS development.

QIS can be manual or computerized. The following section discusses quality in the use of CAD software.

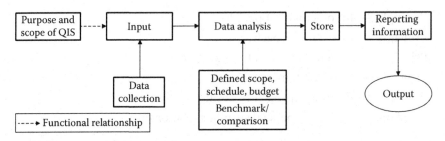

FIGURE 1.23
QIS process.

TABLE 1.17

Questions and Major Response for Development of QIS

Serial Number	Questions	Majority Response
1	Who was the compelling force?	Top level company officer.
2	Were purposes defined?	No.
3	How was information selected?	Issues selected by the user.
4	How was change built into the system?	Did not address change initially. Later incorporated change.
5	Was the system externally benchmarked?	Yes. Companies were compared to external standards.
6	What were the effective ways to analyze performance results?	Comparative charts and training.
7	What were the effective ways to transmit performance results?	Meetings and the Internet with the use of peer pressure.
8	What is quality of measurements?	All companies used filters to improve usefulness and believability.
9	What measurements are made?	Turnaround time; report length; usefulness.
10	How can the QIS be more effective?	Meeting to share information and research reports from their industry.
11	Was training provided?	Training was provided both during the development process and after.
12	What did not work in the development process?	Defining the scope of the QIS to be too expansive.
13	What are the most important problems?	1. Relying on the IT department. 2. Lack of ownership of system by employees. 3. No development process flow model.
14	What are the three most important steps to take?	1. Recognition of project importance. 2. Accurate definition of inputs and outputs. 3. Use of project management tools.
15	Can quality be improved in your current system?	Use more technology such as computers and networks to distribute information.

Source: Gryna, F. (2001), *Quality Planning & Analysis,* Reprinted with permission of The McGraw-Hill Education.

1.10.1 Quality in the Use of CAD Software

With the advent of the computer age, information automation has become a reality. Computers are now being used in design, presentation of documents, estimation, presentation, analysis, planning, and many other applications.

Engineering design is the partial realization of the designer's concept. Engineering drawing is an abstract universal language used to represent the designer's concept to others. The conventional method of representing an engineering drawing is drafting on paper with pen or pencil. Manual drafting is tedious and requires a tremendous amount of patience and time. The invention of computer-aided design (CAD) has had a tremendous effect on the design process, and the computer-aided drafting system has improved drafting efficiency. CAD is a process by which a computer-based program assists in the creation or modification of design. It allows not only passing information more quickly and accurately but also opens the door to integration and automation. A product of the computer era, CAD originated from early comport graphics systems and their development into interactive computer graphics. The 1970s marked the beginning of a new era in CAD—the invention of three-dimensional (3D) solid modeling. Since the invention of computer-aided drafting, there has been tremendous progress in CAD, and it is being regularly used in design drafting. In construction projects, CAD is used by all trades such as architectural, structural, mechanical, electrical, infrastructure, and landscape engineers, and others. Design software has helped engineers to produce and modify drawings in the related field of applications and has made it possible to integrate and visualize their overall effect and put them to use to construct the facility. Its application has helped in producing fully coordinated design drawings and avoiding any conflict during construction.

A CAD system consists of three major parts:

1. Hardware: computer and input/output (I/O) devices
2. Operating system: software
3. Application software: CAD package

Hardware is used to support the software function. Operating system software is the interface between the CAD application software and the hardware. It is important not only for CAD software but for non-CAD software as well. Application software is the heart of the CAD system. CAD consists of programs that do drafting, engineering analysis, 2D (two-dimensional) and 3D drawings, 3D modeling, and many other engineering-related functions.

AutoCAD is the most commonly used software in the construction project/industry. It is the most commonly used software by the construction project professional. It began as a PC-based drafting package running under the MS-DOS environment and gradually evolved into a full-blown CAD system. Its editions are upgraded to meet the current demands of industries. AutoCAD 2010 is the latest version of AutoCAD.

There are numerous off-the-shelf application packages available; it is beyond the scope of this book to list them all, but they are being used in designing systems in construction projects. The following are the most common application software packages used by all the trades:

- General-purpose by all trades
  - AutoCAD
  - Microsoft Office
- Planning, scheduling, and controlling
  - Primavera
  - Microsoft Project
- Structural design
  - STAAD
  - ETABS
- Mechanical design
  - Elite
  - StormCAD, WaterCAD, SewerCAD
- HVAC design
  - HAP
  - APEC
  - Hevacomp
- Electrical design
  - DiaLUX
  - Ecodial
- Landscape
  - DynaSCAPE
- Infrastructure/roads
  - Bentley MX
  - Eagle Point

Quality in use aims at defining the quality attributes that are important to the end user. In 1991, the International Organization for Standardization (ISO) introduced ISO/IEC 9126 (1991): Software evaluation quality characteristics and guidelines for their use. ISO/IEC 9126 is a four-part model:

Part 1: Quality model
Part 2: External metrics
Part 3: Internal metrics
Part 4: Quality in use metrics

There are six characteristics for both external and internal metrics: functionality, reliability, usability, efficiency, maintainability, and portability. They can be further subdivided into subcharacteristics. The quality in use metrics has four characteristics: effectiveness, satisfaction, productivity, and safety.

ISO/IEC 9126 was superseded in 2005 by ISO/IEC 25000: Software Engineering. All the requirements of ISO/IEC 9126 have been taken care of in this standard.

Table 1.18 lists software quality factors to be considered while selecting a software package.

TABLE 1.18

Software Quality Factors

Serial Number	Factor	Description
1	Suitability	Whether the application software is suitable and satisfies the intended use and has all the required parameters
2	Compatibility with the operating system	Application software is usually operating system dependent, therefore attention must be given to the compatibility with the operating system
3	Hardware capability	Whether the RAM and storage memory is sufficient to store the created/modified design data
4	Reliability	Whether the application software shall perform its intended use with required precision
5	Acceptability	Whether the designs produced with application software are accepted by the client
6	Credibility	How credible the product is
7	Usability	How easy is it to learn, operate, and transfer input information by translation or interpreting the same to the specific data format
8	Integrability	Whether the program can be integrated with other application software and how much effort is required for interoperating with other systems
9	Flexibility	Efforts required to modify an operational system
10	Efficiency	Whether the program can perform the amount of computing resources and code required by a program
11	Maintainability	How much effort is required for troubleshooting and fixing an error
12	Testability	How much effort is required to test and ensure that the program is performing its intended function
13	Portability	Efforts required to transfer the program from one hardware to another and to configure the same with new environment
14	Reusability	Whether the program can be used in other applications related to the packaging and scope of the functions that the program performs
15	Safety	How safe is the program to use

1.10.2 Building Information Modeling

Building information modeling (BIM) is an innovative process of generating a digital database for collaboration and managing building data during its life cycle and for preserving the information for reuse and additional industry-specific applications. It is Autodesk's strategy for the application of information technology to the building industry. It helps in better visualization and clash detection and is an excellent tool to develop project staging plans and study phasing and coordination issues during the construction of project life cycle, in preparation of as-built, and also during maintenance of the project.

BIM uses software and processes to digitally develop building data in a manner that is collaborative and integrative. Using BIM as a robust process and not just a tool has transformed the design and construction industry. Different from computer-aided drafting (CAD), BIM leverages digital information about the building (including materials, furnishings, and equipment requirements) in a searchable database that is additive and scalable, enabling design and construction team members to participate in the virtual cocreation and codevelopment of a project design and operational requirements. Accurately described as virtual design and construction, BIM facilitates 3D digital development including coordination, estimating, scheduling, and material selection. BIM also creates a rich data source onto which rules-based project studies such as occupancy simulations, fire and life-safety code reviews, and energy use calculations can be applied. The future of BIM will be leveraged in physical facility operations, where model information will be utilized in continuous facility commissioning for operating and maintaining a building.

The advent of BIM and sharing the parametric geometry and information contained in these files during all phases of design and construction has disrupted the traditional incumbent roles within the construction industry. Currently deployed on large and complicated projects, the use of BIM has become the tool of choice for designers and builders for coordination and scheduling.

1.11 Customer Relationship

Quality is meeting customer requirements and customer satisfaction. Customer expectations are the first step toward customer satisfaction. Customer relationship is interaction with the customer to ensure customer satisfaction and retention of the customer by fulfilling the stated needs/ requirements that the customer wants.

1.11.1 Customer Satisfaction

Customer satisfaction is an indicator that measures the satisfaction of the customers/consumers about the quality of the products/services, acceptance

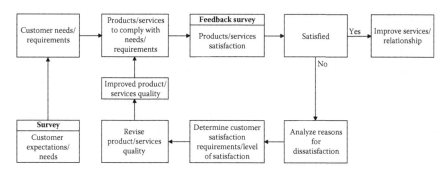

FIGURE 1.24
Customer relationship management process.

of the products/services, and retention of the loyalty toward doing future business that help to manage and improve the business. Customer satisfaction for construction is delivering a project or facility as per defined scope within agreed-upon budget and specified schedule. It is delivering a project or facility that satisfies owner/end user needs and requirements. Customer satisfaction is an important factor in the development of construction process and customer relationship.

In construction, a customer may be defined as the owner of the project or facility and the one that needs the construction facility (end user). Customer satisfaction in the construction industry is how well a contractor meets the customer's (owner's) expectations. In construction projects, the needs to be satisfied include not only those of the client/owner but the expectations of the community into which the completed building/facility will integrate. Figure 1.24 illustrates customer relationship management process.

1.11.2 Customer Relationship in Construction Projects

Construction projects are temporary endeavors, unique and have finite duration. Customer satisfaction enables construction companies to differentiate themselves from their competitors and create sustainable advantage. Construction projects involve numerous stakeholders that closely related and interacted during a given project. The level of stakeholder's satisfaction directly influences the current project and subsequent projects and the level of satisfaction experienced by the stakeholders. For contractors, completing a project in accordance with the defined scope within budget and on time satisfies owner's needs and generates profit. The relationship between customer and contractor is periodic and depends on the duration of the project. In construction, the mutual cooperation, maintaining harmonious working relationship between customer (owner) and contractor, is strongly emphasized. Further, in construction, the project organizations and collaborative relationships are of "one-off" nature; the customer satisfaction survey model is different to those on other industries where the relationship between the client (customer) and supplier is normally a long-term relationship. In construction, each customer has unique needs and expectations; therefore, it is necessary to make a concerted effort to identify customer satisfactions through direct discussions with the owner and participation meetings with the owner and their team prior to the start of the project. Table 1.19 illustrates an example model of customer satisfaction survey in construction projects. Implementation of customer satisfaction survey requirements helps manage and improve customer relationships.

1.12 Supply Chain Management

Supply chain management is a new concept of doing business that has been recognized and practiced by the manufacturing industry for the last two decades. Supply change management has also emerged as new practice in the construction industry.

The emerging technological advances and to maintain competitive advantage in the business environment, it has become essential to plan, organize, and control all supply-chain-related activities at all the levels of product/project life cycle to ensure competitiveness of the end product. It is a process to ensure that the customer gets the right product or service having the right quality and quantity on time.

Supply chain management is a very complex process that has involvement of many participants both within and outside the organization. Each of these participants has influence over the activities to be performed to ensure availability of the product for further processing.

1.12.1 Supply Chain Management Process

Supply chain management is a process to manage and optimize the flow of products, materials, services, and information starting from the creation/inception of need/demand until its installation/usage for specified purpose to achieve a high level of performance for organizational success. It is a process for reducing cost, enhancing quality, and reducing operation time.

Supply chain management process begins from the inception of need and covers all the stages, systems till the installation/fixing/implementation of the product, material, and services. Figure 1.25 illustrates supply chain management process.

TABLE 1.19

Customer Satisfaction Survey in Construction

Serial Number	Stages	Elements	Description
1	Preconstruction	Establish client expectation	Contractor to have one-on-one discussion with owner/client, project manager, consultant (construction supervisor) prior to the start of construction to help 1. Establish customer expectations 2. Determine performance criteria 3. Define team expectations 4. Determine required team members

2	Construction	To know how the work is progressing	Conduct meetings at regular interval to find out whether
			1. Client's expectations are met
			2. Work progress satisfactory
			  a. Schedule
			  b. Material
			  c. Workforce
			3. Specified quality is achieved
			4. Cost variance
			5. Any discrepancies with documents
			6. Staff performance and cooperation
			7. Subcontractor's performance

3	Post-construction	Prepare survey questionnaire and get feedback from owner/client, project manager, consultant	The survey questionnaire to include items related to following factors:
			1. Quality assurance at work
			  1.1 Management and implementation of agreed-upon quality assurance procedures
			  1.2 Quality of material
			2. Workmanship
			  2.1 Quality of work
			3. Skill of personnel
			4. Cooperation of site staff
			  4.1 Agreement about changes/variations
			5. Behavior of site staff
			6. Work supervision
			7. Attendance to defects
			8. Safety at work
			  8.1 Safety measures
			  8.2 Handling of accidents
			  8.3 Attendance to injuries/accidents
			9. Environmental consideration
			  9.1 Cleanliness of site
			  9.2 Handling of hazardous material
			  9.3 Disposal of waste
			10. Subcontractor's performance
			  10.1 Competency of subcontractor(s)

(Continued)

TABLE 1.19 (*Continued*)

Customer Satisfaction Survey in Construction

Serial Number	Stages	Elements	Description
			11. Adherence to schedule
			12. Overrun
			13. Information flow
			14. Handover procedure
			15. Attending snags
			16. Overall satisfaction
			17. Consideration for future projects

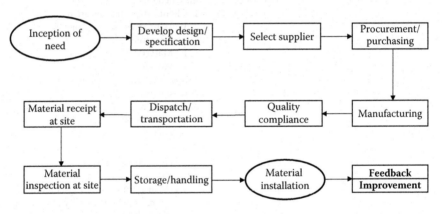

FIGURE 1.25
Supply chain management process.

1.12.2 Supply Chain Stakeholder

Supply chain management has the involvement of many participants. The following are the main stakeholders, participants involved in the supply chain management process:

1. Owner
2. Designer/specification developer
3. Purchasing/procurement department
4. Finance department
5. Manufacturer/supplier
6. Quality department
7. Dispatch section
8. Transportation

TABLE 1.20

Stakeholder's Responsibilities in Supply Chain Management

Serial Number	Stakeholder	Responsibilities
1	Owner	Inception of need
2	Designer	Develop design specifications
3	Procurement manager	Select competitive manufacturer/supplier
4	Finance manager	Arrange finance for procurement
5	Manufacturer/supplier	Produce/supply product as per specifications
6	Quality manager	Ensure the product meets specifications/ product quality
7	Dispatch department	Ensure timely dispatch of material properly identified
8	Transportation department	Ensure economical and short duration process
9	Receiving department	Receive material in good condition
10	Incoming inspection department	Ensure incoming material conforms with specifications
11	Material handling	Proper handling of material
12	Storekeeper	Proper storage of material
13	Installer/user	Installed as specified/recommended. Give feed about the quality of the product

9. Receiving party
10. Incoming inspection
11. Material handling
12. Storage
13. Installation/usage

Table 1.20 illustrates matrix of responsibilities of each of the participants.

1.12.3 Supply Chain Management Processes

The following are the major processes that need to be implemented and monitored in the supply chain process to achieve optimal result:

1. Product specification
2. Supplier selection
3. Purchasing
4. Product quality
5. Product handling
6. Improvement

1.12.3.1 Product Specification

In order to achieve efficient and smooth supply management process, it is essential to develop product design and specifications based on the customers' need/product requirements, goals, and objectives. The following are the major points that the designer has to consider while developing the design and product specifications:

- Suitability to the purpose and objectives
- Performance parameters and compatibility
- Sustainability (environmental, social, economical)
- Availability
- Reliability
- Safety and security

1.12.3.2 Supplier Selection

The selection of supplier mainly depends on the following:

- Experience in supply of similar type of product/services
- Quality management certification
- Technical capability
- Current work load and capacity to supply future requirements
- Resources such as manpower, material
- Financial capability and stability
- Track record in timely execution of past orders
- Product/supply rejection record
- Litigation record

1.12.3.3 Purchasing

Purchasing process starts with identification of potential suppliers who can deliver the needed product/material on time at competitive price, conforming to the product/material quality.

Figure 1.26 illustrates the purchasing process.

1.12.3.4 Product Quality

It is essential that both suppliers and customers are involved to ensure quality compliance of the product/material. The customer must discuss with the supplier about the acceptable level of product/material quality/performance. A quality metrics and method of inspection and quality control should be prepared for acceptance level. Quality management process to meet the

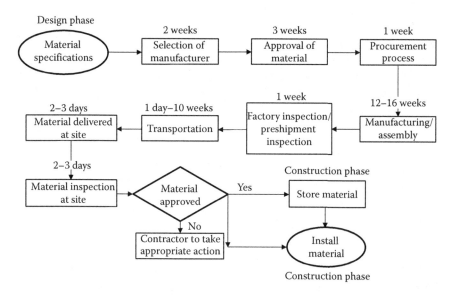

FIGURE 1.26
Supply chain process in construction project.

customer requirements of the specific product/material should be placed in order to ensure smooth supply. Both parties should establish inspection and testing procedure to be mutually agreed upon and to be followed. Supplier and customer should work together to achieve zero-defect policy.

1.12.3.5 Product Handling

Product handling is important to ensure the customer gets the right quantity at the right time. The following points are to be considered for proper handling of the product:

- Proper packing
- Product traceability and verification
- Proper shipping documentation
- Safe and risk-free handling of the product
- Safe loading/unloading
- Shortest and reliable mode of transport to ensure product delivery in time
- Appropriate mode of shipment/transportation
- Maximum/minimum batch size to meet the ordered quantity
- Avoid breakage/damage during handling
- Safety during handling

1.12.3.6 Improvement

For competitive advantage, continual improvement is essential. The supplier has to ensure performance improvement. Customer feedback is necessary to improve the product performance. Supplier has to maintain proper communication with the customer to know the problems or potential problems related to the supplied product and take necessary steps to improve the quality and customer satisfaction.

1.12.4 Supply Chain in Construction Projects

Supply chain management in construction project is managing and optimizing the flow of construction materials, systems, equipment, resources to ensure timely availability of all the construction resources without affecting the progress of works at the site. Figure 1.27 illustrates the supply chain management process in construction projects.

In construction projects, the supply chain management begins from the inception of project. The designer has to consider the following while specifying the products (materials, systems, equipment) for use/installation in the project:

- Quality management system followed by the manufacturer/supplier
- Quality of product
- Reliability of product
- Reliability of manufacturer/supplier
- Durability of product
- Availability of product for entire project requirement
- Price economy/cost efficient

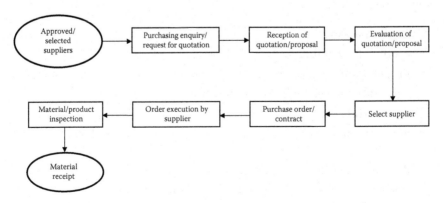

FIGURE 1.27
Purchasing process.

- Sustainability
- Conformance to applicable codes and standards
- Manufacturing time
- Location of the manufacturer/supplier from the project site
- Interchangeability
- Avoid monopolistic product

Product specifications are documented in the construction documents (particular specifications). In certain projects, the documents list the names of recommended manufacturers/suppliers. However, in order of continuous and uninterrupted supply of specified product, the contractor has to consider the following:

- Quality management system followed by the manufacturer/supplier
- Historical rejection/acceptance record
- Reliability of the manufacturer/supplier
- Product certification
- Financial stability
- Proximity to the project site
- Manufacturing/lead time
- Availability of product as per the activity installation/execution schedule
- Manufacturing capacity
- Availability of quantity to meet the project requirements
- Timeliness of delivery
- Location of manufacturer/supplier
- Product cost
- Transportation cost
- Product certification
- Risks in delivery of product
- Responsiveness
- Cooperative and collaborative nature to resolve problem

In major construction projects, the following items are required in bulk quantities:

- Concrete
- Concrete block
- Conduit

- Utility pipes
- Light fixtures
- Electrical devices
- Plumbing fixtures

The contractor can follow just-in-time method to procure these items and avoid large inventory at the project site. The contractor has to select a reliable manufacturer/supplier for these products and to ensure that the manufacturer/supplier is capable of maintaining continuous flow of these items at a short notice. The contractor can sign an agreement with the manufacturer/supplier for an entire project quantity with agreed-upon delivery schedule as per the requirement at site.

In order to ensure supply chain payments are made promptly, the cash flow system should be projected accordingly as the supply chain may be affected due to interruption in payments toward the supply of products.

1.13 Risk Management

Risk management is the process of identifying, assessing, prioritizing different kinds of risks, planning risk mitigation, implementing mitigation plan, and controlling the risks. It is a process of thinking systematically about the possible risks, problems, or disasters before they happen and setting up the procedure that will avoid the risk, or minimize the impact, or cope with the impact. The objectives of project risk management are to increase the probability and impacts of positive events and decrease the probability and impacts of events adverse to the project objectives. Risk is the probability that the occurrence of an event may turn into an undesirable outcome (loss, disaster). It is virtually anything that threatens or limits the ability of an organization to achieve its objectives. It can be unexpected and unpredictable events, which have the potential to damage the functioning of an organization in terms of money or, in the worst-case scenario, may cause the business to close.

1.13.1 Risk Management Process

Risk management process is designed to reduce or eliminate the risk of certain kinds of events happening (occurring) or having an impact on the project. The risk process consists of the following steps:

1. Identify the potential sources of risk on the project.
2. Analyze their impact on the project.
   a. Qualitative
   b. Quantitative

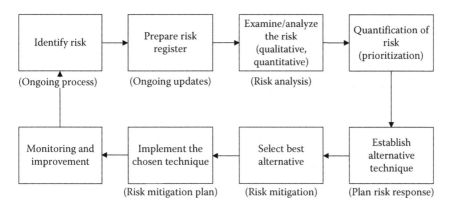

FIGURE 1.28
Risk management cycle. (*Source*: Rumane, A.R. (2013), *Quality Tools for Managing Construction Projects*, CRC Press, Boca Raton, FL. Reprinted with permission from Taylor & Francis Group.)

3. Select those with a significant impact on the project.
    a. Prioritization
4. Determine how the impact of risk can be reduced.
    a. Avoidance
    b. Transfer
    c. Reduction
    d. Retention (acceptance)
5. Select best alternative.
6. Develop and implement mitigation plan.
7. Monitor and control the risks by implementing risk response plan, tracking identified risks, identifying new risks, and evaluating the risk impact.

Figure 1.28 illustrates the risk management cycle (process).

1.13.1.1 Identify Risk

Risk identification involves determining the source and type of risk that may affect the project.
    The following tools and techniques are used to identify risks:

- Benchmarking
- Brainstorming
- Delphi technique
- Interviews
- Past database, historical data from similar projects

- Questionnaires
- Risk breakdown structure
- Workshops

The identified risks are classified into the following:

1. Internal
2. External

These are further divided into the following main categories:

- Management
- Project (contract)
- Technical
- Construction
- Physical
- Logistic
- Health, safety, and environmental
- Statutory/regulatory
- Financial
- Commercial
- Economical
- Political
- Legal
- Natural

Each identified risk is documented in a risk register.

1.13.1.1.1 Risk Register

Risk register is a document recording details of all the identified risks at the beginning of the project and during the life cycle of the project in a format that consists of comprehensive lists of significant risks along with the actions and cost estimated with the identified risks. Risk register is updated every time a new risk is identified or relevant actions are taken. Table 1.21 illustrates an example risk register.

1.13.1.2 Analyze Risk

Risk analysis is the process used to analyze the listed risks. There are two methods of analyzing risks:

1. Qualitative analysis
2. Quantitative analysis

TABLE 1.21

Risk Register

Project Name
Risk Register

Serial Number	Risk Identification Number (Risk ID)	Description of Risk	Owner of Risk	Estimated Likelihood of Risk	Impact	Estimated Severity	Prioritization	List of Activities Influenced	Leading Indicators for Risk	Risk Mitigation Plan	Risk Mitigation Plan on Leading Indicator	Timeline for Mitigation Action	Tracking of Leading Indicators	Date of Review/Update	Forecasting Risk Happenings	Action to be Taken in Future
						SAMPLE FORM										

1.13.1.2.1 Qualitative Analysis

Qualitative analysis is a process that assesses the probability of occurrence (likelihood) of the risk and its impact (consequence).

The following tools and techniques are used for qualitative analysis:

- Failure modes and effects analysis
- Group discussion (workshop)
- Pareto diagram
- Probability and impact assessment
- Probability levels
- Risk categorization

1.13.1.2.2 Quantitative Analysis

Quantitative analysis is a process that quantifies the probability of risk and its impact based on numerical estimation.

The following tools and techniques are used for quantitative analysis:

- Event tree analysis
- Probability analysis
- Sensitivity analysis
- Simulation techniques (Monte Carlo simulation)

Table 1.22 illustrates the risk levels normally assumed for probability impact matrix. The % probability of occurrence shown in the table is indicative. The organization can determine the probability level as per the nature of business.

TABLE 1.22

Risk Probability Levels

Serial Number	Value	Definition	Meaning	% Probability of Occurrence
1	Level 5	Very high (frequent)	• Almost certain that the risk will occur. • Frequency of occurrence is very high.	40–80
2	Level 4	High (likely)	• It is likely to happen. • Frequency of occurrence is less.	20–40
3	Level 3	Moderate (occasional)	• Its occurrence is occasional.	10–20
4	Level 2	Low (unlikely)	• It is unlikely to happen.	5–10
5	Level 1	Very low (rare)	• The probability to occur is rare.	0–5

Source: Rumane, A.R. (2016), *Handbook of Construction Management: Scope, Schedule, and Cost Control*, CRC Press, Boca Raton, FL. Reprinted with permission from Taylor & Francis Group.

Similarly, the risk impact is analyzed for each of the identified risk that may be classified into different levels such as the following:

- Very high
- High
- Substantial
- Possible
- Slight

Based on the impact level, the probability and impact matrix is prepared. Table 1.23 is a sample risk probability and impact matrix.

1.13.1.3 Prioritization

It is the process of prioritizing the list of quantified risks. The results of risk assessment are used to prioritize risks to establish very high to very low ranking. Prioritization of risks depends on the following factors:

1. Probability (occurrence)
2. Impact (consequences)
3. Urgency
4. Proximity
5. Manageability
6. Controllability
7. Responsiveness
8. Variability
9. Ownership ambiguity

The prioritization list helps the project manager to plan actions and assign the resources to mitigate the realization of high value probability.

1.13.1.4 Plan Risk Response

Plan risk response is a process that determines what action (if any) will be taken to address the identified and assessed risks that are listed under Risk Register on prioritization basis. The risk response process is used for developing options and actions to enhance opportunities and reduce the threats to the identified risk activities in the project.

For each identified risk, a response must be identified. The risk owner and project team has to select the risk response for each of the identified risk.

The probability of the risk event occurring and the impacts (threats) is the basis for evaluating the degree to which the response action is to be evolved. Based on the risk probability impact matrix, the risk response strategy for scope, schedule, cost, and quality can be listed as per Table 1.24.

TABLE 1.23

Risk Probability and Impact Matrix

Severity		Probability				
Impact/ Severity		Level 5 Very High (Frequent)	Level 4 High (Likely)	Level 3 Moderate (Occasional)	Level 2 Low (Unlikely)	Level 1 Very Low (Rare)
Level 5	Very High					
Level 4	High					
Level 3	Substantial					
Level 2	Possible					
Level 1	Slight					

Risk Assessment Matrix for: Scope/Schedule/Cost/Quality

TABLE 1.24

Risk Response Strategy

Potential Risk		Probability Level	Impact/Severity Level	Risk Response Strategy
Scope	1			
	2			
Schedule	1			
	2			
		SAMPLE TABLE		
Cost	1			
	2			
Quality	1			
	2			

Generally, risk response strategies for impact (consequences) on the project fall into one of the following categories:

1. Avoidance
2. Transfer
3. Mitigation (reduction)
4. Acceptance (retention)

1.13.1.4.1 *Avoidance*

Avoidance is changing the project scope, objectives, or plan to eliminate the risk or to protect the project objectives from the impact (threat).

1.13.1.4.2 *Transfer*

Transfer is transferring the risk to someone else who will be responsible to manage that risk. Transferring the risk does eliminate the threat, it still exists, however it is owned and managed by an other party.

1.13.1.4.3 *Mitigation*

Mitigation is reduction in the probability and/or impact to an acceptable threshold. It is done by taking a series of control actions.

1.13.1.4.4 *Acceptance*

It is the acceptance of consequences after response actions understanding the risk impact, should it occur.

1.13.1.5 Reduce Risk

Reducing risk involves identifying various steps to reduce the probability and/or impact of the risk. Taking early steps to reduce the probability of risk is more effective and less costly than repairing the damage after the occurrence of the risk.

1.13.1.6 Monitor and Control Risk

It is a systematic process of tracking identified risks, monitoring residual risks, identifying new risks, execution of risk response plan, and evaluating the effectiveness of implementation of actions against established levels of risk in the area of scope, time, cost, and quality throughout the life cycle of the project. It involves timely implementation of risk response to identified risk to ensure the best outcome for a risk to a project.

1.13.2 Risks in Construction Projects

Construction projects have many varying risks. Risk management throughout the life cycle of the project is important and essential to prevent unwanted consequences and effects on the project. Construction projects have involvement of many stakeholders such as project owners, developers, design firms (consultants), contractors, banks, and financial institutions funding the project who are affected by the risk. Each of these parties have involvement with a certain portion of the overall construction project risk; however, the owner has a greater share of risks as the owner is involved from the inception until completion of project and beyond. The owner must take initiatives to develop risk consciousness and awareness among all the parties emphasizing on the importance of explicit consideration of risk at each stage of the project as the owner is ultimately responsible for overall project construction. Traditionally,

1. Owner/client is responsible for the investment/finance risk.
2. Designer (consultant) is responsible for design risk.
3. Contractors and subcontractors are responsible for construction risk.

Construction projects are characterized as very complex projects, where uncertainty comes from various sources. Construction projects involve a cross section of many different participants. They have varying project expectations. Both influence and depend on each other in addition to the "other players" involved in the construction process. The relationships and the contractual groupings of those who are involved are also more complex and contractually varied. Construction projects often require a large amount of materials and physical tools to move or modify these materials. Most items used in construction projects are normally produced by other construction-related industries/manufacturers. Therefore, risk in construction projects is multifaceted. Construction projects inherently contain a high degree of risk in their projection of cost and time as each is unique. No construction project is without any risk. Risk management in construction projects is mainly focused on delivering the project with

1. What was originally accepted (as per **defined scope**)
2. Agreed-upon time (as per **schedule** without any delay)
3. Agreed-upon budget (no overruns to accepted **cost**)

Risk management is an ongoing process. In order to reduce the overall risk in construction projects, the risk assessment (identification, analysis, and evaluation) process must start as early as possible to maximize project benefits. There are a number of risks that can be identified at each stage of the project. Early risk identification can lead to better estimation of the cost in the project budget, whether through contingencies, contractual, or insurance. Risk identification is the most important function in construction projects.

Risk factors in construction projects can be categorized into a number of ways according to the level of details or selected viewpoints. These are categorized based on various risks factors and source of risk. The contractor has to identify related risks affecting the construction, analyze these risks, evaluate the effects on the contract, and evolve the strategy to counter these risks, before bidding for a construction contract. Construction project risks mainly relate to the following:

- Scope and change management
- Schedule/time management
- Budget/cost management
- Quality management
- Resources and manpower management
- Communication management
- Procurement/contract management
- Health, safety, and environmental management

Table 1.25 illustrates typical categories of risks in construction projects.

Quality of construction projects is defined as follows: Construction project quality is fulfillment of owner's needs as per defined scope of works within a budget and specified schedule to satisfy owner's/user's requirements. The phenomenon of these three components can be called the "construction project trilogy." Risk management in all these elements is required to maintain the quality of construction projects. Table 1.26 lists probable risks that occur during the construction phase and its effects on scope, schedule, and cost.

TABLE 1.25

Typical Categories of Risks in Construction Projects

Serial Number	Category	Types
1	Management	Selection of project delivery system
		Selection of project/construction manager
		Selection of designer
		Selection of contractor
2	Contract (project)	Scope/design changes
		Schedule
		Cost
		Conflict resolution
		Delay in changer order negotiations
3	Statutory	Statutory/regulatory delay
4	Technical	Incomplete design
		Incomplete scope of work
		Design changes
		Design mistakes
		Errors and omissions in contract documents
		Incomplete specifications
		Ambiguity in contract documents
		Inconsistency in contract documents
		Inappropriate schedule/plan
		Inappropriate construction method
		Conflict with different trades
		Improper coordination with regulatory authorities
		Inadequate site investigation data
5	Technology	New technology
6	Construction	Delay in mobilization
		Delay in transfer of site
		Different site conditions to the information provided
		Changes in scope of work
		Resource (labor) low productivity
		Equipment/plant productivity
		Insufficient skilled workforce
		Union and labor unrest
		Failure/delay of machinery and equipment
		Quality of material
		Failure/delay of material delivery
		Delay in approval of submittals
		Extensive subcontracting
		Subcontractor's subcontractor
		Failure of project team members to perform as expected
		Information flow breaks

(Continued)

TABLE 1.25 (*Continued*)

Typical Categories of Risks in Construction Projects

Serial Number	Category	Types
7	Physical	Damage to equipment
		Structure collapse
		Damage to stored material
		Leakage of hazardous material
		Theft at site
		Fire at site
8	Logistic	Resources availability
		Spare parts availability
		Consistent fuel supply
		Transportation facility
		Access to work site
		Unfamiliarity with local conditions
9	Health, safety, and environment	Injuries
		Health and safety rules
		Environmental protection rules
		Pollution rules
		Disposal of waste
10	Financial	Inflation
		Recession
		Fluctuations in exchange rate
		Availability of foreign exchange (certain countries)
		Availability of funds
		Delays in payment
		Local taxes
11	Economical	Variation of construction material price
		Sanctions
12	Commercial	Import restrictions
		Custom duties
13	Legal	Permits and licenses
		Professional liability
		Litigation
14	Political	Change in laws and regulations
		Constraints on employment of expatriate workforce
		Use of local agent and firms
		Civil unrest
		War
15	Natural	Flood
		Earthquake
		Cyclone
		Sandstorm
		Landslide
		Heavy rains
		High humidity
		Fire

Source: Rumane, A.R. (2013), *Quality Tools for Managing Construction Projects*, CRC Press, Boca Raton, FL. Reprinted with permission from Taylor & Francis Group.

TABLE 1.26

Potential Risks on Scope, Schedule, and Cost, during Construction Phase, and Its Effects and Mitigation Action

Serial Number	Potential Risk	Probable Effects	Control Measures/Mitigation Action
1.0 Scope			
1.1	Scope/design changes	• Project schedule • Project cost • Claim	• Compress schedule • Resolve change order issues in order not to delay the project
1.2	Different site conditions to the information provided	• Change in Scope of work • Delay in project	• Contractor to investigate site conditions prior to starting the relevant activity
1.3	Inadequate site investigation data	• Additional work • Scope change	• Contractor to investigate site conditions prior to starting the relevant activity
1.4	Conflict in contract documents	• Project delay	• Amicably resolve the issue
1.5	Incomplete design	• Project scope • Project schedule • Project cost	• Raise request for information (RFI) • Resolve issue in accordance with contract documents
1.6	Incomplete scope of work	• Project scope • Project schedule • Project cost	• Raise RFI • Resolve issue in accordance with contract documents
1.7	Design changes	• Project scope • Project schedule • Project cost	• Follow contract documents for change order
1.8	Design mistakes	• Project scope • Project schedule • Project cost	• Raise RFI • Resolve issue in accordance with contract documents
1.9	Errors and omissions in contract documents	• Project scope • Project schedule • Project cost	• Raise RFI • Resolve issue in accordance with contract documents
1.10	Incomplete specifications	• Project scope • Project schedule • Project cost	• Raise RFI • Resolve issue in accordance with contract documents
1.11	Conflict with different trades	• Project delay	• Coordinate with all trades while preparing coordination and composite drawings
1.12	Inappropriate construction method	• Project delay • Claim	• Raise RFI and correct the method statement
1.13	Quality of material	• Project delay	• Locate suppliers having proven record of supplying quality product

(Continued)

TABLE 1.26 (*Continued*)

Potential Risks on Scope, Schedule, and Cost, during Construction Phase, and Its Effects and Mitigation Action

Serial Number	Potential Risk	Probable Effects	Control Measures/Mitigation Action
2.0 Schedule			
2.1	Incompetent sub contractor	• Project delay • Project quality	• Contractor has to monitor the workmanship and work progress.
2.2	Delay in transfer of site	• Project delay	• Contractor to adjust the construction schedule
2.3	Delay in mobilization	• Project delay	• Adjust construction schedule accordingly
2.4	Project schedule	• Project completion	• Compress duration of activities
2.5	Inappropriate schedule/plan	• Project delay	• Contractor to prepare schedule taking into consideration site conditions all the required parameters
2.6	Delay in changer order negotiations	• Project schedule	• Request owner/supervisor/project manager to expedite the negotiations and resolve the issue
2.7	Resource availability (material)	• Project delay	• Contractor to make extensive search
2.8	Resource (labor) low productivity	• Project quality • Project delay	• Contractor to engage competent and skilled labors
2.9	Equipment/plant productivity	• Project delay	• Contractor hire/purchase equipment to meet project productivity requirements
2.10	Insufficient skilled workforce	• Project duration	• Contractor arrange workforce from alternate sources
1.11	Failure/delay of machinery and equipment	• Project delay	• Contractor to plan procurement well in advance
1.12	Failure/delay of material delivery	• Project delay	• Contractor to plan procurement well in advance
2.13	Delay in approval of submittals	• Project delay	• Notify owner/project manager
2.14	Delays in payment	• Project delay • Claim	• Contractor to have contingency plans • Owner to pay as per contract
2.15	Statutory/regulatory delay	• Project delay	• Regular follow-up by the contractor, owner with the regulatory agency

(Continued)

TABLE 1.26 (*Continued*)

Potential Risks on Scope, Schedule, and Cost, during Construction Phase, and Its Effects and Mitigation Action

Serial Number	Potential Risk	Probable Effects	Control Measures/Mitigation Action
3.0 Cost			
3.1	Low bid project cost	• Project quality	• Contractor to try competitive material, improve method statement and higher production rate from its manpower
3.2	Variation in construction material price	• Project quality • Project cost	• Contractor to negotiate with supplier/manufacturer for best price. Contractor to request for change order if applicable as per contract
3.3	Damage to equipment	• Schedule	• Regularly maintain the equipment. Take immediate action to repair damage equipment
3.4	Damage to stored material	• Project delay • Material quality	• Contractor to follow proper storage system
3.5	Structure collapse	• Injuries • Project delays	• Contractor to ensure that formwork and scaffolding is properly installed
3.6	Leakage of hazardous material	• Safety hazards	• Contractor to take necessary protect to avoid leakage. Store in safe area
4.0 General			
4.1	Failure of team members to perform as expected	• Project quality • Project delay	• Select competent candidate. Provide training.
4.2	Change in laws and regulations	• Scope/ specification changes • Variation order	• Contractor to inform owner/ consultant and raise RFI.
4.3	Access to work site	• Extra/ Additional time to access site	• Access road to be planned in coordination with adjacent area and local authority
4.4	Theft at site	• Project delay	• Contractor to monitor access to site. Record entry/exit to the site. Provide fencing around project site.

(*Continued*)

TABLE 1.26 (*Continued*)

Potential Risks on Scope, Schedule, and Cost, during Construction Phase and its Effects and Mitigation Action

Serial Number	Potential Risk	Probable Effects	Control Measures/Mitigation Action
4.5	Fire at site	• Project delay	• Contractor to install temporary firefighting system. Inflammable material to be stored in safe and secured place with necessary safety measures
4.6	Injuries	• Project delay	• Contractor to keep first aid provision at site. Take immediate action to provide medical aid.
4.7	New technology	• Scope change • Schedule • Cost	• Owner/contractor to mutually agree for changes in the contract for better performance of project

Source: Rumane, A.R. (2016), *Handbook of Construction Management: Scope, Schedule, and Cost Control,* CRC Press, Boca Raton, FL. Reprinted with permission from Taylor & Francis Group.

1.14 Six Sigma

1.14.1 Introduction

Six Sigma is, basically, a process quality goal. It is a process quality technique that focuses on reducing variation in the process and preventing deficiencies in the product. In a process that has achieved Six Sigma capability, the variation is small compared to the specification limits.

Sigma is a Greek letter, σ, standing for standard deviation. Standard deviation is a statistical way to describe how much variation exists in a set of data, a group of items, or a process. Standard deviation is the most useful measure of dispersion. Six Sigma means that for a process to be capable at the Six Sigma level, the specification limits should be at least 6σ from the average point. So, the total spread between the upper specification (control) limit and the lower specification (control) limit should be 12σ. With Motorola's Six Sigma program, no more than 3.4 defects per million fall outside the specification limits with a process shift of not more than 1.5σ from the average or mean. Six Sigma started as a defect reduction effort in manufacturing and was then applied to other business processes for the same purpose.

Six Sigma is a measurement of "goodness" using a universal measurement scale. Sigma provides a relative method to measure improvement. Universal means sigma can measure anything from coffee mug defects to missed chances to closing of a sales deal. It simply measures how many times a customer's requirements were not met (a defect), given a million opportunities. Sigma is measured in defects per million opportunities. For example, a level of sigma can indicate how many defective coffee mugs were produced when one million were manufactured. Levels of sigma are associated with improved levels of goodness. To reach a level of Three Sigma, you can only have 66,811 defects, given a million opportunities. A level of Five Sigma only allows 233 defects. Minimizing variation is a key focus of Six Sigma. Variation leads to defects, and defects lead to unhappy customers. To keep customers satisfied, loyal, and coming back, you have to eliminate the sources of variation. Whenever a product is created or a service performed, it needs to be done the same way every time, no matter who is involved. Only then will you truly satisfy the customer. Figure 1.29 illustrates the Six Sigma roadmap.

1.14.2 Six Sigma Methodology

Six Sigma is an overall business improvement methodology that focuses an organization on

- Understanding and managing customer requirements
- Aligning key business process to achieve these requirements
- Utilizing rigorous data analysis to minimize variation in these processes
- Driving rapid and sustainable improvement in the business process by reducing defects, cycle time, impact to the environment, and other undesirable variations
- Timely execution

As a management system, Six Sigma is a high-performance system for executing business strategy. It uses the concept of facts and data to drive better solutions. Six Sigma is a top-down solution to help organizations

- Align their business strategy to critical improvement efforts
- Mobilize teams to attack high-impact projects
- Accelerate improved business results
- Govern efforts to ensure that improvements are sustained

Six Sigma methodology also focuses on

- Leadership principles
- Integrated approach to improvement
- Engaged teams
- Analytic tool
- Hard-coded improvements

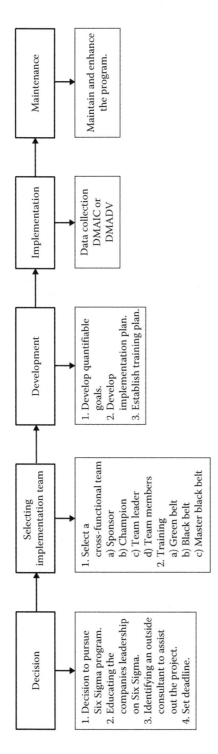

FIGURE 1.29
Six Sigma roadmap.

1.14.2.1 Leadership Principles

The Six Sigma methodology has four leadership principles:

1. Align
2. Mobilize
3. Accelerate
4. Govern

Brief descriptions of these leadership principles follow.

Align: Leadership should ensure that all improvement projects are in line with the organization's strategic goals.

Alignment begins with the leadership team developing a scorecard. This vital tool, the cornerstone of the Six Sigma business improvement campaign, translates strategy into tactical operating terms. The scorecard also defines the metrics an organization can use to determine success. Just as a scoreboard at a sporting event tells you who is winning, the scorecard tells the leadership how well the company is meeting its goals.

Mobilize: Leadership should enable teams to take action by providing clear direction, feasible scope, a definition of success, and rigorous reviews.

Mobilizing sets clear boundaries, lets people go to work, and trains them as required. The key to mobilizing is focus; lack of focused action was one of the downfalls of previous business improvement efforts. True focus means the project is correctly aligned with the organization's scorecard. Mobilized teams have a valid reason for engaging in improvement efforts—they can see benefit for the customer. The project has strategic importance, and they know it. They know exactly what must be done and the criteria they can use to determine success.

Accelerate: Leadership should drive a project to rapid results through tight clock management, training as needed, and shorter deadlines.

More than 70% of all improvement initiatives fail to achieve desired results in time to make a difference. For projects to make an impact, they must achieve results quickly, and that is what acceleration is all about.

The accelerate leadership principle involves three main components:

1. Action learning
2. Clock management
3. Effective planning

Accelerate employs the "action learning" methodology to quickly bridge from "learning" to "doing." Action learning mixes traditional training with direct application. Training is received while working on a real-world project, allowing plenty of opportunity to apply new knowledge. The instructor is not simply a trainer but a coach as well, helping with work on the real-world project. Action learning accelerates improvement over traditional learning methods. It helps in receiving training and also completing a worthwhile project at the same time. In addition to the 4–6 months time frame, accelerate requires teams to set deadlines that are reinforced through rigorous reviews.

Govern: Leadership must visibly sponsor projects and conduct regular and rigorous reviews to make critical midcourse corrections.

Once the leaders select an improvement opportunity, their work is not done. They must remain ultimately responsible for the success of that project. Govern requires leaders to drive for results.

While governing a Six Sigma project, you need

- A regular communications plan and a clear review process
- To actively sponsor teams and their projects
- To encourage proactive dialogue and knowledge sharing in the team and throughout the organization

1.14.2.2 Six Sigma Team

Teamwork is absolutely vital for complex Six Sigma projects. For teams to be effective, they must be engaged—involved, focused, and committed to meeting their goals. Engaged teams must have leadership support. There are four types of teams:

1. Black Belt
2. Green Belt
3. Breakthrough
4. Blitz

These are briefly described next.

1.14.2.2.1 Black Belt

Black Belt teams are led by a Black Belt, and may have Green Belts and functional experts assigned to complex, high-impact process improvement projects or to designing new products, services, or complex processes. Black Belts are internal Six Sigma practitioners, skilled in the application of rigorous statistical methodologies, and they are crucial to the success of Six Sigma. Their additional training and experience provide them with the skills they need to tackle difficult problems. The responsibilities of Black Belts are to

- Function as Team Leader on Black Belt projects
- Integrate their functional discipline with statistical, project, and interpersonal skills
- Serve as internal consultants
- Tackle complex, high-impact improvement opportunities
- Mentor and train Green Belts

1.14.2.2.2 Green Belt

Led by a Green Belt and comprised of nonexperts, Green Belt teams tackle less complex, high-impact process improvement projects. Green Belt teams are often coached by Black Belts or Master Black Belts.

Green Belts are also essential to the success of Six Sigma. They perform many of the same functions as Black Belts, but their work requires less complex analysis. Green Belts are trained in basic problem-solving skills and the statistical tools needed to work effectively as members of process improvement teams.

Green Belt responsibilities include

- Acting as Team Leader on business improvements requiring less complex analysis
- Adding their unique skills and experiences to the team
- Working with the team to come up with inventive solutions
- Performing basic statistical analysis
- Conferring with a Black Belt as questions arise

1.14.2.2.3 Breakthrough

While creating simple processes, sophisticated statistical tools may not be needed. Breakthrough teams are typically used to define low-complexity, new processes.

1.14.2.2.4 Blitz

Blitz teams are put in place to quickly execute improvements produced by other projects. These teams can also implement digitization for efficiency using a new analytic tool set.

For a typical Six Sigma project, four critical roles exist:

1. Sponsor
2. Champion
3. Team Leader
4. Team member

A Sponsor typically

- Remains ultimately accountable for a project's impact
- Provides project resources
- Reviews monthly and quarterly achievements, obstacles, and key actions
- Supports the project Champion by removing barriers as necessary

A Champion typically

- Reviews weekly achievements, obstacles, and key actions
- Meets with the team weekly to discuss progress
- Reacts to changes in critical performance measures as needed
- Supports the Team Leader, removing barriers as necessary
- Helps ensure project alignment

A Team Leader typically

- Leads improvement projects through an assigned, disciplined methodology
- Works with the Champion to develop the Team Charter, review project progress, obtain necessary resources, and remove obstacles
- Identifies and develops key milestones, timelines, and metrics for improvement projects
- Establishes weekly, monthly, and quarterly review plans to monitor team progress
- Supports the work of team members as necessary

Team members typically

- Assist the Team Leader
- Follow a disciplined methodology
- Ensure that the Team Charter and timelines are being met
- Accept and execute assignments
- Add their views, opinions, and ideas

1.14.3 Analytic Tool Sets

The following are the analytic tools used in Six Sigma projects:

1. Ford Global 8D Tool

   D1: Establish the team

   D2: Describe the problem → What problem needs solving?

   → Who should help solve the problem?

   → How do we quantify symptoms?

   D3: Implement and verify containment → How do we contain it?

   D4: Identify and verify root causes → What is the root cause?

   D5: Choose and verify corrective action → What is the permanent corrective action?

D6: Implement and validate permanent corrective action → How do we implement?

D7: Prevent recurrence → How can we prevent this in the future?

D8: Congratulate the team → Who should we reward?

The Ford Global 8D Tool is primarily used to bring performance back to a previous level.

2. DMADV Tool Set Phases

Define → What is important?

Measure → What is needed?

Analyze → How will we fulfill?

Design → How do we build it?

Verify → How do we know it will work?

The DMADV tool is used primarily for the invention and innovation of modified or new products, services, or process. Using this tool set, Black Belts optimize performance before production begins. DMADV is proactive, solving problems before they start. This tool is also called DFSS (Design for Six Sigma).

3. DMAIC Tool

Define → What is important?

Measure → How are we doing?

Analyze → What is wrong?

Improve → What needs to be done?

Control → How do we guarantee performance?

The DMAIC tool refers to a data-driven quality strategy and is used primarily for improvement of an existing product, service, or process.

4. DMADDD Tool

Define → Where must we be leaner?

Measure → What is our baseline?

Analyze → Where can we free capacity and improve yields?

Design → How should we implement?

Digitize → How do we execute?

Drawdown → How do we eliminate parallel paths?

The DMADDD tool is primarily used to drive the cost out of a process by incorporating digitization improvements. These improvements can drive efficiency by identifying nonvalue-added tasks and use simple web-enabled tools to automate certain tasks and improve efficiency. In doing so, employees can be freed to work on more value-added tasks.

Table 1.27 lists the fundamental objectives of DMADV, DMAIC, and DMADDD.

TABLE 1.27

Fundamental Objectives of Six Sigma Tools

Tool	Phase	Fundamental Objective
DMADV		
1	*Define*—What is important?	Define the project goals and customer deliverables (internal and external).
2	*Measure*—What is needed?	Measure and determine customer needs and specifications.
3	*Analyze*—How do we fulfill?	Analyze process options and prioritize based on capabilities to satisfy customer requirements.
4	*Design*—How do we build it?	Design detailed process(es) capable of satisfying customer requirements.
5	*Verify*—How do we know it will work?	Verify design performance capability.
DMAIC		
1	*Define*—What is important?	Define the project goals and customer deliverables (internal and external).
2	*Measure*—How are we doing?	Measure the process to determine current performance.
3	*Analyze*—What is wrong?	Analyze and determine the root cause(es) of the defects.
4	*Improve*—What needs to be done?	Improve the process by permanently removing the defects.
5	*Control*—How do we guarantee performance?	Control the improved process's performance to ensure sustainable results.
DMADDD		
1	*Define*—Where must we be leaner?	Identify potential improvements.
2	*Measure*—What is our baseline?	Analog touch points.
3	*Analyze*—Where can we free capacity and improve yields?	Task elimination and consolidated ops. Value-added/nonvalue-added tasks. Free capacity and yield.
4	*Design*—How should we implement?	Future state vision. Define specific projects. Define drawdown timing. Define commercialization plans.
5	*Drawdown*—How do we eliminate parallel paths?	Commercialize new process. Eliminate parallel path.

1.14.3.1 The DMAIC Process

The majority of the time, Black and Green Belts approach their projects with the DMAIC analytic tool set, driving process performance to never-before-seen levels.

DMAIC has the following fundamental objectives:

1. Define phase: Define the project and customer deliverables.
2. Measure phase: Measure the process performance and determine current performance.
3. Analyze: Collect, analyze, and determine the root causes of variation and process performance.
4. Improve: Improve the process by diminishing defects with alternative remedy.
5. Control: Control improved process performance.

The DMAIC process contains five distinct steps that provide a disciplined approach to improving existing processes and products through the effective integration of project management, problem solving, and statistical tools. Each step has fundamental objectives and a set of key deliverables, so the team member will always know what is expected of him or her and his or her team.

DMAIC stands for the following:

- Define opportunities
- Measure performance
- Analyze opportunity
- Improve performance
- Control performance

1.14.3.1.1 Define Opportunities (What Is Important?)

The objective of this phase is

- To identify and/or validate the improvement opportunities that will achieve the organization's goals and provide the largest payoff, develop the business process, define critical customer requirements, and prepare to function as an effective project team

The key deliverables in this phase include

- Team charter
- Action plan

- Process map
- Quick win opportunities
- Critical customer requirements
- Prepared team

1.14.3.1.2 *Measure Performance (How Are We Doing?)*

The objectives of this phase are

- To identify critical measures that are necessary to evaluate success or failure, meet critical customer requirements, and begin developing a methodology to effectively collect data to measure process performance
- To understand the elements of the Six Sigma calculation and establish baseline sigma for the processes the team is analyzing

The key deliverables in this phase include

- Input, process, and output indicators
- Operational definitions
- Data collection format and plans
- Baseline performance
- Productive team atmosphere

1.14.3.1.3 *Analyze Opportunity (What Is Wrong?)*

The objectives of this phase are

- To stratify and analyze the opportunity to identify a specific problem and define an easily understood problem statement
- To identify and validate the root causes and thus the problem the team is focused on
- To determine true sources of variation and potential failure modes that lead to customer dissatisfaction

The key deliverables in this phase include

- Data analysis
- Validated root causes
- Sources of variation
- Failure modes and effects analysis
- Problem statement
- Potential solutions

1.14.3.1.4 *Improve Performance (What Needs to Be Done?)*

The objectives of this phase are

- To identify, evaluate, and select the right improvement solutions
- To develop a change management approach to assist the organization in adapting to the changes introduced through solution implementation

The key deliverables in this phase include

- Solutions
- Process maps and documentation
- Pilot results
- Implementation milestones
- Improvement impacts and benefits
- Storyboard
- Change plans

1.14.3.1.5 *Control Performance (How Do We Guarantee Performance?)*

The objectives of this phase are

- To understand the importance of planning and executing against the plan and determine the approach to be taken to ensure achievement of the targeted results
- To understand how to disseminate lessons learned, identify replication and standardization opportunities/processes, and develop related plans

The key deliverables in this phase include

- Process control systems
- Standards and procedures
- Training
- Team evaluation
- Change implementation plans
- Potential problem analysis
- Solution results
- Success stories
- Trained associates

- Replication opportunities
- Standardization opportunities

The Six Sigma methodology is not so commonly used in construction projects; however, the DMAIC tool can be applied at various stages in construction projects. These are

1. Detailed design stage—To enhance coordination method in order to reduce repetitive work
2. Construction stage—Preparation of builder's workshop drawings and composite drawings, as it needs much coordination among different trades
3. Construction stage—Preparation of contractor's construction schedule
4. Execution of works

1.14.4 Six Sigma in Construction Projects

The contractor's construction schedule (CCS) is an important document used during the construction phase. It is used to plan, monitor, and control project activities and resources. The document is voluminous and important. It has to be prepared with accuracy in order to follow the work progress without deviation from the milestones set up in the contract documents. Generally, the project interim payment to the contractor is linked to the approval of the CCS. The contractor is not paid unless the CCS is approved by the construction manager/project manager/consultant.

In most cases, contractors experience problems with getting the CCS approved, at the very first submission, from the construction manager/project manager/consultant. It could be rejected if it does not meet the specifications. Therefore, the contractor has to put all effort into collecting relevant data to be fed to develop the CCS.

1.14.4.1 The DMADV Process

The following is an example procedure to develop the CCS using the Six Sigma DMADV analytic tool set. The DMADV method is used primarily for the invention of modified or new products, services, or processes. DMADV stands for

1. Define → What is important?
2. Measure → What is needed?
3. Analyze → How will we fulfill?
4. Design → How do we build?
5. Verify → How do we know it will work?

1.14.4.1.1 Define Phase (What Is Important?)

The objective of this phase is to define the project goals and customer deliverables.

The key deliverables of this phase are

- Establish the goal
- Identify the benefits
- Select project team
- Develop project plan

Goal: Develop CCS using Six Sigma tools.

Benefits: The measurable benefits in adopting this process will result in CCS that will meet all the requirements of the specifications and shall be approved by the construction manager/project manager/consultant at the first submission itself. This will reduce the repetitive work and help implement the schedule right from the early stage of the project.

Selection of team: The team shall consist of

a. Sponsor—project manager
b. Champion—construction manager
c. Team Leader—planning and control manager
d. Team Members—planning engineer, cost engineer, and one representative from each subcontractor

Project plan: Time frame in the form of the Gantt chart shall be prepared to meet the target dates for submitting the CCS.

1.14.4.1.2 Measure Phase (What Is Needed?)

The objective of this phase is to measure and determine customer needs and specifications.

The key deliverable in this phase is

- Identify specification requirements

The following are the requirements listed in most contract documents.

The contractor has to submit the construction schedule in a bar chart time-scaled format to show the sequence and interdependence of activities required for complete performance of all items of work under the contract. The contractor shall use a computerized precedence diagram critical path method (CPM) technique in preparation of CCS. The schedule shall include, but not be limited to, the following:

1. Project site layout.
2. Concise description of the work.

3. Milestones (contractual milestones or constraints).

4. Number of working days.

5. Work breakdown structure (WBS) activities shall consist of all those activities that take time to carry out execution/installation and on which resources are expended.

6. Construction network of project phases (if any), including various subphases.

7. Construction network of the project arrangements (activities) and sequence.

8. Time schedules for various activities in a bar chart format.

9. The minimum work activities to be included in the program shall include items stated in the bill of quantity (BOQ).

10. WBS activities shall consist of all those activities that take time to carry out execution/installation and on which resources are expended.

11. Early and late finish dates.

12. Time schedule for critical path.

13. Schedule text report showing activity, start and finish dates, total float, and relationship with other activities.

14. Summary schedule report showing number of activities, project start, project finish, number of relations, open ends, constraints, and milestone.

15. Total float of each activity.

16. Cost loading.

17. Expected progress cash flow S-curve.

18. Resource-loaded S-curve.

19. Manpower loading.

20. Labor and crew movement and distribution.

21. Resource productivity schedule.

22. The number of hours per shift.

23. Average weekly usage of manpower for each trade.

24. Resource histogram showing the manpower required for different trades per time period for each trade (weekly or monthly).

25. Equipment and machinery loading.

26. Schedule of mobilization and general requirements.

27. Schedule of subcontractors and suppliers' submittal and approval.

28. Schedule of materials submittals and approvals.

29. Schedule of long lead materials.

30. Schedule of procurement.

31. Schedule of shop drawings submittals and approvals.

32. Regulatory/authorities' requirements.

33. Schedule of testing, commissioning, and handover.

34. Expected cash flow for executed work (during progress of work).

1.14.4.1.3 Analyze Phase (How Will We Fulfill?)

The objective of this phase is to analyze process options and prioritize based on capability to satisfy customer requirements.

The key deliverables in this phase are

- Data collection
- Prioritization of data under major variables

Data collection: The objectives of this process are to

1. Identify milestone dates and constraints
2. Identify the project calendar
3. Identify the resource calendar
4. Review contract conditions and technical specifications
5. Identify mobilization requirements
6. Identify project method statement
7. Identify subcontractors/suppliers
8. Identify materials requirements
9. Identify long lead items
10. Identify procurement schedule
11. Identify shop drawing requirements
12. Identify regulatory/authorities' requirements
13. Identify WBS activities using BOQ
14. Relate WBS activities with BOQ and contract drawings
15. Identify zoning/phasing
16. Identify codes for all activities per contract document divisions/ sections per the CSI format
17. Identify the volume of work for each activity
18. Identify the duration/time schedule of each activity
19. Identify early and late finish dates
20. Identify critical activities and its effect on critical path
21. Identify logical relationship

22. Identify sequencing of activities
23. Identify project progress cash flow (work in place)
24. Identify manpower resources with productivity rate
25. Identify equipment and machinery
26. Identify project constraints such as access, logistics, delivery, seasonal, national, safety, existing work flow discontinuity, and proximity of adjacent concurrent work
27. Identify testing, commissioning, and handover requirements
28. Identify special inspection requirements
29. Identify closeout requirements
30. Identify and include items not listed in the specifications but are important for project scheduling
31. Identify suitable software program
32. Identify submittal requirements

Arrangement of data: The generated data can be prioritized in an orderly arrangement under the following major variables:

1. Milestones
2. WBS activities
3. Time schedule
4. General requirements
5. Resources
6. Engineering
7. Cost loading

Figure 1.30 illustrates these variables along with related subvariables arranged in the form of the Ishikawa diagram.

1.14.4.1.4 Design Phase (How Do We Build It?)

The objective of this phase is to design detailed processes capable of satisfying customer requirements.

The key deliverable in this phase is

• Preparation of program using suitable (specified) software program

The project and control manager can prepare the CCS based on the collected data and sequence of activities.

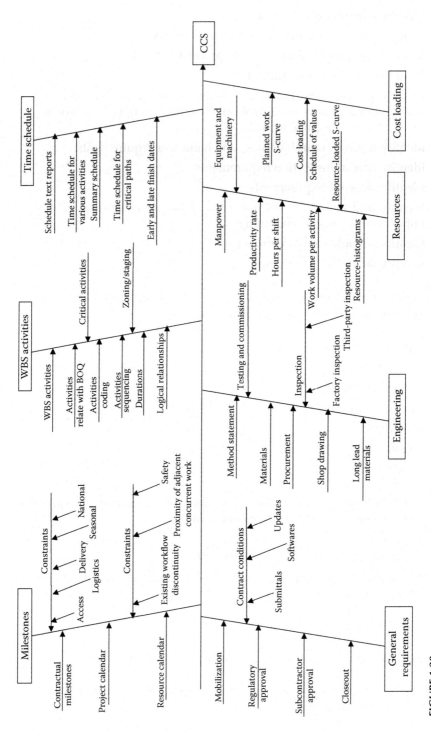

FIGURE 1.30
Ishikawa diagram for CCS data.

1.14.4.1.5 Verify Phase (How Do We Know It Will Work?)
The objective of this phase is to verify design performance capability.

The key deliverables in this phase are

- Review the schedule by the team members to ascertain that all the required elements are included for compliance with specification requirements.
- Submit CCS to construction manager/project manager/consultant.
- Update the schedule as and when required.

1.15 TRIZ

TRIZ is short for *teirija rezhenijia izobretalenksh zadach* ("theory of inventive problem solving"), developed by the Russian scientist Genrich Altshuller. TRIZ provides systematic methods and tools for analysis and innovative problem solving to support the decision-making process.

Continuous and effective quality improvement is critical to the organization's growth, sustainability, and competitiveness. The cost of quality is associated with both chronic and sporadic problems. Engineers are required to identify and analyze the causes and solve these problems by applying various quality improvement tools. Any of these quality tools taken individually does not allow a quality practitioner to carry out the whole problem-solving cycle. These tools are useful for solving a particular phase of the problem and need a combination of various tools and methods to find the solution.

TRIZ is an approach that starts at a point where fresh thinking is needed to develop a new process or redesign an existing one. It focuses on a method for developing ideas to improve a process, get something done, design a new approach, or redesign an existing approach. TRIZ offers a more systematic although still universal approach to problem solving. It has advantages over other problem-solving approaches in terms of time efficiency and low-cost quality improvement solution. The pillar of TRIZ is the realization that contradictions can be methodically resolved through the application of innovative solutions. Altshuller defined an inventive problem as one containing a contradiction. He defined contradiction as a situation where an attempt to improve one feature of the system detracts from another feature.

1.15.1 TRIZ Methodology

Traditional processes for increasing creativity have a major flaw in that their usefulness decreases as the complexity of the problem increases. At times,

TABLE 1.28

Level of Inventiveness

Level	Degree of Inventiveness	% of Solutions	Source of Solution
1	Obvious solution	32	Personal skill
2	Minor improvement	45	Knowledge within existing systems
3	Major improvement	18	Knowledge within the industry
4	New concept	4	Knowledge outside industry and are found in science, not in technology
5	Discovery	1	Outside the confines of scientific knowledge

the trial-and-error method is used in every process, and the number of trials increases with the complexity of the inventive problem. In 1946, Altshuller was determined to improve the inventive process by developing the "science" of creativity, which led to the creation of TRIZ. TRIZ was developed by Altshuller as a result of analysis of many thousands of patents. He reviewed over 200,000 patents looking for problems and how they were solved. He selected 40,000 as representative of inventive solutions, and the rest were classified as direct improvements easily recognized within the system. Altshuller recognized a pattern where some fundamental problems were solved with solutions that were repeatedly used from one patent to another, although the patent subject, applications, and timings varied significantly. He categorized these patterns into five levels of inventiveness. Table 1.28 summarizes Altshuller's findings.

Altshuller noted that, with each succeeding level, the source of the solution required broader knowledge and more solutions to consider before an ideal solution could be found.

TRIZ is a creative thinking process that provides a highly structured approach to generating innovative ideas and solutions for problem solving. It provides tools and methods for use in problem formulation, system analysis, failure analysis, and pattern of system evolution. TRIZ works in contrast to techniques such as brainstorming and aims to create an algorithmic approach to the invention of new systems and refinement of old systems. Using TRIZ requires some training and a good deal of practice.

The TRIZ body of knowledge contains 40 creative principles drawn from the analysis of how complex problems have been solved, such as

- The laws of systems solution
- The algorithm of inventive problem solving
- Substance-field analysis
- 76 standard solutions

1.15.2 Application of TRIZ

Engineers can apply TRIZ for solving the following problems in construction projects:

- Nonavailability of specified material
- Regulatory changes for using certain types of material
- Failure of dewatering system
- Casting of lower grade of concrete to that of specified higher grade
- Collapse of trench during excavation
- Collapse of formwork
- Collapse of roof slab while casting is in progress
- Chiller failure during peak hours in the summer
- Modifying method statement
- Quality auditor can use to develop corrective actions to audit findings

1.15.3 TRIZ Process

Altshuller has recommended four steps to invent new solutions to a problem:

Step 1—Identify the problem
Step 2—Formulate the problem
Step 3—Search for precisely well-solved problem
Step 4—Generate multiple ideas and adopt a solution

The foregoing referred methods are primarily used for low-level problems. To solve more difficult problems, more precise tools are used. These are as follows:

- Algorithm for Inventive Problem Solving
- Separation Principles
- Substance-Field Analysis
- Anticipator Failure Determination
- Direct Product Evaluation

The quality function deployment (QFD) matrix is also used to identify new functions and performance levels to achieve a truly exciting level of quality by eliminating technical bottlenecks at the conceptual stage. QFD may be used to feed data into TRIZ, especially using the "rooftop" to help develop contradictions.

The different schools for TRIZ and individual practitioners have continued to improve and add to the methodology.

1.16 Value Methodology

1.16.1 Value

Value is difficult to define because it is used in a variety of ways. There are seven classes of value that are recognized today:

1. Economic
2. Moral
3. Aesthetic
4. Social
5. Political
6. Religious
7. Judicial

Of all these classes, only economic value can be measured in terms of (hopefully) objective monetary units such as dollars, euros, yen, or dinar. However, economic value is also established through an item's use value (facilities that provide a unit of use, work, or service) and esteem value (properties that make something desirable). We can say that use values cause a product to perform, and esteem values cause it to sell. Use value and esteem value defy precise quantification in monetary terms, so we often resort to multiattribute techniques for evaluating the total value of complex designs and complicated systems or machinery. Accordingly, value can be defined as

$$\text{Value} = \text{Quality}/\text{Cost, or value is quality divided by cost.}$$

It means that, if you can provide better facility to the client with lower price, the value goes up.

1.16.2 Value Engineering

Value engineering (VE) is a technique for evaluating the design of a product/project to assure that essential functions are provided at minimal overall cost to the manufacturer or client/end user. It is a systematic approach to identifying the project's functional objectives with the goal of optimizing design, construction, and future operations. VE is a creative approach to optimizing value. The authors of *Quality in Constructed Projects* by ASCE (p. 239) have described that

> In its broadest sense, project value is determined by the relationship of the worth of the project and its elements to their cost. The following expression summarizes the relationship:
>
> $$\text{Value} \sim \left(\text{Function} + \text{Performance} + \text{Quality}\right)/\left(\text{Cost}\right)$$
>
> The goal of the VE process is to identify alternatives that maximize this value relationship. The owner's requirements are the basis for establishing values for the items in the numerator. For example, an owner who intends to sell a facility within 5 years of its completion may place less value in long-term maintenance characteristics than an owner who intends to keep a facility for decades. Owner requirements directly affect the relative value of aesthetics, reliability, sustainable development, maintainability, operability, construction duration, and other characteristics. Improving the quality of elements related to these characteristics usually increase cost. The goal of VE is to achieve a ratio of quality to cost that is acceptable to and in the best interest of the owner.

1.16.2.1 Objective of Value Engineering

The objective of VE is to provide the required product/project functions at a minimum cost and to achieve equivalent or better performance while maintaining all functional requirements defined by the customer/client. VE necessitates a detailed examination of a product/project function and the cost of each item in addition to a thorough review of product/project specifications. It does this by identifying and eliminating unnecessary cost.

1.16.2.2 Value Engineering Team

VE studies are conducted by a multidisciplinary team that focuses on a clearly defined scope. The VE team consists of specialists from a variety of disciplines (design, manufacturing, marketing, etc., for product evaluation; and owner, design professional, contractor, consultant, and construction management/project management, etc., for construction project) focusing on determining the most effective way to provide high value at an accepted cost to the customer/client end user. While each member of the project team is free to recommend that a VE study be undertaken, it is typically the owner's responsibility to authorize and formally initiate the VE effort.

The success of a VE effort is linked to the experience of the team members. The number and qualifications of VE team members depend on project objectives and the stage of the project life cycle at which VE studies are conducted. VE studies typically involve multiple disciplines. VE team members must be from diverse disciplines and backgrounds and have a range of expertise related to the project's key issues. Each of the team members must have

- Positive attitude and approach
- Technical knowledge related to the product/project
- Higher education
- Professional experience
- Certification from a recognized body as an expert in VE

The VE team is led by a certified value specialist (CVS). SAVE International (The Value Society) manages the certification program for CVS.

1.16.2.3 Approach

The VE approach involves brainstorming such questions as

- Does the item have any design features that are not necessary?
- Can two or more parts be combined together?
- How can we cut down the weight?
- Are these nonstandard parts that can be eliminated?

The key to successful VE is to ask critical questions and seek creative answers. The following are sample questions that should be included in a VE study:

- Are all the functions provided required by the customer?
- Can a less expensive material be used?
- Can the number of different materials used be reduced?
- Can the design be simplified to reduce the type of items?
- Can a product designed for another project be used?
- Are all surfaces and finishes necessary?
- Would product redesign eliminate a quality problem?
- What is the level of technology required to achieve the owner's objectives?

It is important to question everything and not take anything for granted. Creative solutions can be obtained using classical brainstorming or the nominal group technique.

1.16.2.3.1 Classical Brainstorming

Classical brainstorming is the most well-known and often-used technique for idea generation in a short period of time. It is used when creative and original ideas are desired, and is based on the fundamental principles of deferment of judgment and that quantity breeds quality. The following are the rules for successful brainstorming:

1. Criticism is ruled out—no criticism.
2. Freewheeling is welcomed.
3. Quantity is wanted—more ideas are better.
4. Contribution and improvement are sought—participants should speak whatever ideas pop into their minds.
5. Avoid evaluation—no comments on the idea.

A classical brainstorming session has the following basic steps:

1. *Preparation*: During this period, the participants are selected and a preliminary statement of the problem is circulated among them.
2. *Brainstorming*: A warm-up session with simple unrelated problems is conducted, the relevant problem and the rules of brainstorming are presented, and ideas are generated and recorded using agreed-upon checklists/formats and other techniques if necessary.
3. *Evaluation*: The ideas are evaluated relative to the problem.

Generally, a brainstorming group consists of four to seven people, although some suggest a larger group.

1.16.2.3.2 Nominal Group Technique

The nominal group technique (NGT) involves a structural group meeting designed to incorporate individual ideas and judgments into a group consensus. By correctly applying the NGT, it is possible for groups of people (preferably 5–10) to generate alternatives or other ideas for improving the competitiveness of the product/project.

The NGT, when properly applied, draws on the creativity of the individual participants while reducing undesirable effects of most group meetings such as

- The dominance of one or more participants
- The suppression of conflict ideas

The basic format of an NGT session is as follows:

1. Individual silent generation of ideas
2. Individual round-robin feedback and recording the ideas
3. Group's clarification of each idea
4. Individual voting and ranking to prioritize ideas
5. Discussion of group consensus results

The NGT session begins with an explanation of the procedure and a statement of questions, preferably written by the facilitator. The group members are then asked to prepare individual listing of alternatives that they feel are crucial to the survival and health of the organization.

After this phase has been completed, the facilitator calls on each participant, in round-robin fashion, to present one idea from his or her list. Each idea (or opportunity) is then identified in turn and recorded on a flipchart or board by the NGT facilitator, leaving sufficient time between ideas for comments or classification. This process continues until all the opportunities have been recorded, clarified, and displayed for all the members to see. At this point, a voting procedure is used to prioritize the ideas or opportunities. Finally, voting results lead to the development of group consensus on the topic being addressed.

1.16.2.4 Timings of Value Engineering

VE studies benefit project quality at most stages of the project phases. Conducted in the early stages of a project, they tend to provide the greatest benefit, as there is more potential for cost saving. Figure 1.31 illustrates the

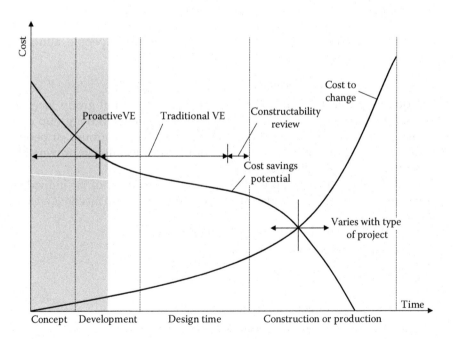

FIGURE 1.31
Effectiveness of VE. (*Source*: Quality in the Constructed Project (2000) by ASCE. Reprinted with permission from ASCE.)

effectiveness of VE. In fact, in recent years, VE has become an aid to owners in the formulation of goals and objectives before most design work begins.

VE studies for construction projects are conducted at one or more of the following project stages:

- Conclusion of concept design or facilities planning
- 30% design completion (preliminary design)
- 60% design completion
- 90% or 100% design completion (detail design)

VE enhances project quality at every stage of the project phase. In construction projects, VE starts during the conceptual phase.

Concept design: VE performed during the conceptual design phase improves the quality of project design. During this phase, VE involves the investigation of alternatives, design concepts with the client/end user, design professional, project or construction manager, and any other project team member involved in this phase. The project team works to identify the client/end user's requirements and to design concepts with the potential to achieve the owner's objectives.

Preliminary design: Major changes resulting from VE studies conducted during this phase can be implemented without significant impacts to the project schedule or budget. The potential for VE-related cost savings is quite substantial during this phase. VE studies conducted during this phase conform to the owner's/end user's goals and objectives.

Detailed engineering: VE studies conducted during this phase (60% completion or more) focus on detail design and construction methods and the constructability of the project. Table 1.29 illustrates areas of VE study by design stage.

Construction: VE recommendations made during the construction phase are called value engineering change proposals. These usually relate to alternatives/substitutes of material, equipment, and construction methods normally proposed by the contractor. Savings made due to VE application during this phase are adjusted per agreed-upon terms of contract.

Figure 1.32 illustrates a schematic for review of a project on target cost basis. Target cost is the amount the owner would like to spend to build the facility. In construction projects, the target-based approach can be applied for the development of the facility from the investment point of view, having a fixed budget allocation. The design is reviewed keeping in mind the budget price of the facility to ensure that the facility shall be functionally qualitative and constructed within the targeted budget. If the total estimated cost is greater than the target cost, then the design must be fed back into the VE process.

TABLE 1.29

Area of VE Area of Value Study by Design Stage

Area of Study	Conceptual	Schematic (Preliminary)	Design Development (Detailed Design)
General Project Budget Layout Criteria and standards	• Design concepts • Program interpretation • Site/facility massing • Access, circulation • Design intentions • Net-to-gross ratios	• Schematic floor plans • Schematic sections • Approach to systems integration • Floor-to-floor height • Functional space	• Floor plans • Sections • Typical details • Integrated systems • Space circulation • Specifications
Structural Foundation Substructure Superstructure	• Performance requirements • Structural bay sizing • Framing systems exploration • Subsurface conditions • Underground concepts • Initial framing review • Structural load criteria	• Schematic basement plan • Selection of foundation systems • Structural system selection • Framing plan outline • Sizing of elements	• Basement floor plan • Key foundation elements details • Floor and roof framing plans • Sizing of major elements • Outline specifications
Architectural Exterior closure Roofing Interior construction Elevators Equipment	• Approach to elevation views to/from building • Roof types and pitch • Interior design • Configuration of key rooms • Organization of circulation schemes • Need and type of vertical circulation • Impact of key equipment on facility and site • Passive solar usage	• Concept elaboration • Selection of wall systems • Schematic elevations • Selection of partitions • Circulation sizing • Basic elevator and vertical transportation concepts • Impact of key equipments on room design	• Elevations • Key elevation details • Key roofing details • Initial finish schedule • Interior construction elements • Integration of structural framing • Key interior elevations • Equipment items
Mechanical HVAC Plumbing Fire protection	• Basic energy concepts • Impact of mechanical concepts on facility • Initial systems selection • Source allocation • Performance requirements for plumbing, HVAC, fire protection	• Mechanical systems selection • Refinement of services and distribution concepts • Input to schematic plans • Energy conservation	• Detailed system selection • Initial system drawings and key details • Distribution and riser diagrams • Outline specification for system design

(Continued)

TABLE 1.29 (*Continued*)

Area of VE Area of Value Study by Design Stage

Area of Study	Conceptual	Schematic (Preliminary)	Design Development (Detailed Design)
Electrical Service and distribution Lighting and power	• Basic power supply • Approach to use of natural and artificial lighting • Performance requirements for lighting • Need for special electrical systems	• Windows/ skylight design and sizing • Selection of lighting and electrical systems • General service, power and distribution concepts	• Detailed systems selection • Distribution diagrams • Key space • Lighting layouts • Outline specification for electrical elements
Site Preparation Utilities Landscaping	• Site selection • Site development criteria • Site forms and massing • Requirements for access • Views to/from facility • Utility supply • Site drawings	• Design concept elaboration • Initial site plan • Schematic planting, grading, paving plans	• Site plan • Planting plan • Typical site details • Outline specifications for site materials

Source: Dell'Isola, A.J. (1997), *Value Engineering: Practical Applications for Design, Construction, Maintenance and Operations.* Reprinted with permission of RSMeans.

1.16.2.5 Stages of Value Engineering Study

Figure 1.33 illustrates the VE process summary.
VE studies typically consist of three sequential stages:

1. Preparation
2. Workshop (execution of the VE job plan)
3. Post workshop

Preparation: This is considered by many to be the most important step of the VE study. During this phase, the owner's need is defined, VE team members are selected, they are informed about the project, and the scope of the study is established.

Workshop: The VE team convenes a workshop that normally has a six-phase job plan:

1. Gathering information
2. Functional analysis
3. Creating alternatives

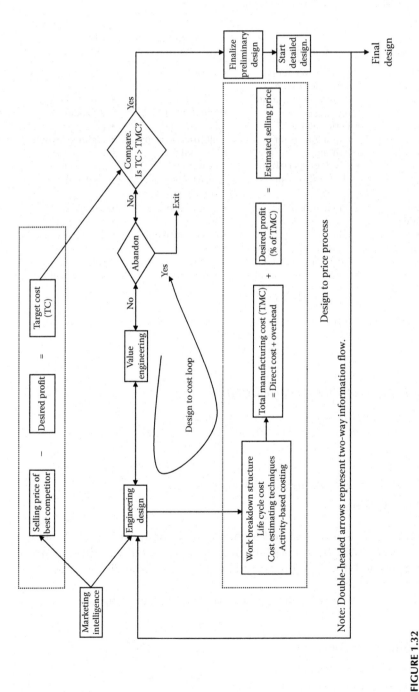

FIGURE 1.32
Schematic for target costing. (*Source:* Sullivan, W.G., Wicks, E.M., and Luxhoj, J.T. (2003), *Engineering Economy,* Reprinted with permission from Pearson Education.)

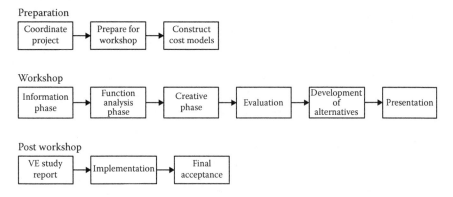

FIGURE 1.33
VE summary process. (*Source*: Quality in the Constructed Project (2000) by ASCE. Reprinted with permission from ASCE.)

4. Evaluation

5. Development

6. Presentation

Post workshop: After the workshop, the VE team prepares a report to support its findings and the implementation program. Its recommendations are reviewed by the owner and the design team before implementation.

Figure 1.34 illustrates the activities performed during the VE study process.

1.16.2.6 Function Analysis System Technique (FAST)

FAST is a technique that specifically illustrates the relationship of all functions within a specific project utilizing a "how–why" logic pattern based on intuitive logic. It is a helpful diagnostic tool, asking questions that generate basic information about project functions. The FAST diagram is especially helpful if the VE team is uncertain of project goals. The following are types of FAST diagrams: classic, customer/user, and technical. Figure 1.35 illustrates a typical FAST concept diagram.

1.16.2.7 The Benefits of Value Engineering

The benefits of VE usually come from improvements in the efficiency of the project delivery system, refinements to specific features, or the development of new approaches to achieving the owner's requirements. The first item most owners look at when they consider VE is potential cost savings. However, VE focuses on improving the relationship of function, performance, and quality to cost—not merely cutting cost.

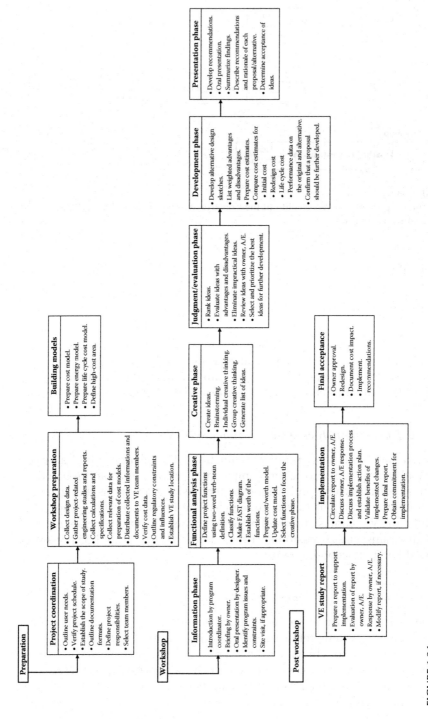

FIGURE 1.34
VE study process activities.

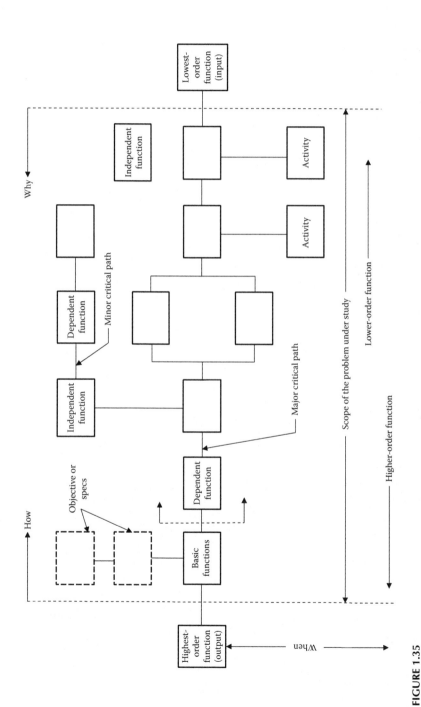

FIGURE 1.35
FAST model "classic." (*Source:* SAVE International-Web site: www.value-eng.org. Reprinted with permission from SAVE International.)

The benefits of VE in construction are

- Reduce project construction cost
- Improve project schedule
- Simplify procedures
- Increase procurement efficiency
- Use resources more effectively
- Decrease operation and maintenance cost

2

Quality Management System

2.1 Introduction

A quality system is a framework for quality management. It embraces the organizational structure, procedure, and processes needed to implement quality management. The quality system should address everything in the organization related to quality of product, services, processes, and operations. The adequacy of the quality system and the quality of products, services, processes, operations, and customer satisfaction are judged by their compliance to specified/relevant standards. Standards have important economic and social repercussions. They are useful to industrial and business organizations of all types, to government and other regulatory bodies, to conformity assessment professionals, suppliers and customers of products and services in both the public and private sectors, and to people in general in their role as customers and users. Standards provide governments with a technical base for health, safety, and environmental legislation.

2.2 Quality Standards

A standard is simply a definition of how something should be. According to Pyzdek (1999):

> Standards are documents used to define acceptable conditions or behaviors and to provide a base line for assuring that conditions or behaviors meet the acceptable criteria. In most cases standards define minimum criteria; world class quality is, by definition, beyond the standard level of performance. Standards can be written or unwritten, voluntary or mandatory. Unwritten quality standards are generally not acceptable. (p. 2)

Pyzdek (1999) further states that quality standards serve the following purpose:

> Standards educate—They set forth ideals or goals for the guidance of manufacturers and users alike. They are invaluable to the manufacturer who wishes to enter a new field and to the naïve purchaser who wants to buy a new product.
>
> Standards simplify—They reduce the number of sizes, the variety of process, the amount of stock, and the paperwork that largely accounts for the overhead costs of making and selling.
>
> Standards conserve—By making possible large-scale production of standard designs, they encourage better tooling, more careful design, and more precise controls, and thereby reduce the production of defective and surplus pieces. Standards also benefit the user through lower costs.
>
> Standards provide a base upon which to certify—They serve as hallmarks of quality which are of inestimable value to the advertiser who points to proven values, and to the buyer who sees the accredited trademark, nameplate, or label. (p. 3)

Standards are used to ensure that a product, system, or service measures up to its specifications and is safe for use. Standards are the key to any conformity assessment activity.

The International Organization for Standardization (ISO) has given the importance of standards as follows:

> Standards make an enormous contribution to most aspects of our lives. Standards ensure desirable characteristics of products and services such as quality, environmental fitness, safety, reliability, efficiency, and interchangeability and at an economical cost.
>
> When products and services meet our expectations, we tend to take this for granted and remain unaware of the role of standards. However, when standards are absent, we soon take notice. We care when products turn out to be of poor quality, do not fit, are incompatible with equipment that we already have, and are unreliable or dangerous. When products, systems, machinery and devices work well and safely, it is often because they meet standards.

Standard setting is one of the first issues in developing a quality assurance system, and increasingly organizations are relying on readily available standards rather than developing their own. Each standard should be

- Clearly written in simple language that is unambiguous
- Convenient in understanding
- Specific in setting out precisely what is expected
- Measurable so that the organization can know whether it is being met

- Achievable, that is, the organization must have the resources available to meet the standard
- Constructible

Chung (1999) defines standards as a reference base that is required to judge the adequacy of a quality system. He further states that a "quality system has to cover all the activities leading to the finished product. Depending on the scope of operation of the organization, these activities include planning, design, development, purchasing, production, inspection, storage, delivery, and after-sales service" (p. 14).

2.3 Standards Organizations

There are many organizations that produce standards; some of the best-known organizations in the quality field are

1. International Organization for Standardization (ISO)
2. International Electrotechnical Commission (IEC)
3. American Society for Quality (ASQ)
4. American National Standards Institute (ANSI)
5. American Society for Testing and Materials (ASTM)
6. American Society of Mechanical Engineers (SME)
7. American Society for Heating, Refrigerating, and Air-Conditioning Engineers (ASHRAE)
8. National Fire Protection Association (NFPA)
9. Institute of Electrical and Electronic Engineers (IEEE)
10. European Committee for Standardization (CEN)
11. European Committee for Electrotechnical Standardization (CENELEC)
12. British Standards Institution (BSI)

Standards produced by these organizations/institutes are recognized worldwide. These standards are referred in the contract documents by the designers to specify products or systems or services to be used in a project. They are also used to specify the installation method to be followed or the fabrication works to be performed during the construction process. Table 2.1 lists some of the most common standards referred to in the particular specifications of building construction projects.

Apart from these, there have been many other national and international quality system standards. These various standards have commonalities and

TABLE 2.1

Most Common Standards Used in Building Construction Projects

Sr. No.	Related Division	Section	Related Standards
1	Concrete	Reinforcement	BS 4449, ASTM A615/A615M
		Cement	ASTM C150, ASTM C295, ASTM C33
		Concrete	ASTM C94, ACI 301, ACI 117
2	Masonry	Concrete Masonry Units	ASTM C140, ASTM C1314, ASTM C270, ASTM C1019
3	Metals	Material Fabrication	ASTM A6/A6M, ASTM A36, ASTM 490, AISC 89
4	Wood and Plastic	Material, Treatment, Panelling	BSEN 942, BSEN 636, BS 1203, BS 1088, BS 4079, BS 1282, BSEN 301, BSEN 302, ASTM E84, AWI Quality Standards
5	Thermal and Moisture Protection	Liquid Waterproofing, Bituminous Waterproofing, Membrane Waterproofing	ASTM C836, D412, D570, D903, ASTM D1187, ASTM D312, ASTM D1227, ASTM D4479
		Building Insulation	ASTM C558, ASTM C578, ASTM C612, ASTM 665
6	Doors and Windows	Steel Doors	SDI 105, SDI 108, SDI 111
		Aluminium Doors	AAMA 101, AAMA 603
		Wooden Doors	NWWDA-ISI –A
		Curtain Wall	AAMA 101, ASTM E283
		Glazing	AAMATIR A7, ANSIZ 971
7	Finishes	Gypsum Plaster	ASTM C11, ASTM E119, ASTM E90, ASTM E413
		Cement Plaster	ASTM C150, ASTM C260, ASTM C897, ASTM C926, ASTM C932, ASTM C1063, ASTM E488, ASTM A641M, ASTM A653M, ASTM C847
		Tiling	ANSI A137.1
		Acoustic Ceiling	ASTM E1264, ASTM E795, ASTM E1264, ASTM C635, ASTM E488, ASTM B633, ASTM A641
		Metallic Ceiling	ASTM B209, ASTM 591, ASTM E1264, ASTM E795, ASTM C635, ASTM C636, ASTM A641, ASTM A653
8	Specialities	Partitions	
		Landscape	
9	Conveying Systems	Elevators, Escalators	EN 81, 95/96/EC, 89/336/EEC, 89/106/EEC, ANSI/ASME/A17.1, BS 5655/BSCP 407
10	Fire Suppression	Fire Fighting System	NFPA 10, NFPA 13, NFPA 14, B 5274
		Fire Fighting Pumps	NFPA 20

(Continued)

TABLE 2.1 (*Continued*)

Most Common Standards Used in Building Construction Projects

Sr. No.	Related Division	Section	Related Standards
11	Plumbing	Plumbing Piping	ISO 2531, BS 4660, BS 4514, ASTM B88, ASTM B 306, ASTM A6.A8, ASTAM 16.22, API 600, ASTM A-74, ASTM A-53, ASTM A-106, ANSI B 16.3, ASTM D1785, ASTM D2729, ASTM D2665, ISO R161
		Hydronic Piping	ASTM B 16.9, ASTM 16.3
		Hydronic Pumps	DIN 24255, EN 733, ISO 2858, ISO 5199, BS 3829, BS 5257, BS 3456, ANSI 316
		Plumbing Equipment	ASTM B16.9, ASTM B16.3, BS 3829, BS 5257, BS 3456, ANI 316
		Water Distribution Pumps	UL 778
12	HVAC	Chillers	ASHRAE 15, ARI, UL 465
		Chiller Piping	ANSI/ASME B-31.1
		Chilled Water Pumps	DIN 24255, EN 733, ISO 2858, ISO 5199, BS 3829, BS 5257, BS 3456, ANSI 316
		Cooling Towers	CTI 201
		Air Handling Unit	EN 1216, ARI
		Fan Coil Units	ARI 410, ARI 440
		Fans and Ventilators	UL 705
13	Automation	Building Automation Systems	
14	Electrical	Conduit, Trunking, Cable Tray, Raceways	BS 4568, BS 46O7, BS 6099, BS 4678, BS 1449, BS 1494, BS 729 (for galv.), BS 21767, BS 31, BS 731
		Wiring Accessories	BS 3676, BS 1363, BS 4343, BS 5420, BS 4662, BS 5733, BS 5419, BS 546, BS 7001, IEC 309
		Wires and Cables	BS 6387, BS 6724, BS 6207, BS 7629, BS 4533, BS 6007, BS 6500, IEC 502, IEC 331, BS 6234, BS 6346, IEC 227, IEC 540
		Bus Duct	IEC 439
		Structured Cabling	ANSI EIA/TIA-568, ANSI EIA/TIA-569, ANSI EIA/TIA 606, ANSI EIA/TIA 607, IEC 11801, EN 50173, IEEE 802.3ab, IEEE 802.3z
		Lighting Fixtures	IEC 598. BS 4533, BS 5042, BS 1853, BS 4782

(Continued)

TABLE 2.1 (*Continued*)

Most Common Standards Used in Building Construction Projects

Sr. No.	Related Division	Section	Related Standards
		Grounding, Lightning Protection	BS 6651, EN 62305. BS 7430
		Main Low Tension Panels, Switch Boards, Distribution Boards, Motor Control Centers	IEC 185, IEC 186, IEC 269, IEC 292, IEC 439, IEC 529, IEC 947, BS 5486, IEC 5750, BS 5420, BS 159, BS 37, IEC 1008
		Motor Starters	IEC 292
		Disconnect Switches	BS 5419, AC 23, BS88, IEC 408
		Emergency Power Supply	ISO 8528, NFPA 37, NFPA 99, NFPA 101, NFPA 110, BS 5514, BS 4999, BS 649, BS 5000
		Automatic Transfer Switch	IEC 947, UL 508, UL 1008, NFPA 70, NFPA 99, NFPA 110, IEEE 446, NEMA Standard ICS 10
		Fire Alarm System	NFPA 70, NFPA 71, NFPA 72, BS 5839
15	Communication	Communication System	ISO/IEC 11801, EN 50173, EIA/TIA 568, EIA/TIA 569

historical linkage. However, in order to facilitate international trade, delegates from 25 countries met in London in 1946 to create a new international organization. The objective of this organization was to facilitate international coordination and unification of industrial standards. The new organization, International Organization for Standardization, ISO, officially began operation on February 23, 1947.

2.4 International Organization for Standardization

ISO is an independent nongovernmental organization with membership of 161 (as of April 2016) national standards bodies/institutes, formed on the basis of one member per country, with a Central Secretariat in Geneva, Switzerland, that coordinates the system.

ISO is the world's largest developer and publisher of international standards. It is a nongovernmental organization that forms a bridge between the public and private sectors. ISO has more than 21,000 international standards. Of all the standards produced by ISO, the ones that are most widely known are the ISO 9000 and ISO 14000 series. ISO 9000 has become an international reference for quality requirements in business-to-business dealings, and ISO 14000 looks to achieve at least as much, if not more, in helping organizations to meet their environmental changes. ISO 9000 and ISO 14000 families are known as "generic management system standards."

The ISO 9000 is a series, or family of standards, primarily concerned with "quality management." This means what the organization does to

- Fulfill the customer's quality requirements
- Fulfill applicable regulatory requirements, while aiming to enhance customer satisfaction
- Achieve continual improvement of its performance in pursuit of the objectives

The ISO 9000 family addresses various aspects of quality management and contains some of ISO's best known standards. The standards provide guidance and tools for companies and organizations who want to ensure that their products and services consistently meet customer's requirements, and that quality is consistently improved.

The ISO 14000 family is primarily concerned with "environmental management." This means to

- Minimize harmful effect on the environment caused by its activities
- Achieve continual improvement with its environmental performance

ISO standards are updated periodically since they were originally published in 1987. ISO 9000 actually comprises several standards.

The 1994 version of the ISO 9000 series was made up of the following standards:

ISO 9000: Quality management and quality assurance standards

ISO 9001: Quality systems—Model for quality assurance in design, development, production, installation, and servicing

ISO 9002: Quality systems—Model for quality assurance in production installation and servicing

ISO 9003: Quality systems—Model for quality assurance in final inspection and test

ISO 9004: Quality management and quality systems element guidelines

ISO 9001:1994 standards had 20 key elements. These standards were simplified in December 2000 and are known as ISO 9000:2000. In the revised version, ISO 9002 and ISO 9003 no longer exist, having been incorporated into ISO 9001:2000.

ISO 9000:2000 consists of the following standards:

ISO 9000: Quality systems: Fundamentals and vocabulary

ISO 9001: Quality management systems: Requirements

ISO 9004: Quality management systems: Guidelines for performance improvement

ISO 9000: 2000 Standards are based on a process model that any enterprise can use. Instead of 20 elements in ISO 9001: 1994, the requirements were grouped into five sections. These are

1. Section 4—Quality management system
2. Section 5—Management responsibility
3. Section 6—Resource management
4. Section 7—Product realization
5. Section 8—Measurement, analysis, and improvement

Table 2.2 illustrates the correlation between ISO 9001:1994 and ISO 9000:2000.

ISO 9000:2000 specifies requirements for a quality management system for any organization that needs to demonstrate its ability to consistently provide products that meet customer and applicable regulatory requirements and to enhance customer satisfaction.

In keeping with the process of updating the standards, certain clauses of ISO 9001:2000 of the quality management system were amended during 2008 in order to improve the quality management system, and accordingly the amended standard is known as ISO 9001:2008. ISO 9001:2008 includes Annex B, which outlines the text changes that have been made to specific clauses.

ISO 9001:2008 was revised by various committees, societies, and institutes, and ISO 9001:2015 was published in September 2015. It has the following clauses:

1. Context of Organization/Quality Management System
2. Leadership
3. Planning for Quality Management System
4. Support
5. Operation
6. Performance Evaluation
7. Improvement

Changes in ISO 9001:2015 are an opportunity to revisit organizational areas that yet need to be improved. An awareness of the upcoming changes in ISO 9001:2015 will enable quality professionals to better prepare for the

TABLE 2.2

Correlation between IS0 9000:1994 and ISO 9000:2000

ISO 9001 Clause Number	ISO 9000:1994 Quality System Element	ISO 9000:2000 Section	ISO 9000:2000 Clauses
4.1	Management Responsibility	5	**Management Responsibility**
		5.1	Management commitment
		5.3	Quality policy
		5.5	Responsibility, Authority, and Control
		5.6	Management Review
4.2	Quality System	4	**Quality Management System**
		4.1	General Requirements
		4.2.1	Documentation requirements (General)
4.3	Contract Review	5&7	**Management Responsibility**
			Product Realization
		5.2	Customer focus
		7.2	Customer-related process
4.4	Design Control	7	**Product Realization**
		7.3	Design and development
4.5	Document and Data Control	4	**Quality Management System**
		4.2.3	Control of documents
4.6	Purchasing	7	**Product Realization**
		7.4.1	Purchasing process
		7.4.2	Purchasing information
		7.4.3	Verification of purchased product
4.7	Control of Customer Supplied Product	7	**Product Realization**
		7.5.4	Customer property
4.8	Product Identification and Traceability	7	**Product Realization**
		7.5.3	Identification and Traceability
4.9	Process Control	5,6,&7	**Management Responsibility**
			Resource Management
			Product Realization
		6.3	Infrastructure
		6.4	Work environment
		7.5.1	Control of production and service provision
		7.5.2	Validation of process for production and service provision
4.10	Inspection and Testing	4,7,&8	**Quality Management System**
			Product Realization

(Continued)

TABLE 2.2 (*Continued*)

Correlation between IS0 9000:1994 and ISO 9000:2000

ISO 9001 Clause Number	ISO 9000:1994 Quality System Element	ISO 9000:2000 Section	ISO 9000:2000 Clauses
			Measurement, Analysis, and Improvement
		7.4.3	Verification of purchased product
		8.2.4	Monitoring and measurement of product
		7.5.3	Identification and Traceability
4.11	Control of Inspection, Measurement, and Test Equipment	7	**Product Realization**
		7.6	Control of monitoring and measuring devices
4.12	Inspection and Test Status	7	**Product Realization**
		7.5.3	Identification and Traceability
4.13	Control of Non Conforming Product	8	**Measurement, Analysis, and Improvement**
		8.3	Control of non conforming product
4.14	Corrective and Preventive Action	8	**Measurement, Analysis, and Improvement**
		8.5.2	Corrective action
		8.5.3	Preventive action
4.15	Handling, Storage, Packing, Preservation and Delivery	7	**Product Realization**
		7.5.5	Preservation of product
4.16	Control of Quality Records	4	**Quality Management System**
		4.2.4	Control records
4.17	Internal Quality Audits	8	**Measurement, Analysis, and Improvement**
		8.2.1	Internal audit
4.18	Training	6	**Resource Management**
		6.2.2	Competence, awareness, and training
4.19	Servicing	7	**Product Realization**
		7.5.1	Control of production and service provision
4.20	Statistical Techniques	8	**Measurement, Analysis, and Improvement**
		8.1	General
		8.2.3	Measurement and monitoring of process
		8.4	Analysis of data

future. The change is to incorporate risk-based thinking into the management system by considering the context of the organization. In other words, all processes are not equal for all organizations with some being more critical than others, resulting in different levels of risk. Table 2.3 lists the clauses of ISO 9001:2015 and ISO 9001:2008 and their correlations.

The tremendous impact of ISO 9001 and ISO 14001 on organizational practices and on trade stimulated the development of other ISO standards and deliverables that adapt the generic management system to specific sectors or aspects. These are

1. Food Safety Management Systems ISO 22000
2. Information Security Management Systems ISO 27001
3. Supply Chain Security Management Systems ISO 28000

ISO 22000:2005, published on September 1, 2005, is related to the safe food supply management system to ensure that food is safe at the time of human consumption. ISO 27001:2005 is related to information security system. ISO 28000:2005 is related to supply management system to help combat threats to safe and smooth flow of international trade.

2.5 ISO 9000 Quality Management System

ISO 9000 quality system standards are a tested framework for taking a systematic approach to managing the business process so that organizations turn out products or services conforming to customer's satisfaction. The typical ISO quality management system is structured on four levels, usually portrayed as a pyramid. Figure 2.1 illustrates this.

On top of the pyramid is the quality policy, which sets out what management requires its staff to do in order to ensure quality management system. Underneath the policy is the quality manual, which details the work to be done. Beneath the quality manual are work instructions or procedures. The number of manuals containing work instructions or procedures is determined by the size and complexity of the organization. The procedures mainly discuss the following:

- What is to be done?
- How is it done?
- How does one know that it has been done properly (e.g., by inspecting, testing, or measuring)?
- What is to be done if there are problems (e.g., failure)?

TABLE 2.3

Correlation between ISO 9001:2015 and ISO 9001:2008

CLAUSE	ISO 9001:2015	Correlation between ISO 9001:2015 to ISO 9001:2008	CLAUSE	ISO 9001:2008
1.0	**Scope**		1.0	**Scope**
	1.1 General			1.1 General
	1.2 All exclusions from ISO 9001:2008 Clause 1.2 removed			1.2 Application
2.0	**Not valid**		2.0	**Normative references**
3.0	**Included into primary standard, ISO 9000**		3.0	**Terms and Definitions**
4.0	**Context of the Organization/Quality Management System**	1.0	4.0	**Quality Management System**
	4.1 Understanding the Organization and its Context	1.1		4.1 General requirements
	4.2 Understanding the needs and expectations of both the parties	1.1		4.2 Documentation requirements
				4.2.1 General
				4.2.2 Quality manual
				4.2.3 Control of documents
				4.2.4 Control of records
	4.3 Determine the scope of Quality Management System	1.2,4.2.2		
	4.4 Quality Management System and its processes	4.0,4.1,4.2.2		

(Continued)

TABLE 2.3 (Continued)

Correlation between ISO 9001:2015 and ISO 9001:2008

CLAUSE	ISO 9001:2015	Correlation between ISO 9001:2015 to ISO 9001:2008	CLAUSE	ISO 9001:2008
5.0	**Leadership**	**5.0,5.5**	**5.0**	**Management Responsibility**
5.1	Leadership and Commitment	5.1	5.1	Management commitment
	5.1.1 Leadership and Commitment to Quality Management System	5.1		
	5.1.2 Customer Focus	5.2	5.2	Customer focus
5.2	Quality policy	5.3	5.3	Quality policy
	5.2.1 Requirements of top management in respect of quality policy		5.4	Planning
	5.2.2 Specific requirements in respect of organization's policy		5.4.1	Quality objective
			5.4.2	Quality management system planning
5.3	Organizational roles, responsibilities and authorities	5.5.1,5.5.2	5.5	Responsibility, authority and communication
			5.5.1	Responsibility and authority
			5.5.2	Management perspective
			5.5.3	Internal communication

(Continued)

TABLE 2.3 (Continued)
Correlation between ISO 9001:2015 and ISO 9001:2008

CLAUSE	ISO 9001:2015	Correlation between ISO 9001:2015 to ISO 9001:2008	CLAUSE	ISO 9001:2008
			5.6	Management review
			5.6.1	General
			5.6.2	Review input
			5.6.3	Review output
6.0	**Planning for Quality Management System**		6.0	**Resource Management**
6.1	Actions to address risks and opportunities	5.4,5.4.2	6.1	Provision of resources
6.1.1	Organization's context when planning for their QMS	5.4.2,8.5.3		
6.1.2	How to address risks and opportunities			
6.2	Quality objectives and planning to achieve them	5.4.1	6.2	Human resources
6.2.1	Enhancement and extension of ISO 9001:2008 requirements		6.2.1	General
6.2.2	Enhancement of ISO 9001:2008 sub-clause 5.4.2		6.2.2	Competence, training, and awareness
6.3	Planning of changes	5.4.2	6.3	Infrastructure
		6.0	6.4	Work environment

(Continued)

TABLE 2.3 (*Continued*)
Correlation between ISO 9001:2015 and ISO 9001:2008

CLAUSE	ISO 9001:2015		Correlation between ISO 9001:2015 to ISO 9001:2008	CLAUSE	ISO 9001:2008
7.0	**Support**		6.0	7.0	**Product Realization**
7.1	Resources		6.0	7.1	Planning of product realization
		7.1.1 General	6.1		
		7.1.2 People	6.1		
		7.1.3 Infrastructure	6.3		
		7.1.4 Environment for the operation of processes	6.4		
		7.1.5 Monitoring and measuring resources	7.6		
		7.1.6 Organizational knowledge	New		
7.2	Competence		5.2,6.2.1,6.2.2	7.2	Customer-related processes
				7.2.1	Determination of requirements related to the product
				7.2.2	Review of requirements related to the product
				7.2.3	Customer communication
7.3	Awareness		6.2.2	7.3	Design and development
				7.3.1	Design and development planning
				7.3.2	Design and development input
				7.3.3	Design and development output
				7.3.4	Design and development review

(*Continued*)

TABLE 2.3 (*Continued*)
Correlation between ISO 9001:2015 and ISO 9001:2008

CLAUSE	ISO 9001:2015	Correlation between ISO 9001:2015 to ISO 9001:2008	CLAUSE	ISO 9001:2008
			7.3.5	Design and development verification
			7.3.6	Design and development validation
			7.3.7	Control of design and development changes
7.4	Communication	5.5.3	7.4	Purchasing
			7.4.1	Purchasing process
			7.4.2	Purchasing information
			7.4.3	Production and service provision
7.5	Documented information	4.2	7.5	Production and service provision
7.5.1	General	4.2.1,4.2.2	7.5.1	Control of production and service provision
7.5.2	Creation and updating	4.2.3,4.2.4	7.5.2	Validation of process for production and service provision
7.5.3	Control and documented information	4.2.3,4.2.4	7.5.3	Identification and traceability
7.5.3.1	Availability of document when needed		7.5.4	Customer property
7.5.3.2	Distribution, access, and retrieve		7.5.5	Preservation of product
			7.6	Control of monitoring and measuring equipment

(*Continued*)

TABLE 2.3 (*Continued*)

Correlation between ISO 9001:2015 and ISO 9001:2008

CLAUSE	ISO 9001:2015		Correlation between ISO 9001:2015 to ISO 9001:2008	CLAUSE	ISO 9001:2008	
8.0	Operation		7.0	8.0	**Measurement, analysis and improvement**	
	8.1	Operational planning and control	7.1		8.1	General
	8.2	Determination of requirements for products and services	7.2		8.2	Monitoring and measurement
		8.2.1 Customer communication	7.2.3		8.2.1	Customer satisfaction
		8.2.2 Determination of requirements related to products and services	7.2.1		8.2.2	Internal audit
		8.2.3 Review of requirements related to products and services	7.2.2		8.2.3	Monitoring and measurement of process
					8.2.4	Monitoring and measurement of product
	8.3	Design and development of products and services	7.3,7.3.1		8.3	Control of nonconforming product
		8.3.1 General	7.3.1			
		8.3.2 Design and development planning	7.3.1			
		8.3.3 Design and development inputs	7.3.2			
		8.3.4 Design and development controls	7.3.4,7.3.5,7.3.6			
		8.3.5 Design and development outputs	7.3.3			
		8.3.6 Design and development changes	7.3.7			

(*Continued*)

TABLE 2.3 (*Continued*)

Correlation between ISO 9001:2015 and ISO 9001:2008

CLAUSE	ISO 9001:2015		Correlation between ISO 9001:2015 to ISO 9001:2008	CLAUSE	ISO 9001:2008	
8.4	Control of externally provided products and services		7.4,7.4.1	8.4	Analysis data	
	8.4.1	General	7.4.1			
	8.4.2	Type and extent of control of external provision	7.4.1,7.4.3			
	8.4.3	Information for external providers	7.4.2			
8.5	Production and service provision		7.3,7.5	8.5	Improvement	
	8.5.1	Control of production and service provision	7.5.1		8.5.1	Continual improvement
	8.5.2	Identification and traceability	7.5.3		8.5.2	Corrective action
	8.5.3	Property belonging to customers or external providers	7.5.4		8.5.3	Prevention action
	8.5.4	Preservation	7.5.5			
	8.5.5	Post-delivery activities	7.5.1			
	8.5.6	Control of changes	7.3.7			
8.6	Release of products and services		8.2.4,7.4.3			
8.7	Control of nonconforming process outputs, products, and services		8.3			

(*Continued*)

TABLE 2.3 (*Continued*)

Correlation between ISO 9001:2015 and ISO 9001:2008

CLAUSE	ISO 9001:2015		Correlation between ISO 9001:2015 to ISO 9001:2008	CLAUSE	ISO 9001:2008
9.0	**Performance evaluation**		New		
9.1	Monitoring, measurement, analysis, and evaluation		8.0		
	9.1.1	General	8.1,8.2.3		
	9.1.2	Customer satisfaction	8.2.1		
	9.1.3	Analysis and evaluation	8.4		
9.2	Internal audit		8.2.2		
	9.2.1	Organization's requirement to carry out internal audit			
	9.2.2	Requirements to how audit programmes must be structured			
9.3	Management review		5.6		
	9.3.1	Review of QMS by top management	5.6.2		
	9.3.2	Specific requirements in respect of management review	5.6.3		
10	**Improvement**		8.5		
10.1	General		8.5.1		
10.2	Nonconformity and corrective action		8.3,8.5.2		
	10.2.1	How the organization is to act when nonconformity is identified			
	10.2.2	Documented information relating to nonconformity	8.3,8.5.2		
10.3	Continual improvement		8.5.1		

Source: Abdul Razzak Rumane (2016). *Handbook of Construction Management: Scope, Schedule, and Cost*, CRC Press, Boca Raton, FL. Reprinted with permission from Taylor & Francis Group.

FIGURE 2.1
QMS pyramid.

The bottom level of hierarchy contains forms and records that are used to capture the history of routine events and activities.

The ISO 9000 quality management system requires documentation that includes a quality manual and quality procedures, as well as work instructions and quality records. All documentation (including quality records) must be controlled according to a document control procedure. The structure of the quality management system depends largely on the management structure in the organization.

ISO 9001:2000 identifies certain minimum requirements that all quality management systems must meet to ensure customer satisfaction. ISO 9001:2000 specifies requirements for quality management systems when an organization

- Needs to demonstrate its ability to consistently provide product that meets customer and applicable regulatory requirements
- Aims to enhance customer satisfaction through the effective application of the system, including processes for continual improvement of the system and the assurance of conformity to customer and applicable regulatory requirements

2.5.1 Quality System Documentation

A quality system has to cover all the activities leading to the final product or service. The quality system depends entirely on the scope of operation of the organization and particular circumstances such as number of employees, type of organization, and physical size of the premises of the organization. The quality manual is the document that identifies and describes the quality management system. The quality management system is based on the guidelines for performance improvement per ISO 9004:2000 and the quality management requirements. ISO 9000:2000 outlines the necessary steps to implement the quality management system. These are

1. Identify the process (activities and necessary elements) needed for quality management system.
2. Determine the sequence and interaction of these processes and how they fit together to accomplish quality goals.
3. Determine how these processes are effectively operated and controlled.
4. Measure, monitor, and analyze these processes and implement action necessary to correct the process and achieve continual requirements.
5. Ensure that all information is available to support the operation and monitoring of the process.
6. Display the most options, thus helping make the right management system.

ISO 9001:2000 requirements fall into the following sections:

1. Quality management system
2. Management responsibility
3. Resource management
4. Product realization
5. Measurement analysis and improvement

In the construction industry, a contractor may be working at any time on a number of projects of varied natures. These projects have their own contract documents to implement project quality, which require a contractor to submit a contractor's quality control plan to ensure that specific requirements of the project are considered to meet client's requirements. Therefore, while preparing a quality management system at a corporate level, the organization has to take into account tailor-made requirements for the projects and accordingly the manual should be prepared.

Table 2.4 lists an example contents of a quality management system manual for building construction organization (contractor) and Table 2.5 lists example contents of a quality management manual for an engineering consultant (design and supervision).

TABLE 2.4

List of Quality Manual Documents-Consultant (Design and Supervision)

Document No.	Title Quality Document	Relevant Clause in 9001:2015	Correlated Clause in ISO 9001:2008	Version/ Revision Date
QC-0	Circulation list			
QC-00	Records of revision			
QC-1.1	Understanding the organization and its context	4.1	1.1	
QC-1.2	Monitoring and review of internal and external issues	4.1	1.1	
QC-2.1	Relevant requirements of stakeholders	4.2	1.1	
QC-2.2	Monitoring and review of stakeholder's information	4.2	1.1	
QC-3	Scope of quality management system	4.4	4.1	
QC-4	Project quality management system	4.4	4.1	
QC-5	Management responsibilities	5.1	5.5.1	
QC-6	Customer focus	5.1.2	5.2	
QC-7.1	Quality Policy (Organization)	5.2	5.3	
QC-7.2	Quality Policy (Project)	5.2.2	5.3	
QC-8	Organizational roles, responsibilities, and authorities (Organization chart)	5.3	5.5.1/2	
QC-9	Preparation and control of project quality plan	6.0	5.4.2	
QC-10.1	Project risk (during design)	6.1	5.4.2,8.5.3	
QC-10.2	Project risk (during construction)	6.1	5.4.2,8.5.3	
QC-11	Project quality objective	6.2	5.4.1	
QC-12.1	Change Management (during design phase)	6.3	5.4.2	
QC-12.2	Change Management (during construction phase)	6.3	5.4.2	
QC-13	Office resources (Human resources, office equipment, design software)	7.1	6.1	
QC-14	Infrastructure	7.1.3	6.3	
QC-15	Work Environment	7.1.4	6.4	
QC-16	Human Resources (Design team, supervision team)	7.2	6.2.1	
QC-17.1	Training in quality system	7.2	6.2.2	

(Continued)

TABLE 2.4 *(Continued)*

List of Quality Manual Documents-Consultant (Design and Supervision)

Document No.	Title Quality Document	Relevant Clause in 9001:2015	Correlated Clause in ISO 9001:2008	Version/ Revision Date
QC-17.2	Training in quality auditing	7.2	6.2.2	
QC-17.3	Training in operational/ technical skills	7.2	6.2.2	
QC-18	Communication internal and external	7.4	5.5.3	
QC-19.1	Control of documents for general application	7.5.2/3	4.2.3	
QC-19.2	Control of documents for specific projects	7.5.2/3	4.2.3	
QC-20	Records updates	7.5.2	4.2.3/4	
QC-21	Control of quality records	7.5.3	4.2.4	
QC-22.1	Planning of engineering design and quality plan	8.1/8.3/8.3.1/2/3	7.1/7.3.1/7.3.7	
QC-22.2	Design development (Design-Bid-Build)	8.1/8.3/8.3.1/2/3	7.1/7.3.1/7.3.7	
QC-22.3	Design development (Design-Build)	8.1/8.3/8.3.1/2/3	7.1/7.3.1/7.3.7	
QC-23	Evaluation of sub consultant and selection	8.4	7.4.1	
QC-23	Communication with sub consultant	8.4.3	7.4.1/2/3	
QC-24	Engineering Design Procedure	8.5	7.5.1/2	
QC-25.1	Construction supervision procedure	8.5	7.5.1/2	
QC-25.2	Project management procedure	8.5	7.5.1/2	
QC-25.3	Construction management procedure	8.5	7.5.1/2	
QC-26.1	Control of nonconforming work (design errors)	8.7	8.3	
QC-27	Project review (Management and Control)	9.1	8.1	
QC-28	Internal quality audits	9.2	8.2.2	
QC-29	Management Review	9.3	5.6	
QC-30.1	Corrective action	10.2.2	8.5.2	
QC-30.2	Preventive action	10.3	8.5.3	
QC-32	Control of client complaints	10.2.2	8.3	

TABLE 2.5

List of Quality Manual Documents-Contractor

Document No.	Title Quality Document	Relevant Clause in 9001:2015	Correlated Clause in ISO 9001:2008	Version/ Revision Date
QC-0	Circulation list			
QC-00	Records of revision			
QC-1.1	Understanding the organization and its context	4.1	1.1	
QC-1.2	Monitoring and review of internal and external issues	4.1	1.1	
QC-2.1	Relevant requirements of stakeholders	4.2	1.1	
QC-2.2	Monitoring and review of stakeholder's information	4.2	1.1	
QC-3	Scope of quality management system	4.3/4	4.0,4.1	
QC-4	Project quality management system	4.4	4.1	
QC-5	Management responsibilities	5.1	5.5.1	
QC-6	Customer focus	5.1.2	5.2	
QC-7.1	Quality Policy (Organization)	5.2	5.3	
QC-7.2	Quality Policy (Project)	5.2.2	5.3	
QC-8	Organizational roles, responsibilities, and authorities (Organization chart)	5.3	5.5.1/2	
QC-9	Preparation and control of project quality plan	6.0	5.4.2	
QC-10	Project risk management	6.1	5.4.2,8.5.3	
QC-11	Project quality objectives	6.2	5.4.1	
QC-12.1	Change Management (Scope)	6.3	5.4.2	
QC-12.2	Change Management (Variation orders. Site work instructions)	6.3	5.4.2	
QC-13	Construction resources (Human resources, Equipment and Machinery)	7.1	6.1	
QC-14	Infrastructure	7.1.3	6.3	
QC-15	Work Environment	7.1.4	6.4	
QC-16	Control of construction, material, measuring and test equipment	7.1.5	7.6	
QC-17	Control of Human Resources	7.2	6.2.1	
QC-18.1	Training and development in quality system	7.2	6.2.2	
QC-18.2	Training in quality auditing	7.2	6.2.2	
QC-18.3	Training in operational/technical skills	7.2	6.2.2	

(Continued)

TABLE 2.5 *(Continued)*

List of Quality Manual Documents-Contractor

Document No.	Title Quality Document	Relevant Clause in 9001:2015	Correlated Clause in ISO 9001:2008	Version/ Revision Date
QC-19	Communication internal and external	7.4	5.5.3	
QC-20.1	Control of documents for general application	7.5.2/3	4..2.3	
QC-20.2	Control of documents for specific projects	7.5.2/3	4..2.3	
QC-21	Records updates	7.5.2	4.2.3/4	
QC-22	Control of quality records	7.5.3	4.2.4	
QC-24	Documents control (logs)	7.5.3.2	4.2.3/4	
QC-25	Project planning and control	8.1	7.1	
QC-26	Project specific requirements	8.2	7.2	
QC-27	Project specific quality control plan	8.2	7.2	
QC-28-1	Tender documents	8.2	7.2	
QC-28.2	Tender review	8.2	7.2.1/2/3	
QC-28.3	Contract Review	8.2.1	7.2.1/2/3	
QC-29	Variation review	8.2.3	7.2.2	
QC-30	Construction processes	8.2.2	7.2.1	
QC-31	Engineering and shop drawings	8.3	7.3.1	
QC-32	Design developments for Design-Build projects	8.3		
QC-33	Evaluation of subcontractors and selection	8.4	7.4.1	
QC-34	Communication with Sub contractors, material suppliers, vendors	8.4.3	7.4.1/2/3	
QC-35.1	Inspection of subcontracted work	8.4.2	7.4.3	
QC-35.2	Incoming material inspection and testing	8.4.3	7.4.3	
QC-36	Installation procedures	8.5	7.5.1/2	
QC-37	Product identification and traceability	8.5.2	7.5.3	
QC-38	Identification of inspection and test status	8.5.2	7.5.3	
QC-39	Control of owner supplied items	8.5.3	7.5.4	
QC-40	Handling and storage	8.5.4	7.5.5	
QC-41	Construction inspection, testing and commissioning	8.6	8.2.4	
QC-42.1	Control of nonconforming work	8.7	8.3	
QC-42.2	Control of nonconforming work	8.7	8.3	

(Continued)

TABLE 2.5 (*Continued*)

List of Quality Manual Documents-Contractor

Document No.	Title Quality Document	Relevant Clause in 9001:2015	Correlated Clause in ISO 9001:2008	Version/ Revision Date
QC-42.1	Project performance review	9.1.1	8.1	
QC-42.2	Project quality assessment and measurement	9.1.2	8.1	
QC-43	Internal quality audits	9.2	8.2.2	
QC-44	Management Review	9.3	5.6	
QC-45	New Technology in construction	10.1	8.5.1	
QC-46.1	Corrective action	10.2.2	8.5.2	
QC-46.2	Preventive action	10.3	8.5.3	
QC-47	Control of client complaints	10.2.2	8.3	

2.6 ISO Certification

Following are the details of ISO certification, registration, and accreditations, as mentioned on the ISO website (2008):

In the context of ISO 9000 or ISO 14000, "certification" refers to the issuing of written assurance (the certificate) by an independent, external body that has audited an organization's management system and verified that it conforms to the requirements specified in the standard.

"Registration" means that the auditing body then records the certification in its client register so the organization's management system has therefore been both certified and registered. Therefore, in the ISO 9000 and ISO 14000 contexts, the difference between the two terms is not significant and both are acceptable for general use.

"Certification" seems to be the term most widely used worldwide, although "registration" is often preferred in North America, and the two are also used interchangeably.

On the contrary, using "accreditation" as an interchangeable alternative for "certification" or "registration" is a mistake, because it means something different.

In the ISO 9000 or ISO 14000 context, accreditation refers to the formal recognition by a specialized body—an accreditation body—that a certification body is competent to carry out ISO 9000 or ISO 14000 certification in specified business sectors.

In simple terms, accreditation is like certification of the certification body. Certificates issued by accredited certification bodies may be perceived on the market as having increased credibility.

Thus, it should be understood that the certification body is a third-party company registered with an established national accreditation board and is authorized to issue a certificate of conformance after evaluating the conformance of an organization's management system to the requirements of appropriate standard.

With the advent of globalization and competitive market, it has become essential to implement a quality management system in an organization and to get it certified from a third party to enhance business opportunities in the international market. The ISO 9000 quality management system is accepted worldwide, and international customers prefer to do business with organizations having ISO certification. An ISO quality management system includes all activities and overall management functions that determine quality policy, objectives, and responsibilities and their implementation.

ISO certification is not compulsory; however, it is required for competitive advantage. Certification can be a useful tool to add credibility by demonstrating that the product or services meet the expectations of the customers. For some industries, certification is a legal or contractual requirement. ISO does not perform certification. ISO certification is valuable to firms because it provides a framework so they can assess where they are, where they would like to be, and what is their standing in the international market. Implementation of an ISO management system in the organization brings in increased effectiveness and efficiency of operations and ensures that the product satisfies customer requirements. ISO 9000 and ISO 14000 concern the way an organization goes about its work and processes. ISO 9000 and ISO 14000 are not product standards.

There are three types of audits that can be done on ISO quality management systems:

1. First-party audit—Audit your own organization (internal audit).
2. Second-party audit—Audit of supplier by the customer.
3. Third-party audit—Totally independent of the customer–supplier relationship. The best certification of a firm is through third party.

ISO 9000 certification audit is done by a certification body that has been accredited or has been officially approved as competent to carry out certification in a specified business sector by a national accreditation body.

Figure 2.2 diagrammatically summarizes the ISO certification process, and Figure 2.3 illustrates an example quality management system certification schedule that is developed taking into consideration overall certification requirements. The duration of each activity and the overall period to obtain certification may vary from company to company depending on the size and nature of the business.

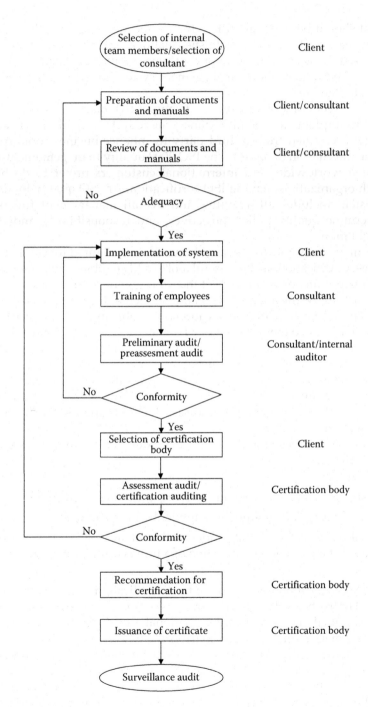

FIGURE 2.2
ISO certification process flow diagram.

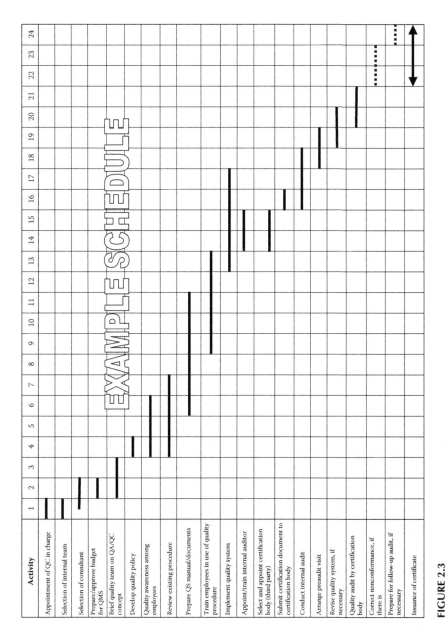

FIGURE 2.3
Quality management system certification schedule.

With the certification of ISO Standards, organizations obtain the following advantages:

- Customer satisfaction and confidence in the organization's products/ services
- Increase in revenues
- Increased market share
- Continuous improvement in organizational process
- Consistency in products/services quality
- Improvement in staff performance
- Effectiveness in the utilization of staff
- Efficient utilization of time, money, and other resources
- Environmental benefits

2.7 ISO 14000 Environmental Management System

ISO 14000 is a series of international standards that have been developed to incorporate environmental aspects into business operations and product standards. ISO 14001 is a specific standard in the series for a management system that incorporates a set of interrelated elements designed to minimize harmful effects on the environment due to the activities performed by an organization, and to achieve continual improvement of its environmental performance. ISO 14001 incorporates quality management system philosophy, terminology, and requirement structure similar to that of ISO 9001 and provides system compatibility.

2.7.1 Benefits of ISO 14000

The following are the benefits of implementing an environmental management system:

- Pollution prevention and waste reduction opportunities
- Cost reduction
- Customer satisfaction
- Compliance with regulatory requirements on environmental considerations
- Reduction in consumption of energy
- Reduction in use of natural resources
- Minimization of environmental liability and risk
- Commitment to social responsibility

2.8 Occupational Health and Safety Assessment Series 18000

The Occupational Health and Safety Assessment Series (OHSAS) 18000 has been developed to help organizations control and minimize occupational health and safety risks. OHSAS 18001 is a specific standard for occupational health and safety management systems designed to eliminate or minimize the risk to employees and other related parties who may be exposed to occupational health and safety risks associated with business activities. OHSAS 18000 is compatible with ISO 9001 and ISO 14001 management systems. OHSAS 18001 represents a progression of a management philosophy from quality management to environmental management to occupational health and safety management.

2.8.1 Benefits of OHSAS Management System

The benefits of implementing an OHSAS management system are

- Reduced accidents and injuries to the employees
- Reduced insurance liability and risk
- Decreased costs due to personal injury and production downtime
- Reduced worker compensation insurance costs
- Ease of managing safety risks
- Enhanced employee safety awareness

2.9 Quality Audit

2.9.1 Introduction

A quality audit is formal or methodical examining, reviewing, and investigating of an existing system to determine whether agreed upon requirements are being met. An audit is a systematic, independent, and documented process to verify or evaluate and report the degree of compliance to the agreed upon quality criteria, or the specification or contract requirements of the product, services, or project. There are mainly five types of audits:

1. Product audit
2. Process audit
3. System audit
4. Compliance audit
5. Adequacy audit

To achieve competitive advantage, effective quality improvement is critical for an organization's growth. Quality audit serves the purpose of examining the effectiveness of the management-directed control programs. The audit of existing quality system that is being implemented provides the factual information to developing long-term organizational strategies for quality.

2.9.2 Categories of Auditing

Audits are mainly classifieds as follows:

First party—Audit your own organization (internal audit)

Second party—Customer audits the supplier (external audit)

Third party—Audits performed by independent audit organization (independent audit)

Third-party audits may result in independent certification of a product, process, or system such as ISO 9000 quality management system certification. Third-party certification enhances organizations' image in the business circle. Quality audit provides feedback to the management on the adequacy implementation and effectiveness of the quality system.

2.9.3 Quality Auditing Process

Figure 2.4 illustrates the quality auditing process.

2.9.4 Quality Auditing in Construction Projects

Quality audit system in construction projects is a system or method to measure and evaluate the quality of workmanship, finishes, and performance of the constructed works based on approved standards and quality requirements to the satisfaction of client/end-user. In order to achieve competitive advantage and organization's growth, it is essential that the organization has a well-established quality system and effective quality improvement procedure. Quality assessment/measurement needs to be continuous, ongoing, and performed in a timely manner. In construction projects, assessment/measurement is carried out by using checklists, project monitoring and controlling system to ensure meeting the owner needs, and is mainly performed during the following phases:

1. Design phase
2. Construction phase
3. Testing, commissioning and handover phase

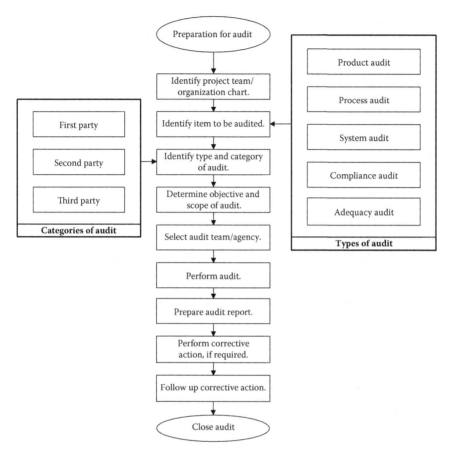

FIGURE 2.4
Quality auditing process.

Quality assessment/measurement system in construction projects can be classified as

1. Internal quality system (internal audit)
2. External quality audit

Internal quality system is a process in which quality audit is performed by the own organization.

2.9.4.1 Quality Audit by Designer

Quality audit during the design phase is normally carried out by the design team as per the quality management system before it is sent for review by the owner (project manager). The designer (consultant) also submits the design to regulatory authorities to obtain their approval for compliance to regulatory requirements prior to submission of drawings (documents) to the owner/client.

2.9.4.2 Quality Audit by Contractor

The contractor has to check the executed works prior to submission of check-list to the supervision engineer (consultant).

2.9.4.3 External Quality System

External assessment in construction projects can be divided into

1. Quality inspection/checking by owner-appointed supervision engineer (consultant).
2. Independent testing agency appointed by the owner (consultant as per contract requirements) to inspect/test mainly the following items/works:
    I. Concrete strength testing
    II. Reinforcement testing
    III. Elevator works
    IV. Chiller testing
    V. Diesel generator testing
    VI. Electrical switch boards
3. Quality compliance audit by regulatory authorities. There are certain countries where the design documents are to be submitted to the regulatory authorities for their approval.

Following are the main four steps needed to establish quality assessment/ measurement process in construction projects:

1. Identify the component/items/products/systems to be reviewed and checked in each of the trades (architectural, structural, mechanical, HVAC, electrical, low voltage, landscape, external, etc.)
2. Identify the frequency of checking (timing)
3. Identify the persons/agency authorized to check and approve
4. Identify the indicators/criteria for performance monitoring

Table 2.6 lists items to be verified, checked internally by the designer before submission to regulatory authorities and, subsequently, submission to the owner (project manager).

TABLE 2.6

Check List for Design Drawings

Serial Number	Items to Be Checked
1	Whether design meets owner requirements (TOR)
2	Whether designs were prepared using authenticated and approved software
3	Whether design calculation sheets are included in the set of documents
4	Whether design is fully coordinated for conflict between different trades
5	Whether design has taken into consideration relevant collected data requirements
6	Whether reviewer's comments are responded
7	Whether regulatory approval is obtained and comments, if any, incorporated and all review comments responded
8	Whether design has environmental compatibility
9	Whether energy efficiency measures are considered
10	Whether design constructability is considered
11	Whether design matches with property limits
12	Whether legends matches with layout
13	Whether design drawings are properly numbered
14	Whether design drawings have owner logo, designer logo as per standard format
15	Whether the design format of different trades have uniformity
16	Whether project name and contract reference is shown on the drawing

Apart from project quality check/inspection, it is also required to assess the following activities on a regular basis to track project progress and its compliance to contract documents:

1. Schedule
2. Budget
3. Resources
4. Project team performance
5. Risk assessment
6. Safety

Monitoring and control of schedule, budget, and resources (project progress) is performed on a regular basis by project monitoring and control team members. Also team members from the supervision team as well as from the contractor side are assigned to carry out risk assessment and safety assessment. Performance evaluation is done by the line manager in coordination with the corporate office requirements.

2.9.5 Auditing versus Inspection

Table 2.7 lists the major differences between audit and inspection.

TABLE 2.7

Audit versus Inspection

Audit	Inspection
• Audit is a function of QA Process • Audit is done during the product is being built or during the process • Audits often go deeper than the inspection. Audits are typically concerned with a wider range of requirements • Audits are external to the process so the process can continue whether the auditor is present or find discrepancies during the audit (exceptions include safety issues or regulatory audits, which can result in process being stopped) • Audits are less frequent than inspections	• Inspection is a function of QC Process • Inspection is done after completion of a particular activity or product or process • Inspections generally focus on smaller number of characteristics • Inspections are typically a mandatory step in the process and included as such in process instructions and other workflow documentation (routing lists, and so on) • Inspections are typically more frequent than audits

Similarities
• Audits and inspections can both be performed on products, projects, processes, or systems • Audits and inspections both typically assess compliance with some standard.

2.9.6 Risk in Auditing

The following are the risks in quality auditing:

- The scope and objective of audits are not clearly defined.
- The auditing person/agency does not have expertise or knowledge about the type of audit to be performed.
- Auditor does not attempt to or cannot check all the transactions that need to be audited.
- Auditor does not identify and assess the risk.
- Auditing procedure does not cover all the broad areas.
- Auditor's understanding of the entity and the environment including internal control systems in which it operates.
- Assessment of risk.

2.10 Quality Cost

2.10.1 Introduction

Quality has an impact on the costs of products and services. The cost of poor quality is the annual monetary loss of products and processes that are not achieving their quality objective.

According to Gryna (2001), "the concept of quality costs emerged during the 1950s, and different people assigned different meaning to the term. Some people equated quality costs with the costs of attaining quality; some people equated the term with the extra costs incurred because of poor quality" (p. 19). He further states that

> The cost of poor quality is the annual monetary loss of products and processes that are not achieving their quality objectives. The main components of the cost of low quality are
>
> 1. Cost of nonconformities.
> 2. Cost of inefficient processes.
> 3. Cost of loss opportunities of sales revenue. (p. 20)

Juran and Godfrey (1999) also state that "the term *quality costs* has different meanings to different people. Some equate quality costs with the cost of poor quality (mainly the costs of finding and correcting defective work); others equate the term with the costs to attain quality; still others use the term to mean the costs of running the quality department" (p. 8.2).

2.10.2 Categories of Costs

Costs of poor quality are those associated with providing poor-quality products or services. These are costs that would not be incurred if things were done right from the start and at every stage thereafter, in order to achieve the quality objective. There are four categories of costs:

1. Prevention costs (the costs incurred to keep failure and appraisal costs to minimum).
2. Appraisal costs (the costs incurred to determine the degree of conformance to quality requirements).
3. Internal failure costs (the costs associated with defects found before the customer receives the product or service. It also consists of cost of failure to meet customer satisfaction and needs and cost of inefficient processes).
4. External failure costs (the costs associated with defects found after the customer receives the product or service. It also includes lost opportunity for sales revenue).

Figure 2.5 illustrates the categories of cost of quality.

Thomas Pyzdek (1999) has detailed these costs as follows:

> 1. *Prevention costs*: costs incurred to prevent the occurrence of nonconformances in the future. Examples of prevention costs include
>   - Quality planning
>   - Process control planning
>   - Design review
>   - Quality training
>   - Gage design

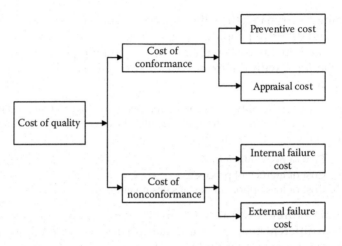

FIGURE 2.5
Categories of cost of quality.

2. *Appraisal costs*: costs incurred in measuring and controlling concurrent production to assure conformance to requirements. Examples of appraisal costs include
   * Receiving inspection
   * Laboratory acceptance testing
   * In-process inspection
   * Outside endorsements (e.g., UL approval)
   * Calibration
   * Inspection and test equipment
   * Field testing
3. *Internal failure costs*: costs generated before a product is shipped as a result of nonconformance to equipment. Examples of internal failure costs include
   * Scrap
   * Rework
   * Process troubleshooting
   * Vendor-caused scrap or work
   * Material review board activity
   * Reinspection or retest
   * Downgrading
4. *External failure costs*: costs generated after a product is shipped as a result of nonconformance to requirement. Examples of external failure costs include
   * Processing of customer complaints
   * Service
   * Unplanned field repair
   * Recalls
   * Processing of returned materials
   * Warranty (p. 148)

These cost categories allow the use of quality cost data for a variety of purposes. Quality costs can be used for measurement of progress, for analyzing the problem, or for budgeting. By analyzing the relative size of the cost categories, the company can determine if its resources are properly allocated.

CII (construction industry institute) product no. EM-4A (1994) states that "the cost of quality is the penalty paid for an imperfect world. It is the costs of all the extra work we do beyond merely doing a task correctly the first time to meet the requirements and expectations. The simple formula for defining the cost of quality is

$$\text{Cost of Quality} = \text{Cost of Prevention and Appraisal}$$

$$+ \text{Cost of Deviation Correction"} \qquad \text{(p. 13)}.$$

It has further elaborated the components of costs of quality as follows:

Prevention and appraisal: all measures taken to assure that requirements are met, such as quality control systems, inspection, work checking, design review, constructability or maintainability review, shop inspection, and auditing

Deviation correction: work done more than once because it did not meet requirements the first time

2.10.3 Reasons for Poor Quality

According to the survey carried out by the CII, the primary reasons for poor quality are mainly due to poor management and are illustrated in Figure 2.6.

Quality has an impact on the costs of products and services. The cost of poor quality is the annual monetary loss of products and processes that are not achieving their quality objective. The main components of the cost of quality are

1. Cost of conformance
2. Cost of nonconformance

Table 2.8 illustrates elements of cost of quality.

Table 2.9 illustrates different elements of categories of cost of quality during design of the project.

Chung (1999) has quoted Robert (1991): "quality does not cost—it pays" (p. 9). Figure 2.7 summarizes the quality-related costs expressed as a percentage of

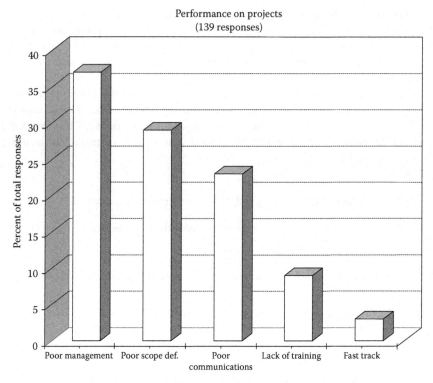

FIGURE 2.6
Primary reasons for poor quality. (*Source*: CII Source document 79. Reprinted with permission of CII, University of Texas.)

total construction costs. He further states that through the implementation of a proactive quality system that costs about 1% of the project value (the prevention cost), the expenditure as a result of repair, and so forth (failure cost) drops from 10% to 2%, representing a saving of 7%. These categories of costs may represent an increase of cost in one area and a reduction of cost in another.

The CII has made the following recommendations to reduce the rework:

1. Reduce the number of design changes
2. Implement a quality management program
3. Adopt the standard set of quality-related terminology
4. Develop and implement system to establish a database
5. Implement a QPMS

TABLE 2.8

Cost of Quality

Cost of Compliance	Cost of Noncompliance
• Quality planning	• Scrap
• Process control planning	• Rework
• Quality training	• Additional material/inventory cost
• Quality audit	• Expedition
• Design review	• Reinspection or retest
• Product design validation	• Downgrading
• Work procedure	• Maintenance service
• Method statement	• Field repairs
• Process validation	• Customer complaints
• Prevention action	• Product recalls
• Corrective action	• Warranty
• In-process inspection	• Rectification of returned material
• Field testing	• Loss of business
• Third-party inspection	• Damaged reputation
• Receiving inspection	
• Laboratory acceptance testing	
• Outside endorsement	
• Calibration of equipment	

Source: Abdul Razzak Rumane (2013), *Quality Tools for Managing Construction Projects*, CRC Press, Boca Raton, FL. Reprinted with permission from Taylor & Francis Group.

TABLE 2.9

Cost of Quality during Design Stage

Prevention Cost	Appraisal Cost
• Conduct technical meetings for proper coordination	• Review of design drawings
• Follow quality system	• Review of specifications
• Meeting submission schedule	• Review of contract documents to ensure meeting owner's needs, quality standards, constructability, and functionality
• Training of project team members	• Review for regulatory requirements, codes
• Update of software used for design	
Internal Cost	**External Cost**
• Redesign/redraw to meet fully coordinated design	• Incorporate design review comments by client/project manager
• Rewrite specifications/documents to meet requirements of all other trades	• Incorporate specifications/documents review comments by client/project manager
	• Incorporate comments by regulatory authority(ies)
	• Resolve RFI (request for information) during construction

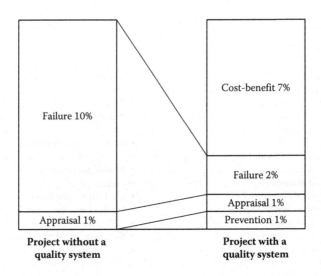

FIGURE 2.7
Implementation of quality management (Roberts, 1991). (*Source*: Chung, H.W. (1999), *Understanding Quality Assurance in Construction*. Reprinted with permission from Taylor & Francis Group.)

2.10.4 Quality Cost in Construction

Quality of construction is defined as including the following:

1. Scope of work
2. Time
3. Budget

Cost of quality refers to the total cost incurred during the entire life cycle of construction project in preventing nonconformance to owner requirements (defined scope). There are certain hidden costs that may not directly affect the overall cost of the project; however, it may cost the consultant/designer to complete the design within the stipulated schedule to meet owner requirements and conformance to all the regulatory codes/standards, and for the contractor to construct the project within the stipulated schedule, meeting all the contract requirements. Rejection/nonapproval of executed/installed works by the supervisor due to noncompliance with specifications will cause the contractor loss in terms of

- Material
- Manpower
- Time

The contractor shall have to rework or rectify the work, which will need additional resources and will need extra time to do the work as specified.

This may disturb the contractor's work schedule and affect execution of other activities. The contractor has to emphasize the "Zero Defect" policy, particularly for concrete works. To avoid rejection of works, the contractor has to take the following measures:

1. Execution of works per approved shop drawings using approved material
2. Following approved method of statement or manufacturer's recommended method of installation
3. Conducting continuous inspection during construction/installation process
4. Employ trained workforce properly
5. Maintaining good workmanship
6. Identifying and correcting deficiencies before submitting the checklist for inspection and approval of work
7. Coordinating requirements of other trades, for example, if any opening is required in the concrete beam for crossing of services pipe

Timely completion of a project is one of the objectives to be achieved. To avoid delay, proper planning and scheduling of construction activities are necessary. Since construction projects have the involvement of many participants, it is essential that the requirements of all the participants are fully coordinated. This will ensure execution of activities as planned resulting in timely completion of the project.

Normally, the construction budget is fixed at the inception of the project; therefore, it is necessary to avoid variations during the construction process as it may take time to get approval of an additional budget resulting in time extension to the project. Quality costs related to construction projects are summarized as follows.

2.10.4.1 Prevention Costs

- Preventive action
- Training
- Work procedures
- Method statement
- Calibration of instruments/equipment

2.10.4.2 Appraisal Costs

- Design review/preparation of shop drawings
- Preparation of composite/coordination drawings

- On-site material inspection/test
- Off-site material inspection/test
- Prechecklist inspection

2.10.4.3 Internal Failure Costs

- Rework
- Rectification
- Rejection of checklist
- Corrective action

2.10.4.4 External Failure Costs

- Breakdown of installed system
- Repairs
- Maintenance
- Warranty

Quality costs during the design phases are different from those of the construction phase. Costs of quality during design phases are mainly to ensure development of project design and documents to ensure conformance to the client's requirements/TOR (Terms of Reference)/Matrix of Owner's Requirements. Quality costs related to design development/contract documents of construction projects can be summarized as follows:

Prevention Costs
- Conducting technical meetings for proper coordination
- Following quality system
- Meeting submission schedule
- Training of project team members
- Updating of software used for design

Appraisal Costs
- Review of design drawings
- Review of specifications
- Review of contract documents to ensure meeting owner's needs, quality standards, constructability, and functionality
- Review for regulatory requirements, codes

Internal Failure Costs
- Redesign/redraw to meet requirements of other trades
- Redesign/redraw to meet fully coordinated design
- Rewrite specifications/documents to meet requirements of all other trades

External Failure Costs
- Incorporate design review comments by client/project manager
- Incorporate specifications/documents review comments by client/project manager
- Incorporate comments by regulatory authority(ies)
- Resolve RFI (request for information) during construction

2.10.5 Quality Performance Management System

Quality performance management system (QPMS) is a product of the CII. QPMS is one of the tools available for a TQM project and is a good implementation tool for a project to utilize in a TQM environment. It is a management tool developed by the CII to give management the information necessary to identify quality improvement opportunities.

QPMS focuses on reducing the cost of quality in four ways:

1. It provides a process that facilitates awareness of individual and group quality performance (how well we do things right) by measuring these costs in dollars.
2. It arms managers with information on quality costs and activities that enable proactive decisions affecting quality outcome.
3. It provides a database for estimating quality performance on future projects.
4. If and when widely accepted, the data should provide benchmarking information throughout the industry. (Benchmarking is a point of reference by which the performance is judged or measured.)

According to the CII, the QPMS has been developed as a management tool to meet the following criteria. It must

1. Be capable of tracking quality-related costs involved in the design and construction of engineered projects and answer the following four questions:

   What quality management activities and deviation costs are involved?

   When were the quality management activities and deviation costs incurred?

Why did the deviations occur (i.e., their root causes)?

How did the rework relate to the quality management?

2. Provide valuable cost-of-quality information to establish baseline and identify opportunities for improvement, without providing either too much or too little detail

3. Be adaptable to various types and aspects of design and construction projects

4. Be easily implementable by owners, designers, and contractors

5. Be cost-effective

6. Be compatible with existing cost systems used by management

Thus, it can be summarized that with implementation of quality management system, costs of quality is reduced and ultimately results in savings.

2.11 Integrated Quality Management

The integrated quality management system (IQMS) is the integration and proper coordination of functional elements of quality to achieve efficiency and effectiveness in implementation and maintaining an organization's quality management system to meet customer requirements and satisfaction. IQMS consists of any element or activity that has an effect on quality. Customer satisfaction is the goal of quality objectives.

During the past three decades, many programs have been implemented for organizational improvements. In the 1980s, programs such as statistical process control, various quality tools, and total quality management were implemented. In the 1990s, ISO 9000 came into being, which resulted in improved productivity, cost reduction, improved time, improved quality, and customer satisfaction.

With globalization and competition, it became necessary for organizations to continuously improve to achieve the highest performance and a competitive advantage.

In the 1980s, the major challenge facing most organizations was to improve quality. In the 1990s, it was to improve faster by restructuring and reengineering all operations.

In today's global competitive environment, organizations are facing many challenges due to an increase in customer demand for higher performance requirements at a competitive cost. They are finding that their survival in the competitive market is increasingly in doubt. To achieve a competitive advantage, effective quality improvement is critical.

Processes and systems are essential for the performance and expansion of any organization. ISO 9000 is an excellent tool to develop a strong foundation

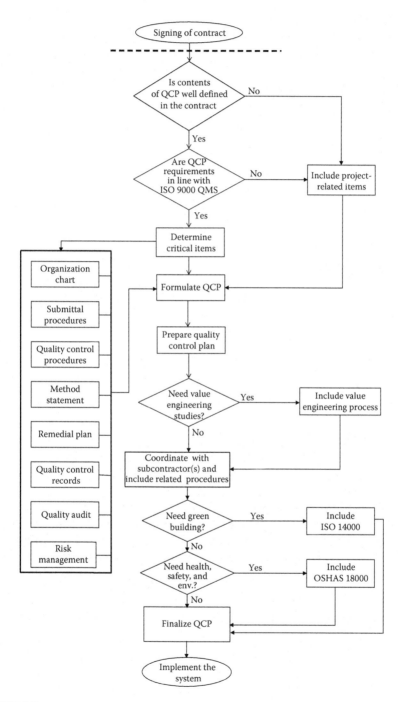

FIGURE 2.8
Logic flow diagram for development of IQMSIs.

for good processes and systems. The ISO 9000 quality management system is accepted worldwide, and ISO 9000 certification has global recognition.

An IQMS is developed by merging recommendations and specifications from ISO 9000 (Quality Management System), ISO 14000 (Environmental Management System), and OHSAS 18000 (Occupational Health and Safety Management), together with other contract documents. If an organization has a certified Quality Management System (ISO 9000), it can build an IQMS system by adding environmental, health, safety, and other requirements of management system standards.

The benefits of implementing an IQMS are

- Reduced duplication and, therefore, cost
- Improved resource allocation
- Standardized process
- Elimination of conflicting responsibilities and relationship
- Consistency
- Improved communication
- Reduced risk and increased profitability
- Facilitated training development
- Simplified document maintenance
- Reduced record keeping
- Ease of managing legal and other requirements

Construction projects are unique and nonrepetitive in nature, and have their own quality requirements that can be developed by integration of project specifications and an organization's quality management system. Normally, quality management system manuals consist of procedures to develop project quality control plans, taking into consideration contract specifications. This plan is called the contractor's quality control plan (CQCP). Certain projects specify which value engineering studies should be undertaken during the construction phase. The contractor is required to include these while developing the CQCP. This plan can be termed an Integrated Quality Management System (IQMS) for construction projects. The contractor has to implement a quality system to ensure that the construction is carried out in accordance with the specification details and approved CQCP. Figure 2.8 illustrates the logic flow diagram for development of IQMS for construction projects.

3

Construction Projects

3.1 Project Definition

The authors of *A Guide to the Project Management Body of Knowledge* (PMBOK 2000) define the word *project* in terms of its distinctive characteristics: "A project is a temporary endeavor undertaken to create a unique product or service." "Temporary" means that every project has a definite beginning and a definite end. "Unique" means that the product or service is different in some distinguishing way from all similar products or services.

It further states that projects are often critical components of the performing organization business strategy. Examples of projects include

- Developing a new product or service
- Effecting a change in structure, staffing, or style of an organization
- Designing a new transportation vehicle/aircraft
- Developing or acquiring a new or modified information system
- Running a campaign for political office
- Implementing a new business procedure or process
- Constructing a building or facility

The duration of a project is finite; projects are not ongoing efforts, and the project ceases when its declared objectives have been attained. Among other shared characteristics, projects are

1. Performed by people
2. Constrained by limited resources
3. Planned, executed, and controlled

Pyzdek (1999) defined "project" as

1. A plan or proposal; a scheme
2. An undertaking requiring concrete effort

The "plan" is defined as

1. A scheme, program, or method worked beforehand for the accomplishment of an objective; a plan of attack
2. A proposed or tentative projective or course of action
3. A systematic arrangement of important parts (p. 48)

According to Kerzner (2001), a project can be considered to be a set of activities and tasks that

- Have a specified objective to be completed within certain specifications
- Have defined start and end dates
- Have funding limits (if applicable)
- Consume human and nonhuman resources (i.e., money, people, and equipment)
- Are multifunctional (i.e., cut across several lines) (p. 2)

Based on various definitions, the project can be defined as follows: "A project is a plan or program performed by the people with assigned resources to achieve an objective within a finite duration."

3.2 Construction Projects

Construction has a history of several thousand years. The first shelters were built from stone or mud and the materials collected from the forests to provide protection against cold, wind, rain, and snow. These buildings were primarily for residential purposes, although some may have had some commercial function.

During the New Stone Age, people introduced dried bricks, wall construction, metal working, and irrigation. Gradually, people developed the skills to construct villages and cities, and considerable skills in building were acquired. This can be seen from the great civilizations in different parts of the world—some 4000–5000 years ago. During the early period of Greek settlement, which was about 2000 BCE, the buildings were made of mud using timber frames. Later, temples and theaters were built from marble. Some 1500–2000 years ago, Rome became the leading center of world culture, which extended to construction.

Marcus Vitruvius Pollo, the first-century military and civil engineer, penned in Rome the world's first major treatise on architecture and construction.

It dealt with building materials, the styles and design of building types, the construction process, building physics, astronomy, and building machines.

During the Middle Ages (476–1492), improvements occurred in agriculture and artisanal productivity and exploration, and as a consequence, the broadening of commerce took place and in the late Middle Ages, building construction became a major industry. Craftsmen were given training and education in order to develop skills and to raise their status. At this time, guilds came up to identify true craftsmen and set standards for quality.

The fifteenth century brought a "renaissance" or renewal in architecture, building, and science. Significant changes occurred during the seventeenth century and thereafter due to the increasing transformation of construction and urban habitat.

The scientific revolution of the seventeenth and the eighteenth centuries gave birth to the great Industrial Revolution of the eighteenth century. After some delay, construction followed these developments in the nineteenth century.

The first half of the twentieth century witnessed the construction industry becoming an important sector throughout the world, employing many workers. During this period, skyscrapers, long-span dams, shells, and bridges were developed to satisfy new requirements and marked the continuing progress of construction techniques. The provision of services such as heating, air-conditioning, electrical lighting, water mains, and elevators in buildings became common. The twentieth century has seen the transformation of the construction and building industry into a major economic sector. During the second half of the twentieth century, the construction industry began to industrialize, introducing mechanization, prefabrication, and system building. The design of building services systems changed considerably in the last 20 years of the twentieth century. It became the responsibility of designers to follow health, safety, and environmental regulations while designing any building.

Building and commercial—traditional architect and engineer (A&E) type—construction projects account for an estimated 25% of the annual construction volume. Building construction is a labor-intensive endeavor. Every construction project has some elements that are unique. No two construction or research and development (R&D) projects are alike. Though it is clear that many building projects are more routine than R&D projects, some degree of customization is a characteristic of the projects.

Construction projects involve a cross section of many different participants. These both influence and depend on each other in addition to the "other players" involved in the construction process. Figure 3.1 illustrates the concept of the traditional construction project organization.

Traditional construction projects involve three main groups:

1. Owners—A person or an organization that initiates and sanctions a project. He/she outlines the needs of the facility and is responsible for arranging the financial resources for creation of the facility.

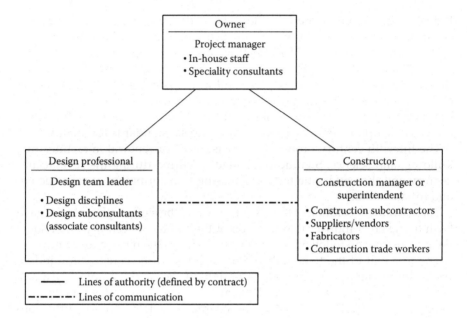

FIGURE 3.1
Traditional Construction Project Organization. (*Source*: Quality in the Constructed Project (2000) by ASCE. Reprinted with permission from ASCE.)

2. Designers (A&E)—This group consists of one or more architects or engineers and consultants. They are the owner's appointed entities accountable for converting the owner's conception and need into a specific facility with detailed directions through drawings and specifications adhering to the economic objectives. They are responsible for the design of the project and, in certain cases, its supervision.

3. Contractors—A construction firm engaged by the owner to complete the specific facility by providing the necessary staff, workforce, materials, equipment, tools, and other accessories to the satisfaction of the owner/end user in compliance with the contract documents. The contractor is responsible for implementing the project activities and for achieving the owner's objectives.

Construction projects are executed based on a predetermined set of goals and objectives. With traditional construction projects, the owner heads the team, designating a project manager. The project manager is a person/member of the owner's staff or independently hired person/firm with overall or principal responsibility for the management of the project as a whole.

Oberlender (2000) states that the working environment and culture of a construction project is unique compared to most working conditions. A typical construction project consists of a group of people, normally from several organizations, that are hired and assigned to a project to build the facility. Due to the relatively short life of a construction project, these people may view the construction project as accomplishing short-term tasks. However, the project manager of the construction team must instill in the team the concept that building a long-term relationship is more important in career advancement than trying to accomplish short-term tasks.

In certain cases, owners engage a professional firm, called a construction manager, trained in the management of construction processes, to assist in developing bid documents, and overseeing and coordinating the project for the owner. The basic construction management concept is that the owner assigns a contract to a firm that is knowledgeable and capable of coordinating all the aspects of the project to meet the intended use of the project by the owner. In the construction management type of construction projects, the consultants (architect/engineer) prepare complete design drawings and contract documents; then the project is put for competitive bid and the contract is awarded to the competitive bidder (contractor). Next, the owner hires a third party (construction manager) to oversee and coordinate the construction.

The authors of "Quality in the Constructed Project" by the American Society of Civil Engineers (ASCE 2000) have categorized two types of construction managers: agency construction managers (ACM) and construction managers-at-risk (CM-at-risk). An ACM functions wholly within the policies, procedures, and practices of the owner's organization. A CM-at-risk typically contracts with the owner in two stages. During the first stage, CM-at-risks act as consultants or even design professionals, and when the design is completed they become involved in the completion of the construction work.

There are numerous types of construction projects:

- Process-type projects
  - Liquid chemical plants
  - Liquid/solid plants
  - Solid process plants
  - Petrochemical plants
  - Petroleum refineries
- Nonprocess-type projects
  - Power plants
  - Manufacturing plants
  - Support facilities
  - Miscellaneous (R&D) projects
  - Civil construction projects
  - Commercial/A&E projects

Civil construction projects and commercial/A&E projects can further be categorized into four somewhat arbitrary but generally accepted major types of construction:

1. Residential construction

2. Building construction (institutional and commercial)

3. Industrial construction

4. Heavy engineering construction

Residential construction: Residential construction includes single-family homes, multiunit town houses, garden, apartments, high-rise apartments, and villas.

Building construction: Building construction includes structures ranging from small retail stores to urban redevelopment complexes, from grade schools to new universities, hospitals, commercial office towers, theaters, government buildings, recreation centers, warehouses, and neighborhood centers.

Industrial construction: Industrial construction includes petroleum refineries, petroleum plants, power plants, heavy manufacturing plants, and other facilities essential to our utilities and basic industries.

Heavy engineering construction: Heavy engineering construction includes dams and tunnels, bridges, railways, airports, highways and urban rapid transit system, ports and harbors, water treatment and distribution, sewage and storm water collection, treatment and disposal system, power lines, and communication network.

Table 3.1 illustrates types of construction projects.

3.3 Construction and Manufacturing

Construction has unique problems compared to manufacturing. A few of these are listed as follows:

- Construction is a custom rather than a routine, repetitive business and differs from manufacturing.

- Quality in manufacturing passes through series of processes. The output is monitored by inspection and testing at various stages of production.

- Construction is different from both that of mass production and batch (lot) production manufacturing.

TABLE 3.1

Types of Construction Projects

1	**Process-Type Projects**		
1.1	Liquid chemical plants		
1.2	Liquid/solid plants		
1.3	Solid process plants		
1.4	Petrochemical plants		
1.5	Petroleum refineries		
2	**Nonprocess-Type Projects**		
2.1	Power plants		
2.2	Manufacturing plants		
2.3	Support facilities		
2.4	Miscellaneous (R&D) projects		
2.5	Civil construction projects	**Categories of Civil Construction Projects and Commercial A/E Projects**	Residential construction	Family homes, multiunit town houses, garden, apartments, condominiums, high-rise apartments, villas
2.6	Commercial A/E projects		Building construction (institutional and commercial)	Schools, universities, hospitals, commercial office complexes, shopping malls, banks, theaters, stadiums, government buildings, warehouses, recreation centers, amusement parks, holiday resorts, neighborhood centers
			Industrial construction	Petroleum refineries, petroleum plants, power plants, heavy manufacturing plants, steel mills, chemical processing plants
			Heavy engineering	Dams, tunnels, bridges, highways, railways, airports, urban rapid transit system, ports, harbors, power lines and communication network
			Environmental	Water treatment and clean water distribution, sanitary and sewage system, waste management

Source: Abdul Razzak Rumane (2013), *Quality Tools for Managing Construction Projects*, CRC Press, Boca Raton, FL. Reprinted with permission from Taylor & Francis Group.

- In construction projects, the scenario is not the same as that of manufacturing. If anything goes wrong, the nonconforming work is very difficult to rectify and remedial action is sometimes not possible. Quality costs play an important role in construction projects.
- In construction, an activity may be repeated at various stages, but it is done only one time for a specific work. Therefore, it has to be right from the onset.
- In manufacturing, the buyer does not enter the scene until the product comes into being, whereas in construction the buyer is involved from beginning to end. Even during the construction phase, it is likely that certain modifications may take place.
- The owner is deeply involved in the construction process, while the purchaser of manufactured goods is not. Buyers of the usual manufactured products seldom have access to the plant where they are made, nor do they deal directly with factory managers.
- Most projects or their individual work phases are of relatively short duration. One consequence is that management teams and, possibly, the workforce must be assembled quickly and cannot be shaken out or restructured before the project or work phase is completed.
- To a great extent, each project has to be designed and built to serve a specific need and therefore it is necessary to make certain modifications in the system process to fit the particular conditions of each construction project and its specific problems.
- The location of construction projects varies widely. In a manufacturing plant, a given operation is assigned to and carried out in one place. In contrast, specialized construction crews progress from location to location.
- Operations are commonly conducted out of doors and are subject to all the interruptions and variation in conditions and the other difficulties that rain, snow, heat, and cold can introduce.
- The final product is usually of unique design and differs from workstation to workstation so that no fixed arrangement of equipment or aids such as jigs and fixtures are possible as is in the case of manufacturing.
- Construction is a preliminary step leading to a completed facility; the layout and arrangements may make access for construction difficult and permanent provisions for safety impossible.
- Construction often needs highly skilled craftsmen rather than unskilled workers; individual crews, whether union or nonunion, usually do specialized operations.

- Construction involves installation and integration of various materials, equipment, systems, or other components to complete the facility.
- Construction focuses mainly on overall performance of the project or facility in which a product(s) or a system(s) is a part and assembled/installed to achieve the objectives.
- Construction projects work against defined scope, schedules, and budget to achieve the specified result.
- Performance of construction projects can be evaluated only after it is completed and put into use/operation.

3.4 Tools for Construction Projects

3.4.1 Introduction

Quality tools are the charts, check sheets, diagrams, graphs, techniques, and methods that are used to create an idea, engender planning, analyze the cause, analyze the process, foster evaluation, and create a wide variety of situations for continuous quality improvement. Applications of tools enhance chances of success and help maintain consistency, accuracy, and increase efficiency and process improvement.

There are several types of tools, techniques, methods, in practice, which are used as quality improvement tools and have variety of applications in manufacturing and process industry. However, all of these tools are not used in construction projects due to the nature of construction projects that are customized and nonrepetitive. Some of these quality management tools that are most commonly used in construction industry are listed under the following broader categories:

1. The quality classic tools
2. Management and planning tools
3. Process analysis tools
4. Process improvement tools
5. Innovation and creative tools
6. Lean tools
7. Cost of quality
8. Quality function deployment
9. Six Sigma
10. Triz

3.4.2 Design Tools

3.4.2.1 Quality Function Deployment

Quality function deployment (QFD) is a technique for translating customer requirements into technical requirements. It was developed in Japan by Dr. Yoji Akao in the 1960s to transfer the concepts of quality control from the manufacturing process into the new product development process. QFD is referred to as the "voice of the customer," which helps in identifying and developing customer requirements through each stage of product or service development. It is a development process that utilizes a comprehensive matrix involving project team members.

QFD is being applied virtually in every industry and business—from aerospace, communication, and software to transportation, manufacturing, services, and construction. QFD helps in constructing one or more matrices containing information related to others. The assembly of several matrices showing correlation with one another is called "the house of quality" and is the most recognized form of QFD. The house of quality is made up of the following major components:

1. WHAT
2. HOW
3. Correlation matrix (Roof)—technical requirements
4. Interrelationship matrix
5. Target value
6. Competitive evaluation

Figure 3.2 illustrates the basic house of quality.

The WHAT is the first step in developing the house of quality. It is a structured set of needs/requirements ranked in terms of priority and the levels of importance being specified quantitatively. It is generated by using questions such as:

- What types of finishes are needed for the building?
- What type of air-conditioning system is required for the building?
- What type of communication system is required for the building?
- What type of flooring material is required?
- Does the building need a security system?

The HOW is the second step in which project team members translate the requirements (WHAT) into technical design characteristics (specifications) and are listed across the columns of the matrix.

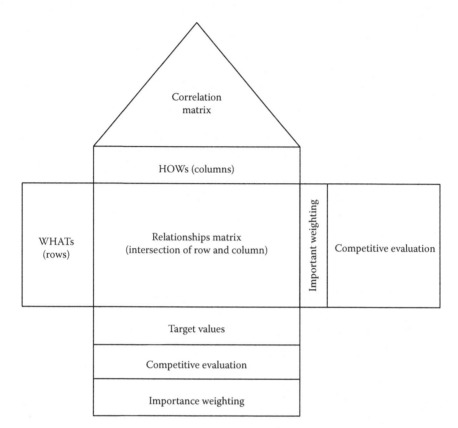

FIGURE 3.2
The house of quality. (*Source*: Pyzdek, Thomas (1999), *Quality Handbook*. Reprinted with permission from Quality America Inc.)

The correlation matrix identifies the technical interaction or physical relationship among the technical specifications. The interrelationship matrix illustrates team members' perceptions of the interrelationship between owner's requirements and technical specifications.

The bottom part allows for technical comparison between possible alternatives, target values for each technical design characteristic, and performance measurement.

The right side of the house of quality matrix is used for planning. It illustrates customer perceptions observed in the market survey.

The QFD technique can be used to translate the owner's need/requirements into developing a set of technical requirements during conceptual design.

Figure 3.3 illustrates the house of quality for a smart building system.

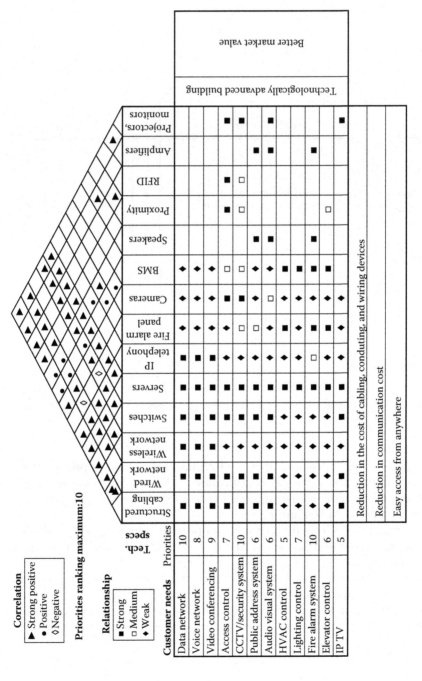

FIGURE 3.3
House of quality for smart building system.

3.4.2.2 Six Sigma DMADV Tool in Design of Construction Projects

3.4.2.2.1 Define Phase

What is important? (Define the project goals and customer deliverables.) The key deliverables of this phase are

- Establish the goal
- Identify the benefits
- Select project team
- Develop project plan
- Develop project charter

3.4.2.2.1.1 Goal Develop Construction project design using Six Sigma tool to meet owner's needs and satisfy the project quality requirements.

3.4.2.2.1.2 Benefits The measurable benefits of adopting this process will

- Minimize design errors
- Minimize omissions
- Reduce design rework
- Reduce risk and liabilities
- Provide construction project (building) design that will meet owner's needs without rejection of documents by the owner's review team (Project Manager, Construction Manager) at the first submission itself and without affecting the construction process, thus reducing external failure cost.
- Not affect the construction process, thus reducing external failure cost
- Reduce construction overruns
- Increase customer satisfaction
- Improve reputation
- Increase profits

3.4.2.2.1.3 Selection of Team The team shall consist of the following:

1. Sponsor—project manager
2. Champion:
   - Design manager (civil/structural)
   - Quality manager

3. Team leader—principal engineer (civil/structural)
4. Team members—structural design engineer, CAD technician, quality control engineer, quantity surveyor, cost engineer, planner, owner's representative, and end user

Similar teams can be organized for other trades such as architectural, HVAC, mechanical (P&FF), electrical including low-voltage systems, landscape, and external works.

Figure 3.4 illustrates the project design organization chart.

3.4.2.2.1.4 Project Plan   Time frame in the form of Gantt chart shall be prepared to meet the target dates for

- Data collection and analysis
- Development of concept design
- Submission and approval of concept design for client review and comments
- Development of schematic design
- Submission and approval of schematic design for client review and comments
- Development of detail design (design development)
- Regulatory/authority approvals
- Preparation of contract documents
- Submission of detail design and documents for client review and comments
- Submission of the project design and documents (final)
- Review and coordination meetings (both internal and with client)
- Monitoring of design schedule
- Monitoring and controlling resources

3.4.2.2.1.5 Project Charter   In this example, the project objective is to create construction project (building) design within stipulated schedule that

- Will meet owner's need
- Will meet relevant standards, codes, regulatory requirements
- Will be approved by the owner/owner's representative at the first submission without any major comments on the design and specification documents
- Will have all documents accurate and shall not affect construction quality

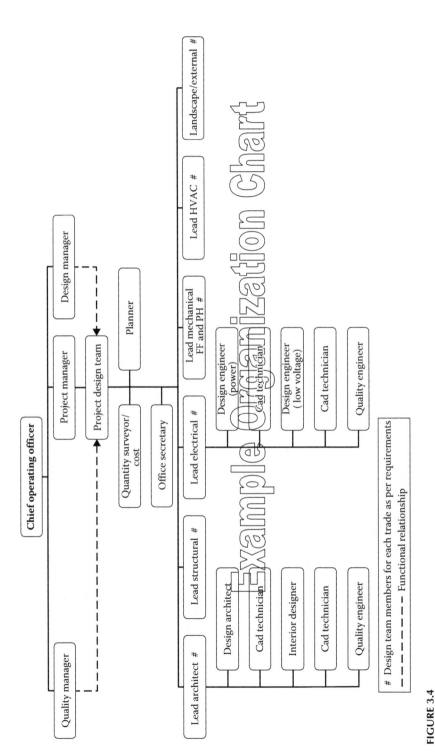

FIGURE 3.4
Project design organization chart.

3.4.2.2.2 Measure Phase

What is needed? (Measure and determine customer needs and specifications.)
The key deliverables in this phase are

1. Matrix of owner's requirements (Terms of Reference—TOR)-scope of work
2. Codes and standards to be followed
3. Regulatory requirements
4. Sustainability (environmental, social, economical)
5. Implementation of LEED requirements
6. Energy conservation requirements
7. Maintainability
8. Constructability
9. HSE requirements
10. Fire protection requirements
11. Project schedule
12. Project budget
13. Design review procedure by the owner/client
14. Other disciplines' requirements for coordination purpose
15. Procedure to incorporate changes/revisions requested by the client/owner
16. Number of drawings to be produced
17. Duration of completion of project design and submission of documents at different stages/phases of the project
18. Quality system requirements within the organization or portion of organization
19. Any special condition by the client

3.4.2.2.3 Analyze Phase

How will we fulfill? (Analyze process options and prioritize based on capability to satisfy customer requirements.)
The key deliverables in this phase are

- Data collection
- Prioritization of data under major variables
- Compliance to organization's procedures and guidance

3.4.2.2.3.1 Data Collection   Data shall be collected at different phases of construction project life cycle, for design considerations, to meet requirements for development of the following phases:

1. Concept design
2. Schematic/preliminary design
3. Detailed design/design development

The following data shall be collected to develop construction project design at different phases:

- Identification of need by the owner
- Identify number of floors
- Identify building usage
- Identify technical and functional capability requirements
- Soil profile and laboratory test of soil
- Identify topography of the project site
- Identify wind load, seismic load, dead load, and live load
- Identify existing services passing through the project site
- Identify existing roads, structure surrounding the project site
- Identify environmental compatibility requirements
- Identify energy conservation requirements
- Identify sustainability requirements
- Identify regulatory/authority requirements
- Identify codes and standards to be followed
- Identify social responsibility requirements
- Identify health and safety features
- Identify fire protection requirements
- Identify aesthetics requirements
- Identify zoning requirements
- Identify project constraints
- Identify ease for constructability
- Identify critical activities during construction
- Identify method statement requirements
- Identify requirements of other disciplines for coordination purpose
- Identification of team members
- Project time schedule

- Financial implications and resources for the project
- Identify cost-effectiveness of the project
- Identify 3D information areas
- Identify suitable software program
- Identify number of drawings to be produced
- Identify milestone for development of each phase of design

3.4.2.2.3.2 Arrangement of Data   The generated data shall be prioritized in an orderly arrangement under the following major variables during designing of concept, schematic, and design development phases:

- Owner's need
- Regulatory compliance
- Sustainability
- Safety
- Constructability

3.4.2.2.4 Design Phase

How we build it? (Design detailed process(es) capable of satisfying customer requirements.)
  The key deliverables in this phase are

- Development of concept design
- Development of schematic design
- Development of detail design

3.4.2.2.4.1 Development of Concept Design   While developing the concept design, the designer shall consider the following:

- Project goals/owner need
- Number of floors
- Usage of building
- Technical and functional capability
- Regulatory requirements
- Authorities requirements
- Environmental compatibility
- Constructability

- Health and safety
- Cost-effectiveness over the entire life cycle of project

3.4.2.2.4.2 Development of Schematic Design   While developing the schematic design, the designer shall consider the following:

- Concept design deliverables
- Site location in relation to the existing environment
- Building structure
- Floor grade and system
- Tentative size of columns, beams
- Stairs
- Roof
- Authorities' requirements
- Energy conservation issues
- Available resources
- Environmental issues
- Sustainability
- Requirements of all stake holders
- Optimized life-cycle cost (value engineering)
- Constructability
- Functional/aesthetic aspect
- Services requirements
- Project schedule
- Project budget
- Preliminary contract documents (outline specifications)

3.4.2.2.4.3 Development of Detail Design   While developing the detail design, the designer shall consider the following:

- Schematic design deliverables
- Authorities' requirements
- Energy conservation issues
- Environmental issues
- Sustainability
- Requirements of all stake holders
- Environmental compatibility

- Available resources
- Number of floors
- Property limits/surrounding areas
- Excavation
- Dewatering
- Shoring
- Backfilling
- Substructure
- Design of foundation based on field and laboratory test of soil investigation
- Subsurface profiles and subsurface conditions, and subsurface drainage
- Coefficient of sliding on foundation
- Degree of difficulty for excavation
- Method of protection of below-grade concrete members against impact of soil and groundwater
- Geotechnical design parameters
- Design load such as dead load, live load, and seismic load
- Grade and type of concrete
- Type of footings
- Type of foundation
- Energy-efficient foundation
- Size of bars for reinforcement and the characteristic strength of bars
- Clear cover for reinforcement
- Reinforcement bar schedule, stirrup spacing
- Superstructure
- Columns
- Walls
- Stairs
- Beams
- Slab
- Parapet wall
- Height of each floor
- Beam size and height of beam
- Location of columns in coordination with architectural requirements
- Openings for services
- Deflection which may cause fatigue of structural elements, crack or failure of fixtures, fittings or partitions, or discomfort of occupants

- Movement and forces due to temperature
- Equipment vibration criteria
- Expansion joints
- Insulation
- Concrete tanks (water storage)
- Services requirements (shafts, pits)
- Shafts and pits for conveying system
- Building services to fit in the building
- Coordination with other trades and conflict resolution
- Calculations required as per contract requirements

3.4.2.2.4.4 Preparation of Contract Documents   Contract documents shall be prepared as per MasterFormat® (2014 edition).

3.4.2.2.5 Verify Phase

How do we know it will work? (Verify design performance capability.)
The key deliverables in this phase are

- Review and check the design for quality assurance using thorough itemized review check lists to ensure that design drawings fully meet the owner's objectives/goal.
- Review and check contract documents (design drawings, specifications, and contract documents).
- Check for accuracy.
- Check calculations.
- Review studies and reports.
- Review discipline requirements.
- Review interdiscipline requirements and conflict.
- Review constructability.
- Management review.

After verification, the documents can be released for submission.
In case of any comments from the client/client's representative, the design shall be reviewed and modified accordingly.

3.4.2.3 Design of Experiments

Design of experiments (DOE) is a tool that can be used in a variety of experimental situations. DOE allows for multiple input factors (assumptions) to be manipulated/evaluated determining their effect on a desired output (result).

It looks at interrelationship between many variables at one time. It identifies important interactions that may be missed when experimenting with one factor at a time. All possible combinations can be investigated/evaluated, or only a portion of the possible combination. A well-performed experiment may provide answers to questions such as

- What are the key factors in a process?
- At what settings would the process deliver acceptable performance?
- What are the key main and interaction effects on the process?
- What settings would bring about less variation in the output?
- What settings and process will result in acceptable product and customer satisfaction?

In construction projects, the concept of DOE can be termed as development and study of several alternatives. The study of alternatives and evaluation of their impacts can vary significantly, depending on the size and complexity of the project. DOE in construction projects can be used to evaluate

- Conceptual alternatives
- Functional alternatives
- Scheduling alternatives
- Cost alternatives
- Environmental impacts

3.4.3 Project Control

The principle of PDCA cycle model can be used to describe monitoring and controlling construction process activities as follows:

Plan: Determine the information needs, data collection methods, and frequency of data collection.

Do: Collect status information needs, record status, and report progress.

Check: Compare actual performance (status) with planned performance (baseline) and analyze issues.

Act: Take corrective or preventive action, update project plan, update project documents and implement approved changes.

Figure 3.5 illustrates project monitoring and controlling process cycle.

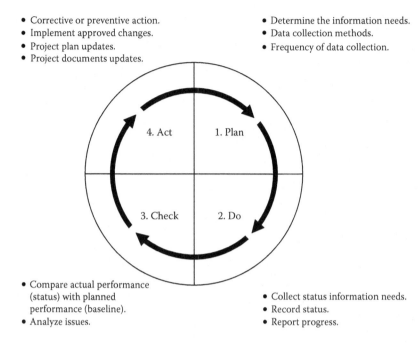

- Corrective or preventive action.
- Implement approved changes.
- Project plan updates.
- Project documents updates.

- Determine the information needs.
- Data collection methods.
- Frequency of data collection.

- Compare actual performance (status) with planned performance (baseline).
- Analyze issues.

- Collect status information needs.
- Record status.
- Report progress.

FIGURE 3.5
Project monitoring and controlling process cycle.

3.4.4 Lean Tools

Lean is a systematic approach and commitment to customer value that identifies and eliminates waste by making work flow across the entire value stream, at the pull of the customer, with an unwavering commitment to the continuous pursuit of perfection. Lean tools

- Create stability/predictability on the projects
- Remove waste in construction design and construction practice, thus improving productivity and reducing time and cost to build the project
- Result in safer projects

3.4.4.1 Concurrent Engineering

Product life cycle begins with need and extends through concept design, preliminary design, detail design, production or construction, product use, phase out, and disposal. Concurrent engineering is defined as systematic approach to create a product design that simultaneously considers all the

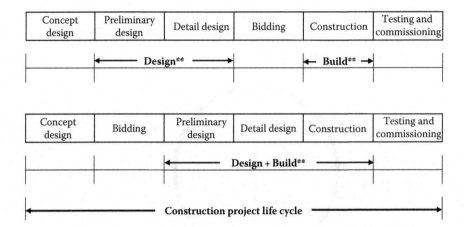

FIGURE 3.6
Concurrent engineering for construction life cycle.

elements of product life cycle, thus reducing the product life-cycle time. It is used to expedite the development and launch of a new product. In construction projects, construction can simultaneously start while the design is under development.

Figure 3.6 illustrates concurrent engineering for construction project life cycle.

3.4.4.2 Value Stream Mapping

This tool is used to establish flow of material or information and eliminating waste and adding value. Value stream mapping is used to identify areas for improvement.

Figure 3.7 illustrates value stream mapping diagram for emergency power system.

3.4.4.3 5S for Construction Projects

5S is a systematic approach for improving quality and safety by organizing a workplace. It is a methodology that advocates

- What should be kept
- Where it should be kept
- How it should be kept

5S is a Japanese concept of housekeeping having reference to five Japanese words starting with letter S. Table 3.2 shows these words with their English equivalent words and what these words stand for.

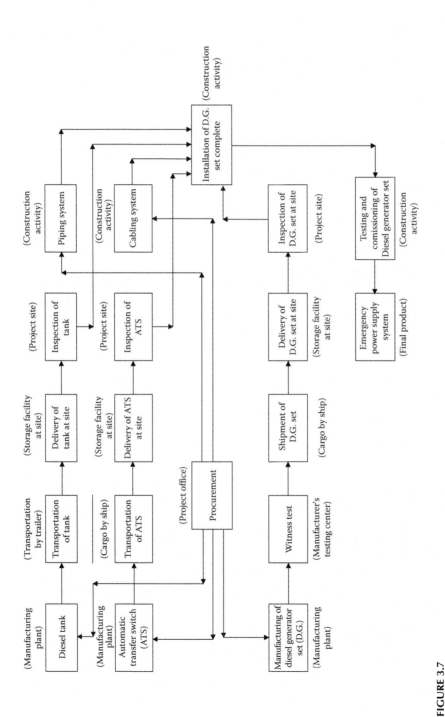

FIGURE 3.7
Value stream mapping for emergency power system.

TABLE 3.2

5S Concept

Japanese	English Equivalent	Stands For
Seiri	Sort	• Keeping only essential items • Sort out necessary from unnecessary and discard the unnecessary
Seiton	Set in order/simplify	• Keep items in a systematic/orderly manner to make easy traceability
Seiso	Shine/sweep	• Maintain cleanliness by arranging things in dirt-free and tidy status, making things always ready to use
Seiketsu	Standardize	• Keep work area organized for operating in a consistent and standardized fashion
Shitsuke	Sustain	• Maintain what has been accomplished

Construction projects are temporary workplaces. Construction project materials are stored in both covered as well as in open areas. The contractor has to keep all the material in the available space in neat and clean ready-to-use condition. The area should be safe and risk-free for free movement of the work personnel. The contractor has to consider site safety while arranging the storage area.

The contractor is responsible for providing all the resources to build the project/facility. These resources are mainly the following:

1. Manpower
2. Construction equipment, machinery, and tools
3. Material to be used/installed in the project
4. Consumables

The following points have to be considered while planning the layout for storage of construction material:

1. Construction documents specify the minimum number of construction equipment, machinery, and tools to be made available at the construction site during the construction process.
2. Materials to be used/installed in the project are documented in the bill of quantity (BOQ) or bill of material (BOM), and contract drawings and documents.
3. Consumables are required by the contractor to fabricate/install/assemble the equipment/panel/material.

Table 3.3 shows an example of how a 5S program can be developed for construction projects.

TABLE 3.3

5S for Construction Projects

Sr. No.	5S	Related Action
1	Sort	• Determine what is to be kept in open and what under shed
		• Allocate area for each type of construction equipment and machinery
		• Allocate area for electrical tools
		• Allocate area for hand tools
		• Allocate area for construction material/equipment to be used/installed in the project
		• Allocate area for hazardous, inflammable material
		• Allocate area for chemicals, paints
		• Allocate area for spare part for maintenance
2	Set in order	• Keep/arrange equipment in such a way that their maneuvering/movement shall be easy
		• Vehicles to be parked in the yard in such a way that frequently used vehicles are parked near the gate
		• Frequently used equipment/machinery to be located near the work place
		• Set boundaries for different type of equipment and machinery
		• Identify and arrange tools for easy access
		• Identify and store material/equipment as per relevant division/section of contract documents
		• Identify and store material in accordance with their usage as per construction schedule
		• Determine items which need special conditions
		• Mark/tag the items/material
		• Display route map and location
		• Put the material in sequence as per their use
		• Frequently used consumables to be kept near work place
		• Label on the drawer with list of contents
		• Keep shuttering material at one place
		• Determine inventory level of consumable items

(Continued)

TABLE 3.3 (*Continued*)

5S for Construction Projects

Sr. No.	5S	Related Action

3	Sweeping	• Clean site on daily basis by removing
		   • Cut pieces of reinforced bars
		   • Cut pieces of plywood
		   • Left out concrete
		   • Cut pieces of pipes
		   • Cut pieces of cables and wires
		   • Used welding rods
		• Clean equipment and vehicles
		• Check electrical tools after return by the technician
		• Attend to breakdown report

4	Standardize	• Standardize the store by allocating separate areas for material used by different divisions/sections • Standardize area for long lead items • Determine regular schedule for cleaning the work place • Make available standard tool kit/box for a group of technicians • Keep everyone informed of their responsibilities and related area where the things are to be placed and are available • Standardize the store for consumable items • Inform suppliers/vendors in advance the place for delivery of material
5	Sustain	• Follow the system till the end of project

The 5S program helps

- Reduce time to search equipment, tools, material, and consumables
- Improve activity timing
- Increase space for storage
- Improve safety
- Organize the workplace

3.4.5 Improvement Tools

3.4.5.1 PDCA

P-D-C-A is mainly used for continuous improvement. It consists of a four-step model for carrying changes. PDCA cycle model can be developed as a process improvement tool to reduce cost of quality.

Figure 3.8 illustrates PDCA cycle for preparation of shop drawings.

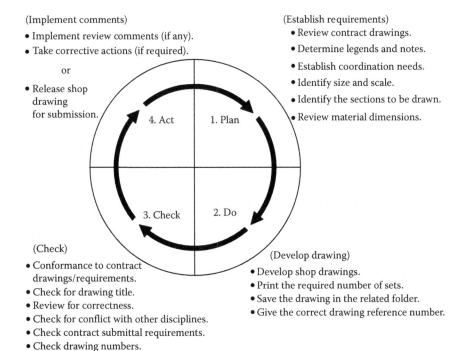

(Implement comments)
- Implement review comments (if any).
- Take corrective actions (if required).

or

- Release shop drawing for submission.

(Establish requirements)
- Review contract drawings.
- Determine legends and notes.
- Establish coordination needs.
- Identify size and scale.
- Identify the sections to be drawn.
- Review material dimensions.

4. Act 1. Plan

3. Check 2. Do

(Check)
- Conformance to contract drawings/requirements.
- Check for drawing title.
- Review for correctness.
- Check for conflict with other disciplines.
- Check contract submittal requirements.
- Check drawing numbers.
- Check for size and scale.

(Develop drawing)
- Develop shop drawings.
- Print the required number of sets.
- Save the drawing in the related folder.
- Give the correct drawing reference number.

FIGURE 3.8
PDCA cycle for preparation of shop drawing.

3.4.5.2 Kaizan

It is used for continually improving through small changes to eliminate waste from manufacturing process by combining the collective talent of every employee of the company.

The concept of Kaizan began in the 1950s, which according to Massaki Imai, the father of Kaizan Strategy, is the most important concept of Japanese management—the key for Japanese success. The spirit of Kaizan is all about achieving improvement by taking small steps instead of drastic, rigorous changes. In Japanese, Kaizan means "small incremental continuous improvement." Kaizan is a philosophy that focuses both on the process and the results. In Kaizan management, the results are continuous and there is no dramatic visibility. It has involvement of everyone and is group activity. Kaizan management is dedicated to the improvement of

- Design
- Productivity
- Efficiency
- Quality
- Business excellence

The Kaizan method and techniques are valuable instruments that are used to achieve competitive advantage.

The Kaizan systems focus on

1. Total quality management (TQM)
2. Totally productive maintenance
3. Suggestion system
4. Just-in-time production system
5. Organization politics management
6. Activities in small groups

The concept of Kaizan is based on the principles of TQM and various philosophies of Quality Gurus.

3.4.5.2.1 Total Quality Management

TQM is a companywide approach through involvement of top management by formulating management policies/philosophies to improve the product/process quality for customer satisfaction and achieving sustainable competitive advantage. In construction projects, TQM is a collaborative and collective effort by all the participants (owner, designer/consultant, and contractor).

3.4.5.2.2 Total Productive Maintenance

In construction projects, activity relates to achieving qualitative, competitive, and economical project, fulfilling the defined scope to the satisfaction of owner/client/end user.

3.4.5.2.3 Suggestion System

Construction projects have involvement of mainly three parties:

1. Owner
2. Designer/consultant
3. Contractor

Coordination among all the parties is required to achieve owner need.

3.4.5.2.4 Just in Time

It is used to reduce inventory levels, improve cash flow, and reduce storage space requirements for material. For example:

1. Concrete block can be received at site just before start of block work and can be stacked near the work area where masonry work is in progress.
2. Chiller can be received at site and directly placed on the chiller foundation without storing in the storage yard.

3.4.5.2.5 Organization Politics Management

TQM is a management approach. The management has to establish clear mid- and long-term vision and strategies. The company has to set standards and continually improving these standards as there is always room for improvement.

3.4.5.2.6 Activities in Small Groups

In construction projects, the activities (scope of work) should be broken down into a group of smaller subgroups/subsystems and then into small, well-defined activities. This breakdown is referred to as work breakdown structure (WBS).

As regards the contracting system, the works should be outsourced to a specialist in a particular discipline or area.

3.4.5.3 Mistake Proofing

Mistake proofing is used to eliminate the opportunity for error by detecting the potential source of error. Mistakes are generally categorized as follows:

1. Information
2. Mismanagement
3. Omission
4. Selection

Table 3.4 illustrates mistake proofing chart for eliminating design error.

3.4.5.4 Corrective Action

A corrective action deals with nonconformity or a problem that has occurred by reworking or restoring or correcting any affected product/outputs to the acceptable situation where it should have been in the first place. Immediate action should be taken to first correct and damage done by the problem and to prevent any additional impact from the problem. Figure 3.9 illustrates the corrective action process.

TABLE 3.4

Mistake Proofing for Eliminating Design Errors

Serial Number	Items	Points to Be Considered to Avoid Mistakes
1	Information	1. Terms of reference (TOR) 2. Client's preferred requirements matrix 3. Data collection 4. Regulatory requirements 5. Codes and standards 6. Historical data 7. Organizational requirements
2	Mismanagement	1. Compare production with actual requirements 2. Interdisciplinary coordination 3. Application of different codes and standards 4. Drawing size of different trades/specialist consultants
3	Omission	1. Review and check design with TOR 2. Review and check design with client requirements 3. Review and check design with regulatory requirements 4. Review and check design with codes and standards 5. Check for all required documents
4	Selection	1. Qualified team members 2. Available material 3. Installation methods

Source: Abdul Razzak Rumane (2013), *Quality Tools for Managing Construction Projects,* CRC Press, Boca Raton, FL. Reprinted with permission from Taylor & Francis Group.

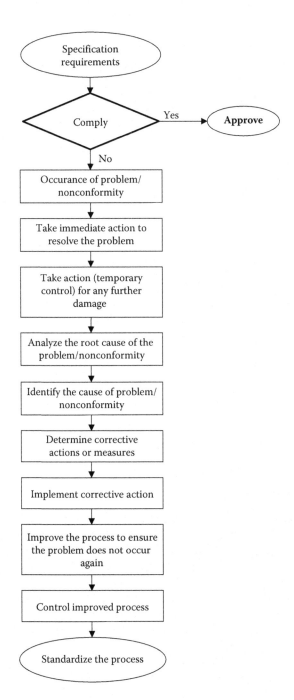

FIGURE 3.9
Corrective action process.

3.4.5.5 Preventive Action

It is an action taken to prevent the occurrence of a problem in an area/product that has not already experienced the problem. In order to identify the risk/nonconformity that need preventive action, adequate monitoring and controls must be in place in the quality system to assure that potential problems are identified and eliminated before they occur.

In order to prevent occurrence of any problem or nonconformity, the following actions are necessary:

- Process performance monitoring and control
- Process planning
- Process review
- Statistical analysis and observance of process trend
- Risk assessment
- Implementation of quality management system
- Customer feedback and priorities for improvement
- Monitoring changes in regulatory requirements
- Latest technology
- Employ qualified personnel having experience in similar types of products, projects, processes, and services

3.5 Systems Engineering

3.5.1 Introduction

Systems are pervasive throughout the universe in which we live. This world can be divided into the natural world and the human-made world. Systems appeared first in natural forms and subsequently with the appearance of human beings. Systems were created based on components, attributes, and relationships.

Systems engineering and analysis, when coupled with new emerging technologies, reveal unexpected opportunities for bringing new improved systems and products into being that will be more competitive in the world economy. Product competitiveness is desired by both commercial and public-sector producers worldwide to meet consumer expectations. These technologies and processes can be applied to construction projects. The systems engineering approach to construction projects helps us understand the entire process of project management in order to manage its activities at different levels of various phases to achieve economical and competitive results. The cost-effectiveness of the resulting technical activities can be enhanced by giving more attention to what they are to do, before addressing what they are composed of. To ensure economic competitiveness regarding the product, engineering must become more closely associated with economics and economic facilities. This is best accomplished through the life-cycle approach to engineering.

Experience in recent decades indicates that properly coordinated and functioning human-made systems will result in a minimum of undesirable side effects through the application of this integrated, life-cycle-oriented "systems" approach. The consequences of not applying systems engineering in the design and development and/or reengineering of systems have been disruptive and costly.

The systems approach is a technique, which represents a broad-based systematic approach to problems that may be interdisciplinary. It is particularly useful when problems are affected by many factors, and it entails the creation of a problem model that corresponds as closely as possible to reality. The systems approach stresses the need for the engineer to look for all the relevant factors, influences, and components of the environment that surround the problem. The systems approach corresponds to a comprehensive attack on a problem and to an interest in, and commitment to, formulating a problem in the widest and fullest manner that can be professionally handled.

3.5.2 System Definition

There are many definitions of *system*. One dictionary definition calls it "a group or combination of interrelated, independent or interacting elements forming a collective entity." A system is an assembly of components or elements having a functional relationship to achieve a common objective for useful purpose. A system is composed of components, attributes, and relationships. These are described as follows:

1. Components are the operating parts of the system consisting of input, process, and output. Each system component may assume a variety of values to describe a system state, as set by some control action and one or more restrictions.

2. Attributes are the properties or discernible manifestations of the components of a system. These attributes characterize the system.

3. Relationships are the links between components and attributes.

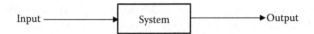

FIGURE 3.10
Black Box.

The properties and behavior of each component of the set have an effect on the properties and behavior of the set as a whole and depend on the properties and behavior of at least one other component on the list. The components of the system cannot be divided into independent subsets. A system is more than the sum of its components and parts. Not every set of items, facts, methods, or procedures is a system. To qualify the system, it should have a functional relationship, interaction between many components, and useful purpose. The purposeful action performed by a system is its function. A basic behavioral concept of a system is that it is a device that accepts one or more inputs and generates from them one or more outputs. This simple behavioral approach to systems is generally known as the Black Box and is represented schematically in Figure 3.10. The Black Box system phenomenon establishes the functional relationship between system inputs and outputs.

Every system is made up of components and components that can be broken down into similar components. If two hierarchical levels are involved in a given system, the lower is conveniently called a subsystem. The designation of system, subsystem, and components are relative because the system at one level in the hierarchy is the component at another level. Everything that remains outside the boundaries of the system is considered to be environmental. Material, energy, and/or information that pass through the boundaries are called "inputs" to the system. In reverse, material, energy, and/or information that pass from the system to the environment are called outputs.

Accordingly, a system is an assembly of components or elements having a functional relationship to achieve a common objective for a useful purpose.

3.5.3 Systems Engineering

INCOSE (International Council on Systems Engineering) defines systems engineering as follows:

> Systems Engineering is an interdisciplinary approach and means to enable the realization of successful systems. It focuses on defining customer needs and required functionality early in the development cycle, documenting cycle, documenting requirements, and then

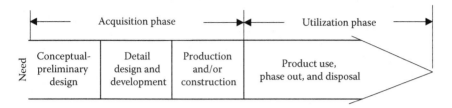

FIGURE 3.11
The product life cycle. (*Source*: Blanchard, B.S. and Fabrycky, W.J. (1998), *Systems Engineering and Analysis*. Reprinted with permission from Pearson Education.)

proceeding with design synthesis and system validation while considering the complete problem:

Operations	Cost & Schedule
Performance	Training & Support
Test	Disposal
Manufacturing	

Systems Engineering integrates all the disciples and specialty groups into a team effort forming a structural development process that proceeds from concept to production to operation. Systems Engineering considers both the business and the technical needs of all customers with the goal of providing a quality product that meets the user needs.

The system life-cycle process is illustrated in Figure 3.11 and is fundamental to the application of systems engineering.

The life cycle begins with the identification of need and extends through conceptual and preliminary design, detail design, and development, production and/or construction, product use, phase-out, and disposal. The program phases are classified as acquisition and utilization to recognize procedure and customer activities. This classification represents a generic approach. Sometimes, the acquiring process may involve both the customer and the producer (or contractor), whereas acquiring may include a combination of contractor and consumer (or ultimate user) activities.

In general, engineering has focused mainly on product performance as the main objective rather than on development of overall system of which the product is a part. Application of a systems engineering process leads to reduction in the cost of design development, production/construction, and operation, and hence results in reduction in life-cycle cost of the product; thus, the product becomes more competitive and economical. Systems engineering provides the basis for a structural and logical approach. The need for systems engineering increases with the size of projects. Application areas of systems engineering are illustrated in Figure 3.12.

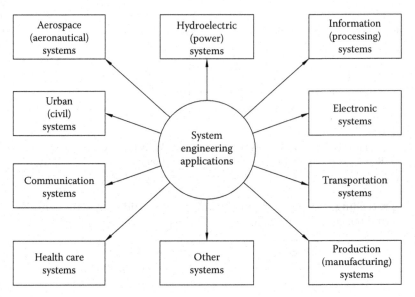

FIGURE 3.12
Application areas for systems engineering. (*Source*: Blanchard, B.S. and Fabrycky, W.J. (1998), *Systems Engineering and Analysis*. Reprinted with permission from Pearson Education.)

3.6 Construction Project Life Cycle

Most construction projects are custom-oriented, having a specific need and a customized design. It is always the owner's desire that his or her project should be unique and better. Further, it is the owner's goal and objective that the facility is completed on time. Expected completion time is important from both financial and acquisition of the facility by the owner/user.

The system life cycle is fundamental to the application of systems engineering. Detailed presentations of the elaborate technological activities and interaction that must be integrated over the system life cycle are shown in Figure 3.13. This figure summarizes major technical functions performed during the acquisition and utilization process of the system life cycle.

A systems engineering approach to construction projects helps in understanding the entire process of project management and in managing and controlling its activities at different levels of various phases to ensure timely completion of the project with economical use of resources to make the construction project most qualitative, competitive, and economical.

Systems engineering starts from the complexity of the large-scale problem as a whole and moves toward the structural analysis and partitioning process until the questions of interest are answered. This process of decomposition is called a work breakdown structure (WBS). The WBS is a hierarchical

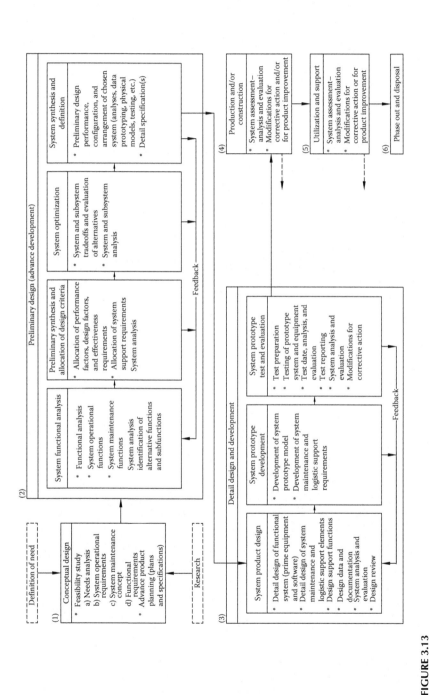

FIGURE 3.13

The system life cycle process. (*Source*: Blanchard, B.S. and Fabrycky, W.J. (1998), *Systems Engineering and Analysis*. Reprinted with permission from Pearson Education.)

representation of system levels. Being a family tree, the WBS consists of a number of levels, starting with the complete system at level 1 at the top and progressing downward through as many levels as necessary to obtain elements that can be conveniently managed.

The benefits of systems engineering applications are

- Reduction in the cost of system design and development, production/construction, system operation and support, system retirement, and material disposal
- Reduction in system acquisition time
- More visibility and reduction in the risks associated with the design decision-making process

Shtub et al. (1994) have divided the project into five phases, as illustrated in Figure 3.14.

Representative construction project life cycle, as per Morris, has four stages (phases) for construction project and is illustrated in Figure 3.15.

Though it is difficult to generalize project life cycle to system life cycle, considering that there are innumerable processes that make up the construction process, the technologies and processes as applied to systems engineering can also be applied to construction projects. The number of phases shall depend on the complexity of the project.

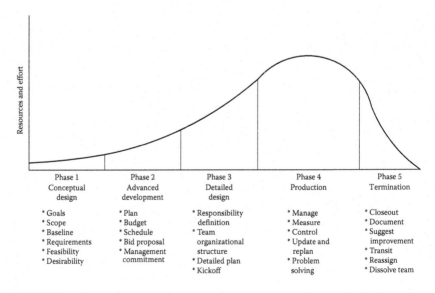

FIGURE 3.14
Project life cycle. (*Source*: Shtub, A., Bard, J.F., and Globerson, S. (1994), *Project Management*. Reprinted with permission from Pearson Education.)

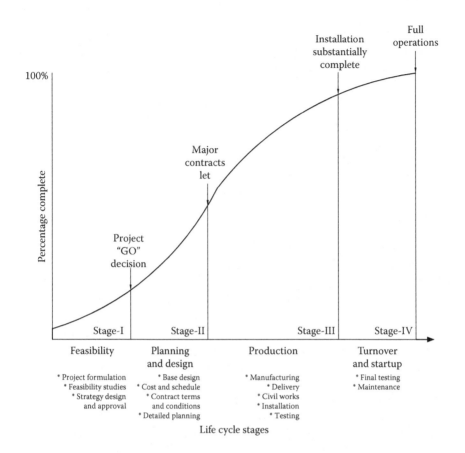

FIGURE 3.15
Representative construction project life cycle; per Morris. (*Source*: PMBOK® Guide 2000 Edition. Reprinted with permission from Project Management Institute (PMI).)

As per Construction-Extension-PMBOK® Guide-Third Edition: "Most Construction projects can be viewed in Five Phases, although they are sometimes shortened to four, each one of these phases can be treated like a project itself, with all of the process groups operating as they do for the overall project. These phases are concept, planning (and development), detail design, construction, and start-up and turnover." Duration of each phase may vary from project to project. Based on the concept of project life cycle shown in Figures 3.13 through 3.15, it is possible to evolve a comprehensive life cycle for construction projects, which may have five of the most common phases. These are as follows:

1. Conceptual design
2. Preliminary design

3. Detailed design

4. Construction

5. Testing, commissioning, and handover

However, in major projects, preparation of construction (contract) documents, its review and approval, and the process of bidding/tendering take considerable time. Thus, for a major construction project, it is ideal to divide the construction project life cycle into seven phases. These are

1. Conceptual design

2. Schematic design

3. Detail design

4. Construction (contract) document

5. Bidding and tendering

6. Construction

7. Testing, commissioning, and handover

Each phase can further be subdivided into the WBS principle to reach a level of complexity where each element/activity can be treated as a single unit that can be conveniently managed. WBS represents a systematic and logical breakdown of the project phase into its components (activities). It is constructed by dividing the project into major elements with each of these being divided into subelements. This is done until a breakdown is achieved in terms of manageable units of work for which responsibility can be defined. WBS involves envisioning the project as a hierarchy of goal, objectives, activities, subactivities, and work packages. The hierarchical decomposition of activities continues until the entire project is displayed as a network of separately identified and nonoverlapping activities. Each activity will be single purposed, of a specific time duration, and manageable; its time and cost estimates easily derived, deliverables clearly understood, and responsibility for its completion clearly assigned. The work breakdown structure helps in

- Effective planning by dividing the work into manageable elements, which can be planned, budgeted, and controlled
- Assignment of responsibility for work elements to project personnel and outside agencies
- Development of control and information system

WBS facilitates the planning, budgeting, scheduling, and control activities for the project manager and its team. By application of WBS phenomenon, the construction phases are further divided into various activities. Division of these phases will improve the control and planning of the construction project at every stage before a new phase starts. The components/activities of construction project life-cycle phases divided on WBS principle are listed as follows:

1. Conceptual design
   - Identification of need
   - Feasibility
   - Identification of alternatives
   - Identification of project team
   - Development of concept design
   - Prepare time schedule
   - Estimate project cost
   - Quality management
   - Estimate resources
   - Identify/manage risk
   - Finalize concept design
2. Schematic design
   - Identification of schematic design requirements
   - Identification of project team
   - Development of schematic design
   - Regulatory/authorities' approval
   - Contract terms and conditions
   - Estimate preliminary schedule
   - Estimate project cost
   - Quality management
   - Estimate resources
   - Identification/management of risk
   - Perform value engineering study
   - Finalize schematic design
3. Design development
   - Identification of design development requirements
   - Identification of project team
   - Develop detail design of the works
   - Regulatory/authorities' approval
   - Prepare contract documents and specifications
   - Prepare schedule
   - Estimate cost
   - Quality management

- Estimate resources
- Identification/management of risk
- Finalize detail design

4. Construction documents
   - Identification of construction documents requirements
   - Identification of project team
   - Develop construction documents
   - Prepare project schedule
   - Estimate cost/budget
   - Quality management
   - Estimate resources
   - Identification/management of risk
   - Finalize construction documents

5. Bidding and tendering
   - Organize tender documents
   - Identification of project team
   - Identification of bidders
   - Management of tender documents
   - Identification/management of risk
   - Select contractor
   - Award contract

6. Construction
   - Mobilization
   - Identification of project team
   - Planning and scheduling
   - Execution of works
   - Management of resources/procurement
   - Monitoring and control
   - Quality management
   - Risk management
   - Contract management
   - Site safety
   - Inspection

7. Testing, commissioning, and handover
   - Develop testing and commissioning plan
   - Identification of project team

TABLE 3.5

Construction Project Life-Cycle (Design-Bid-Build) Phases

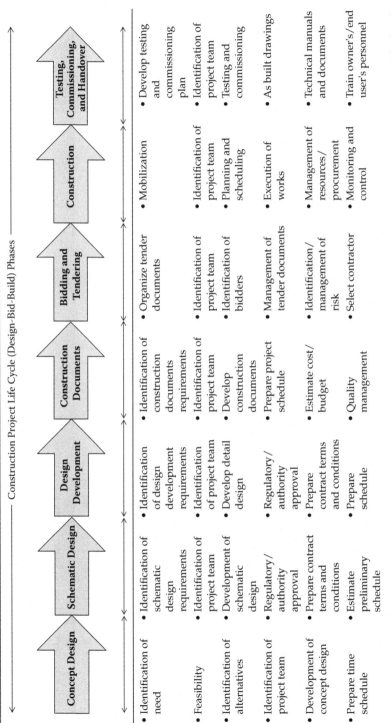

Construction Project Life Cycle (Design-Bid-Build) Phases

Concept Design	Schematic Design	Design Development	Construction Documents	Bidding and Tendering	Construction	Testing, Commissioning, and Handover
• Identification of need	• Identification of schematic design requirements	• Identification of design development requirements	• Identification of construction documents requirements	• Organize tender documents	• Mobilization	• Develop testing and commissioning plan
• Feasibility	• Identification of project team	• Identification of project team	• Identification of project team	• Identification of project team	• Identification of project team	• Identification of project team
• Identification of alternatives	• Development of schematic design	• Develop detail design	• Develop construction documents	• Identification of bidders	• Planning and scheduling	• Testing and commissioning
• Identification of project team	• Regulatory/authority approval	• Regulatory/authority approval	• Prepare project schedule	• Management of tender documents	• Execution of works	• As built drawings
• Development of concept design	• Prepare contract terms and conditions	• Prepare contract terms and conditions	• Estimate cost/budget	• Identification/management of risk	• Management of resources/procurement	• Technical manuals and documents
• Prepare time schedule	• Estimate preliminary schedule	• Prepare schedule	• Quality management	• Select contractor	• Monitoring and control	• Train owner's/end user's personnel

(Continued)

TABLE 3.5 (*Continued*)

Construction Project Life-Cycle (Design-Bid-Build) Phases

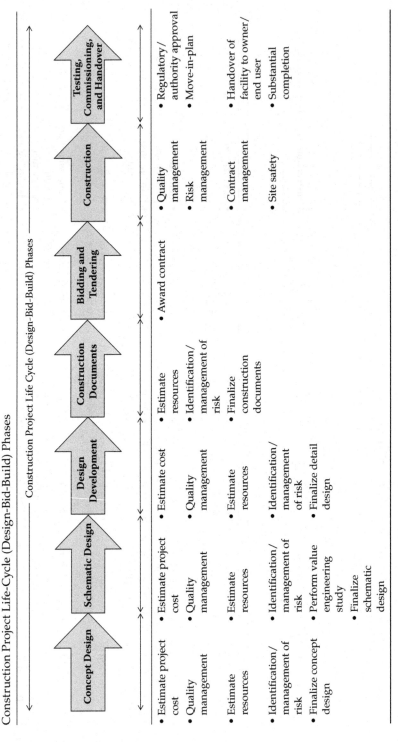

Concept Design	Schematic Design	Design Development	Construction Documents	Bidding and Tendering	Construction	Testing, Commissioning, and Handover
• Estimate project cost	• Estimate project cost	• Estimate cost	• Estimate resources	• Award contract	• Quality management	• Regulatory/authority approval
• Quality management	• Quality management	• Quality management	• Identification/management of risk		• Risk management	• Move-in-plan
• Estimate resources	• Estimate resources	• Estimate resources	• Finalize construction documents		• Contract management	• Handover of facility to owner/end user
• Identification/management of risk	• Identification/management of risk	• Identification/management of risk			• Site safety	• Substantial completion
• Finalize concept design	• Perform value engineering study	• Finalize detail design				
	• Finalize schematic design					

- Testing and commissioning
- As-built drawings
- Technical manual and documents
- Train owner's/end user's personnel
- Regulatory/authority approval
- Move-in-plan
- Handover of facility to owner/end user
- Substantial completion

Table 3.5 illustrates the subdivided activities/components of the construction project life cycle.

These activities may not be strictly sequential; however, the breakdown allows implementation of project management functions more effectively at different stages.

4

Quality in Construction Projects

4.1 Quality in Construction Projects

Construction projects are mainly capital investment projects. They are customized and nonrepetitive in nature. Construction projects have become more complex and technical, and the relationships and the contractual grouping of those who are involved are also more complex and contractually varied. The products used in construction projects are expensive, complex, immovable, and long-lived. Generally, a construction project comprises building materials (civil), electromechanical items, finishing items, and equipment. These are normally produced by other construction-related industries/manufacturers. These industries produce products as per their own quality management practices complying with certain quality standards or against specific requirements for a particular project. Owners of construction projects or their representatives have no direct control over these companies unless they themselves, their representatives, or appointed contractors commit to buying their product for use in their facility. These organizations may have their own quality management program. In manufacturing or service industries, quality management of all in-house manufactured products is performed by the manufacturer's own team or is under the control of the same organization that has jurisdiction over its manufacturing plants at different locations. Quality management of vendor-supplied items/products is carried out as stipulated in the purchasing contract as per the quality control specification of the buyer.

Construction projects are constantly increasing in technological complexity. Electromechanical services constitute between 25% and 35% of the total cost of a building project, depending on what type of technologically advanced services are required for the project. Figure 4.1 illustrates typical values of various trades of a major building construction project. In this project, the electromechanical work constitutes approximately 36% of the total project value, which shows the increasing technological complexity of building construction projects.

In addition, the requirements of construction clients are on the increase and, as a result, construction products (buildings) must meet varied performance standards (climate, rate of deterioration, maintenance, and so on).

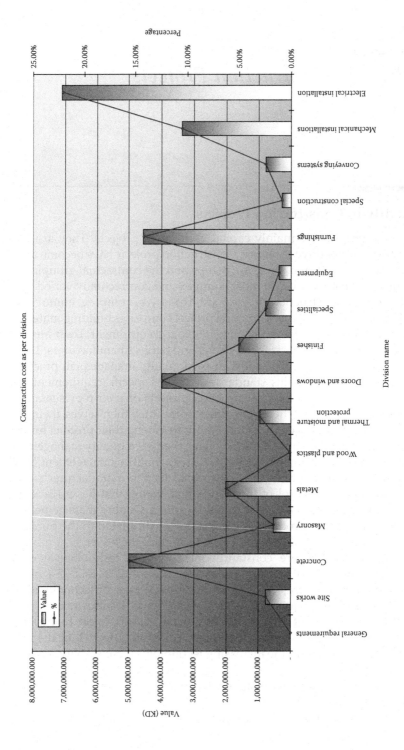

FIGURE 4.1
Divisional values of building construction project.

Therefore, to ensure the adequacy of client brief, which addresses the numerous complex client/user needs, it is now necessary to evaluate the requirements in terms of activities and their interrelationships.

Quality management in construction projects is different from that in manufacturing. Quality in construction projects encompasses not only the quality of products and equipment used in the construction but the total management approach to completing the facility as per the scope of works to customer/owner satisfaction within the budget and in accordance with the specified schedule to meet the owner's defined purpose. The nature of the contracts between the parties plays a dominant part in the quality system required from the project, and the responsibility for fulfilling them must therefore be specified in the project documents. The documents include plans, specifications, schedules, bill of quantities (BOQ), and so on. Quality control in construction typically involves ensuring compliance with minimum standards of material and workmanship in order to ensure the performance of the facility according to the design. These minimum standards are contained in the specification documents. For the purpose of ensuring compliance, random samples and statistical methods are commonly used as the basis for accepting or rejecting work completed and batches of materials. Rejection of a batch is based on nonconformance or violation of the relevant design specifications.

The survey of quality of construction by Federation Internationale des Ingenieurs-Conseils (FIDIC, the international association of consulting engineers) confirmed that failure to achieve appropriate quality of construction is a problem worldwide. Lack of quality in construction is manifested in poor or nonsustainable workmanship, unsafe structure, delays, cost overruns, and disputes in construction contracts.

Defects or failure in construction facilities can result in very large costs. Even with minor defects, reconstruction may be required and facility operation impaired.

Chung (1999) has described the quality of construction as follows:

> The quality of building work is difficult, and often impossible, to quantify since a lot of construction practices cannot be assessed in numerical terms. The framework of reference is commonly the appearance of final product. "How good is good enough?" is often a matter of personal judgment and consequently a subject of contention. In fact, a building is of good quality if it will function as intended for its design life. As the true quality of the building will not be revealed until many years after completion, the notion of quality can only be interpreted in terms of the design attributes. So far as the builder is concerned, it is fair to judge the quality of his work by the degree of compliance with the stipulations in the contract, not only the technical specifications but also the contract sum and the contract period. His client cannot but be satisfied if the contract is executed as specified, within budget and on time. Therefore, a quality product of building construction is one that meets all contractual requirements (including statutory regulations) at optimum cost and time. (p. 4)

About quality in construction, CII Source Document 79 (1992) describes that

> Quality has many meanings; however, for projects, conformance to established requirements has relevance and clarity. While simple, this definition cannot stand alone. Another term is needed for the term *requirements*. Requirements are contractually established characteristics of a product, process, or service. A characteristic is a physical or chemical property, a dimension, a temperature, a pressure, or any other specification used to define the nature of product, process or service.
>
> The requirements are initially set by client/customer (ordinarily the user/operator of the facility) and are then translated during the preplanning phase into a conceptual design and estimate developed into a project scope and more fully defined. During the Design phase, the requirements are translated into specific design documents (drawings, plans, specification, purchase orders, and the like). Procurement of fabricated items often proceeds concurrently with design. The products of design and procurement reach the construction site for erection and installation during the construction phase. (p. 5)

An implicit assumption in the traditional quality control practices is the notion of an acceptable quality level, which is an allowable fraction of defective items. Materials obtained from suppliers or work performed percentage should be within the acceptable quality level. Problems with materials or goods are corrected after delivery of the product. In contrast to this traditional approach of quality control is the goal of total quality control. In this system, no defective items are allowed anywhere in the construction process. While the zero defects goal can never be permanently obtained, it provides a goal so that an organization is never satisfied with its quality control program even if defects are reduced by substantial amounts year after year. This concept and approach to quality control was first developed in manufacturing firms in Japan and Europe but has since spread to many construction companies. Total quality control is a commitment to quality expressed in all parts of an organization and typically involves many elements. Design reviews to ensure safe and effective construction procedures are a major element. Other elements include extensive training for personnel, shifting the responsibility for detecting defects from quality control inspectors to workers, and continually maintaining equipment. Worker involvement in improved quality control is often formalized in quality circles in which groups of workers meet regularly to make suggestions for quality improvement. Material suppliers are also required to ensure zero defects in delivered goods. Initially, all materials from a supplier are inspected and batches of goods with any defective items are returned. Suppliers with good records can be certified, and such suppliers will not be subject to complete inspection subsequently.

Total quality management (TQM) is an organization-wide effort centered on quality to improve performance that involves everyone and permeates

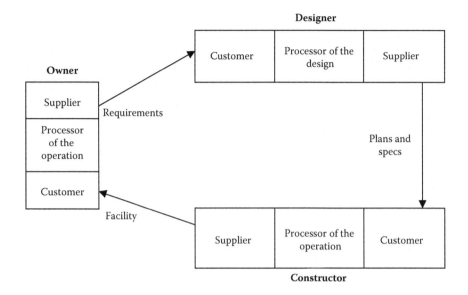

FIGURE 4.2
Juran's triple role concept applied to construction. (*Source*: CII Publication 10-4. Reprinted with permission of CII, University of Texas.)

every aspect of an organization to make quality a primary strategic objective. It is a way of managing an organization to ensure the satisfaction at every stage of the needs and expectations of both internal and external customers.

In case of construction projects, an organizational framework is established and implemented mainly by three parties: owner, designer/consultant, and contractor. Project quality is the result of aggressive and systematic application of quality control and quality assurance (QA). Figure 4.2 illustrates Juran's triple concept applied to construction.

Construction projects being unique and nonrepetitive in nature need specified attention to maintain the quality. Each project has to be designed and built to serve a specific need. TQM in construction projects typically involves ensuring compliance with minimum standards of material and workmanship in order to ensure the performance of the facility according to the design. TQM in a construction project is a cooperative form of doing the business that relies on the talents and capabilities of both labor and management to continually improve quality. The important factor in construction projects is to complete the facility as per the scope of works to customer/owner satisfaction within the budget and to complete the work within the specified schedule to meet the owner's defined purpose. Figure 4.3 shows the various elements that influence the quality of construction.

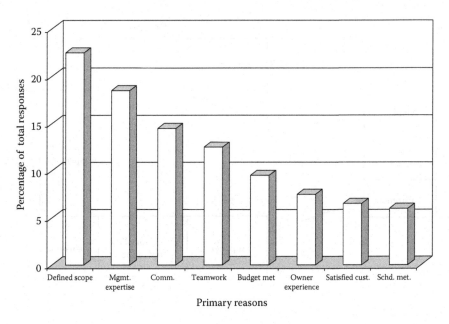

FIGURE 4.3
Primary reasons for quality. (*Source*: CII Document 79. Reprinted with permission of CII, University of Texas.)

Oberlender (2000) has observed:

> Quality in construction is achieved by the people who take pride in their work and have the necessary skills and experience to do the work. The actual quality of construction depends largely upon the control of construction itself, which is the principle responsibility of the contractor. ... What is referred to today as "quality control," which is a part of a quality assurance program, is a function that has for years been recognized as the inspection and testing of materials and workmanship to see that the work meets the requirements of the drawings and specifications. (p. 278)

Crosby's quality definition is "conformance to requirements" and that of Oakland is "meeting the requirements." Juran's philosophy of quality is "fitness for use or purpose."

Based on the philosophies of quality gurus, quality of construction projects can be evolved as follows:

1. Properly defined scope of work.
2. Owner, project manager, design team leader, consultant, and constructor's manager are responsible to implement quality.

3. Continuous improvement can be achieved at different levels as follows:
    a. Owner—Specify the latest needs.
    b. Designer—Specification should include the latest quality materials, products, and equipment.
    c. Constructor—Use the latest construction equipment to build the facility.
4. Establishment of performance measures
    a. Owner
        i. To review and ensure that the designer has prepared the contract documents that satisfy his or her needs.
        ii. To check the progress of work to ensure compliance with the contract documents.
    b. Consultant
        i. As a consultant designer, to include the owner's requirements explicitly and clearly define them in the contract documents.
        ii. As a supervision consultant, supervise contractor's work as per contract documents and the specified standards.
    c. Contractor—To construct the facility as specified and use the materials, products, and equipment that satisfy the specified requirements.
5. Team approach—Every member of the project team should know that TQM is a collaborative effort, and everybody should participate in all the functional areas to improve the quality of the project work. They should know that it is a collective effort by all the participants.
6. Training and education—Both the consultant and contractor should have customized training plans for their management, engineers, supervisors, office staff, technicians, and laborers.
7. Establish leadership—Organizational leadership should be established to achieve the specified quality. Encourage and help the staff and laborers to understand the quality to be achieved for the project.

These definitions when applied to construction projects relate to the contract specifications or owner/end user requirements to be formulated in such a way that construction of the facility is suitable for the owner's use or meets the owner's requirements. Quality in construction is achieved through the complex interaction of many participants in the facilities development process.

The quality plan for construction projects is part of the overall project documentation consisting of the following:

1. Well-defined specification for all the materials, products, components, and equipment to be used to construct the facility
2. Detailed construction drawings

3. Detailed work procedure
4. Details of the quality standards and codes to be compiled
5. Cost of the project
6. Manpower and other resources to be used for the project
7. Project completion schedule

Figure 4.4 illustrates functional relationships between various participants.

Table 4.1 identifies the key quality assurance activities that would take place during the life cycle of a typical construction project. These activities are performed by all the participants of the construction projects at various phases/stages of the project.

Participation involvement of all three parties at different levels of construction phases is required to develop quality system and application of quality tools and techniques. With the application of various quality principles, tools, and methods by all the participants at different stages of construction project, rework can be reduced, resulting in savings in the project cost and making the project qualitative and economical. This will ensure completion

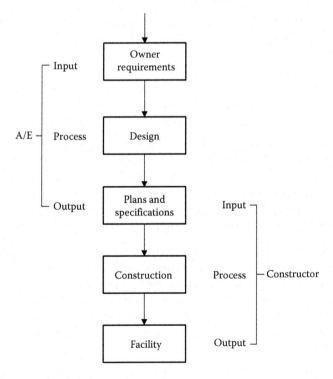

FIGURE 4.4
Juran's triple role-functional relationship. (*Source*: CII source Document No. 51. Reprinted with permission of CII, University of Texas.)

TABLE 4.1

Key Quality Assurance Activities on a Typical Construction Project

Client	Design Consultant	Contractor
Establish project brief/ objectives/ specification (include QA conditions)	Carry out tender review, prepare outline PQP, and submit	
Accept outline PQP		
Place contract	Set up project team, carry out contract review, prepare PQP	
Approve PQP	Submit for approval	
Approve DQP	Prepare DQP, if appropriate	
Approve key drawings	Prepare drawings	
Approve calculations	Prepare calculations, carry out design reviews, prepare detailed specifications	
Monitor design consultant's activities by audit	Issue enquires for construction work including QA conditions	Carry out tender review, prepare QA submission
	Carry out bid appraisal, assess QA submission	
	Place contract with QA conditions	Carry out contract review, set up site team
	Approve PQP	Prepare PQP and submit for approval
	Approve shop drawings and other documents	Place subcontracts including QA condition where appropriate to work package; include requirement for documentation submissions, approvals, and records

(Continued)

TABLE 4.1 (*Continued*)

Key Quality Assurance Activities on a Typical Construction Project

Client	Design Consultant	Contractor
		Receive DQPs from subcontractors for approval prior to work commencing
	Approve DQPs	Prepare DQPs for own work, if required
	Project manage contract; carry out progress meeting; monitor work by inspection test and review of documentation	Place "hold points," etc., on DQPs to monitor work packages; approve DQPs
		Monitor off-site work against DQPs
	Carry out audits to agreed schedules	
		Carry out goods inward inspection to agreed procedure
		Control work on site against PQP. DQPs inspection checklist, etc.
		Carry out audits on- and off-site to agreed audit schedule
		Generate records as construction proceeds
		Mark up drawing to as-built state
		Prepare handover packages and submit
Accept documentation package ⟵	Check handover package and submit with design records ⟵ ⎤

Source: Thorpe, B., Sumner, P., and Duncan, J. (1996), *Quality Assurance in Construction*. Reprinted with permission from Taylor & Francis Group.

Note: This flowchart identifies the key quality assurance activities that would take place during the life cycle of a typical construction project, from project initiation to handover.

TABLE 4.2

Principles of Quality in Construction Projects

Principle	Construction Projects' Quality Principle
Principle 1	Owner, consultant, contractor are fully responsible for application of quality management system to meet defined scope of work in the contract documents
Principle 2	Consultant is responsible to provide owner's requirements explicitly and clearly defined in the contract documents
Principle 3	Contractor should study all the documents during tendering/bidding stage and submit his proposal taking into consideration all the requirements specified in the contract documents
Principle 4	Method of payments (work progress, material, equipment, etc.) to be clearly defined in the contract documents. Rate analysis of BOQ or Bill of Materials item to be agreed before signing of contract
Principle 5	Contract documents should include a clause to settle the dispute arising during construction stage
Principle 6	Contractor shall follow an agreed-upon quality assurance and quality control plan. Consultant shall be responsible to oversee the compliance with contract documents and specified standards
Principle 7	Contractor shall follow the submittal procedure specified in the contract documents
Principle 8	Contractor is responsible to construct the facility as specified and use the material, products, equipment, and methods that satisfy the specified requirements
Principle 9	Each member of project team should participate in all the functional areas to continuously improve quality of project
Principle 10	Contractor is responsible to provide all the resources, manpower, material, equipment, etc. to build the facility as per specifications
Principle 11	Contractor to build the facility as stipulated in the contract documents, plan, specifications within budget and on schedule to meet the owner's objectives
Principle 12	Contractor should perform the works as per agreed-upon construction program and hand over the project as per contracted schedule

Source: Rumane, A.R. (2013), *Quality Tools for Managing Construction Projects,* CRC Press, Boca Raton, FL. Reprinted with permission from Taylor & Francis Group.

of construction and making the project most qualitative, competitive, and economical. Table 4.2 illustrates quality principles of construction projects.

In order to process the construction project in an effective and efficient manner and to improve control and planning, construction projects are divided into various phases. In traditional thinking, there are five phases of a construction project life cycle; however, for major projects there are seven phases, which are further broken into various activities. These are conceptual design, preliminary design, design development, construction document, bidding and tendering, construction, and commissioning and handing over.

Participation of all three parties at different levels during construction is required to develop a quality system and apply quality tools and techniques.

With the application of various quality principles, tools, and methods by all the participants at different stages of construction, rework can be reduced, resulting in savings in the project cost and making the project qualitative and economical. This will ensure completion of a construction project and make the project qualitative, competitive, and economical in a way that will meet the owner's needs and specification requirements.

There are several types of project delivery system that these parties are involved in at different levels contracts. Table 4.3 illustrates different categories of project delivery systems.

TABLE 4.3

Categories of Project Delivery Systems

Serial Number	Category	Classification	Subclassification
1	Traditional system (separated and cooperative)	Design–bid–build	Design–bid–build
		Variant of traditional system	Sequential method
			Accelerated method
2	Integrated system	Design–build	Design–build
		Design–build	Joint venture (architect and contractor)
		Variant of design–build system	Package deal
		Variant of design–build system	Turnkey method (engineering, procurement, construction)
		Variant of design–build system (turnkey)	Build–operate–transfer (BOT)
			Build–own–operate–transfer (BOOT)
			Build–transfer–operate
			Design–build–operate–maintain
		Variant of design–build system (funding option)	Lease–develop–operate
			Wraparound (public–private partnership)
		Variant of design–build system	Build–own–operate (BOT)
			Buy–build–operate
3	Management-oriented system	Management contracting	Project manager (program management)
		Construction management	Agency construction manager
			Construction manager-at-risk
4	Integrated project delivery system	Integrated form of contract	

Source: Rumane, A.R. (2013), *Quality Tools for Managing Construction Projects*, CRC Press, Boca Raton, FL. Reprinted with permission from Taylor & Francis Group.

4.1.1 Design–Bid–Build

In this method, the owner contracts design professionals to prepare detailed design and contract documents. These are used to receive competitive bids from the contractors. A design–bid–build contract has a well-defined scope of work. This method involves three steps:

1. Preparation of complete detailed design and contract documents for tendering
2. Receiving bids from prequalified contractors
3. Award of contract to successful bidder

In this method, two separate contracts are awarded, one to the designer/ consultant and one to the contractor. In this type of contract structure, design responsibility is primarily that of the architect or engineer employed by the client and the contractor is primarily responsible for construction only. In most cases, the owner contracts the designer/consultant to supervise the construction process. These types of contracts are lump-sum, fixed-priced contracts. Any variation, or change, during the construction needs prior approval from the owner. Since a complete design is prepared before construction, the owner knows the cost of the project, time of completion of the project, and the configuration of the project. The client, through the architect or engineer, retains control of design during construction. This type of contracting system requires considerable time; each step must be completed before starting the next step. Figure 4.5 illustrates the design–bid–build type of contract relationship.

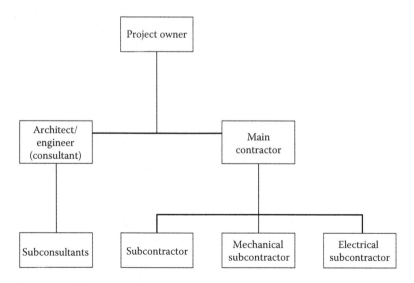

FIGURE 4.5
Design-bid-build (traditional contracting system).

4.1.2 Design–Build

In a design–build contract, the owner contracts a single firm to design and build the project. In this type of contracting system, the contractor is appointed based on an outline design or design brief to understand the project owner's intent. The owner has to clearly define his or her needs and scope of work before the signing of the contract. It is imperative that the project definition be understood by the contractor to avoid any conflict, as the contractor is responsible for detailed design and construction of the project. A design/build type of contract is often used to shorten the time required to complete a project. Since the contract with the design/build firm is awarded before starting any design or construction, a cost plus contract or reimbursable arrangement is normally used instead of lump-sum, fixed-cost arrangement. This type of contract requires extensive involvement on the part of the owner during the entire life cycle of the project. He or she has to be involved in making decisions during the selection of design alternatives and the monitoring of costs, schedules, and quality during construction and, therefore, the owner has to maintain/hire a team of qualified professionals to perform these activities. Design/build contracts are used for relatively straightforward work, where no significant risk or change is anticipated and when the owner is able to specify precisely what is required. Figure 4.6 illustrates the design/build type of contract relationship.

4.1.3 Build–Own–Operate–Transfer

This type of method is generally used by governments to develop public infrastructure by involving the private sector in financing, designing, operating, and managing the facility for a specified period and then

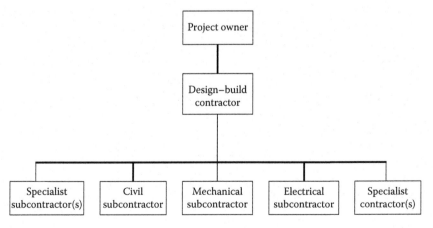

FIGURE 4.6
Design–build contracting system.

transferring the facility to the same government free of charge. The terms BOOT and BOT are used synonymously.

Examples of BOT projects include

Airports

Bridges

Motorways/toll roads

Parking facilities

Tunnels

Certain countries allow the private sector to develop commercial and recreational facilities on government land through the BOT scheme.

4.1.4 The Turnkey Contract

As the name suggests, these are the types of contracts where, upon completion, one turns a key in the door and finds everything working to full operating standards. In this type of method, the owner employs a single firm to undertake design, procurement, construction, and commissioning of the entire work. The firm is also involved in management of the project during the entire process of the contract. The client is responsible for preparation of their statement of requirements, which becomes the strict responsibility of the contractor to deliver. This type of contract is used mainly for the process type of project and is sometimes called engineering, procurement, and construction (EPC).

There are two general types of owners: single-builder owners and multiple-builder owners. Single-builder owners are organizations that do not have a need for projects on a repetitive basis, normally have a limited project staff, and contract all design and construction activities to outside organizations. They usually handle projects with a design/bid/build or construction management contract. Multiple-builder owners are generally large organizations that have a continual need for projects and generally have a staff assigned to project work. They typically handle small-sized, short-duration projects by design/bid/build. For a project in which they desire extensive involvement, a design/build, construction management, or an owner/agent contract arrangement is often used.

4.1.5 Project Manager

A project manager contract is used when the owner decides to turn over the entire project management to a professional project manager. In the project manager type of contract, the project manager is the owner's representative and is directly responsible to the owner. The project manager is responsible for planning, monitoring, and managing the project. In its broadest sense, the project manager has responsibility for all the phases of the project from inception of the project until the completion and handing over of the project to

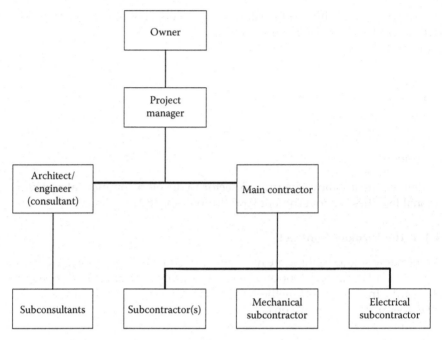

FIGURE 4.7
Project manager contracting system.

the owner/end user. The project manager is involved in giving advice to the owner and is responsible for the appointment of design professionals, consultants, supervision firm, and selection of the contractor who will implement the project. Figure 4.7 illustrates the project manager type of contract relationship.

4.1.6 Construction Manager

In this method, owner contracts a construction management firm to coordinate the project for the owner and provide construction management services. There are two general forms of construction management:

1. Agency construction management (Agency CM)
2. Construction management-at-risk (CM-at-Risk)

The Agency CM type is a management process type of contract system having four-party arrangement involving owner, designer, construction management firm, and the contractor. The construction manager provides advice to the owner regarding cost, time, safety, and about the quality of materials/products/systems to be used on the project. The Agency CM firm performs no design or construction but assists the owner in selecting design firm(s), and contractor(s) to build the project.

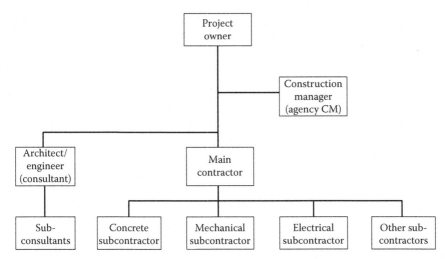

FIGURE 4.8
Agency construction management contractual relationship.

Agency CM could be implemented in conjunction with any type of project delivery system.

The basic concept of construction management type of contract is that the firm is knowledgeable and capable of coordinating all aspects of the project to meet the intended use of the project by the owner. Agency construction manager acts as a principal agent to advise the owner/client, whereas construction manager-at-risk is responsible for on-site performance and actually performs some of the project works. CM-at-Risk type of contract has two stages. The first stage encompasses preconstruction services and during second stage the CM-at-Risk is responsible for performing the construction work. CM-at-Risk project delivery system is also known as construction manager/general contractor (CM/GC).

Figure 4.8 illustrates contractual relationship for Agency CM, and Figure 4.9 illustrates contractual relationship for CM-at-Risk.

4.1.7 Integrated Project Delivery

In this delivery method, the owner, designer, and contractor are contractually required to collaborate among themselves so that the risk responsibility and liability for the project delivery are collectively managed and appropriately shared. Figure 4.10 illustrates contractual relationship between various parties.

All the foregoing project delivery systems follow generic life cycle phases of a construction project; however, the involvement/participation of various parties differs depending on the type of delivery and contracting system adapted for a particular project.

In case of the design/build type of deliverable system, the contractor is contracted right from the early stage of the construction project and is responsible

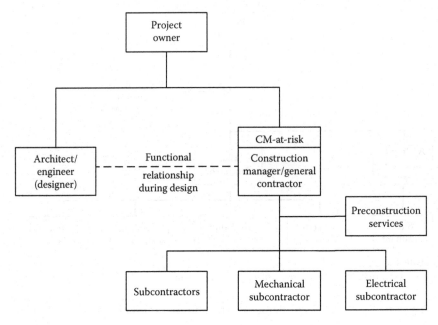

FIGURE 4.9
Construction manager at risk contractual relationship (CM-at-Risk).

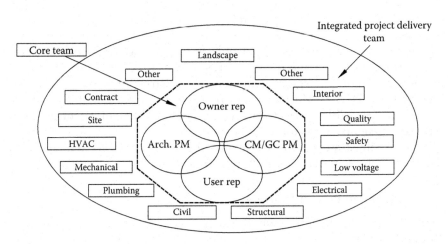

FIGURE 4.10
Integrated project delivery system.

for design development of the project. Figure 4.11 shows the typical logic flow diagram for the design/bid/build type of construction project, and Figure 4.12 shows the diagram for the design/build type of contracting system. Details of activities performed during the various phases of the design/bid/build type of contract delivery system are discussed in related sections.

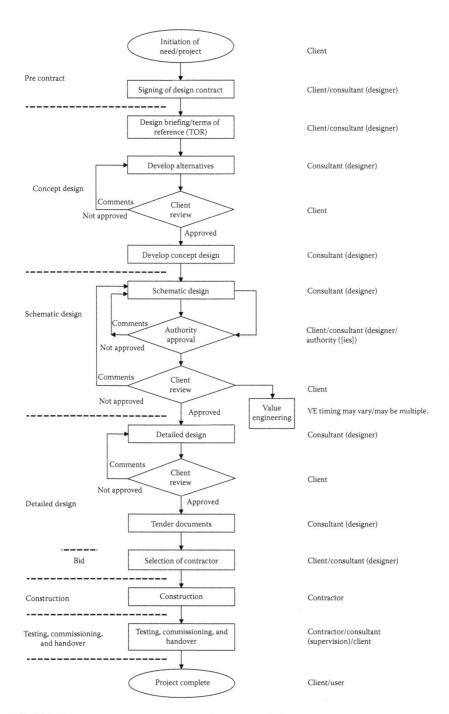

FIGURE 4.11
Logic flow diagram for construction projects-design-bid-build system.

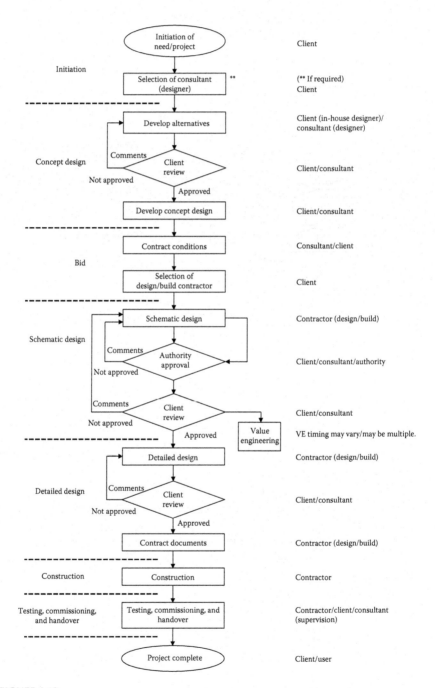

FIGURE 4.12
Logic flow diagram for construction projects-design-build system.

4.2 Conceptual Design

Conceptual design is the first phase of the construction project life cycle. The conceptual design is initiated once the need is recognized. In this phase, the idea is conceived and given an initial assessment. Conceptual design, or the design development phase, is often viewed as most critical to achieving outstanding project performance. During the conceptual phase, the environment is examined, forecasts are prepared, objectives and alternatives are evaluated, and the first examination of the technical performance, cost, and time objectives of the project is made. The conceptual phase includes

Identification of need by the owner, and establishment of main goals

Feasibility study, which is based on owner's objectives

Identification of project team by selecting other members and allocation of responsibilities

Identification of alternatives

Financial implications, resources, based on estimation of life cycle cost of the favorable alternative

Time schedule

Development of concept design

The most significant impacts on the quality in a project occur during the conceptual phase. This is the time when specifications, statement of work, contractual agreements, and initial design are developed. Initial planning has the greatest impact on a project because it requires the commitment of processes, resources schedules, and budgets. A small error that is allowed to stay in the plan is magnified several times through subsequent documents that are second or third in the hierarchy.

Figure 4.13 shows the major activities in the conceptual design phase.

4.2.1 Identification of Need

Most construction projects begin with recognition of the new facility. The owner of the facility could be an individual, a public/private sector company, or a governmental agency. The need for the project is created by the owner and is linked to the financial resources available to develop the facility. The owner's needs are quite simple and are based on the following:

To have best use of the money, that is, to have maximum profit or services at a reasonable cost

On time completion, that is, to meet the owner's/user's schedule

Completion within budget, that is, to meet the investment plan for the facility

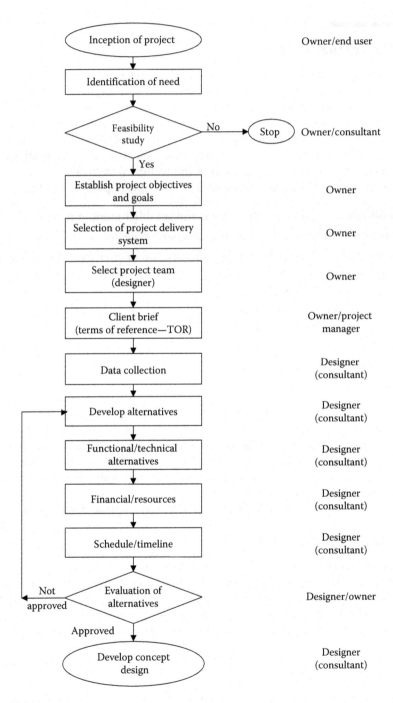

FIGURE 4.13
Logic flow in the conceptual design phase.

The conceptual design is initiated once the owner's need is recognized. Therefore, it is important that the owner define the requirements and objectives of the potential project clearly at the start of the formulation of design. The need statement is an expression of an unfulfilled requirement. It provides a specifically focused requirement that can be addressed as a way of providing a solution.

The owner's need must be well defined, indicating the minimum requirements of quality and performance, an approved main budget, and required completion date. Sometimes, the project budget is fixed and, therefore, the quality of the building system, materials, and completion of the project needs to be balanced with the budget.

Figure 4.14 illustrates a preliminary appraisal and the steps in the project identification.

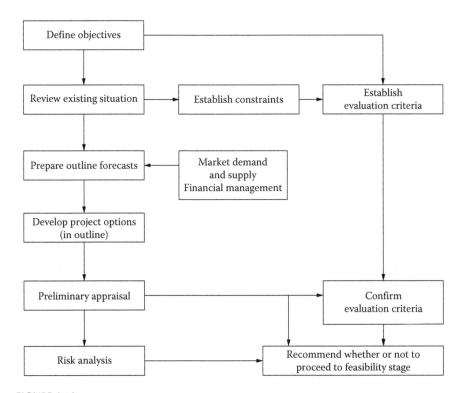

FIGURE 4.14
Steps in project identification. (*Source*: Corrie, R.K. (1991), *Engineering Management Project Evaluation*, Reprinted with permission from Thomas Telford Publishing, UK.)

4.2.2 Feasibility

Once the owner's need is identified, the traditional approach is pursued through a feasibility study or an economical appraisal of owner needs or benefits, also taking into account the many relevant moral, social, environmental, and technical constraints. The feasibility study takes its starting point from the output of the project identification need.

The authors of *Civil Engineering Procedure* (Institution of Civil Engineers 1996) have listed some of the investigations to be carried out for a major project. These are as follows:

- Outline design
- Studies of novel requirements and risks
- Public consultation
- Geotechnical study of site, sources of materials, storage areas, and access routes
- Environmental impact analysis
- Health and safety studies
- Testing for contaminated land and requirements for the disposal of waste
- Estimates of capital and operating costs
- A master program of work, expenditure, and financing
- Assessment of funding (p. 22)

Depending on the circumstances, the feasibility study may be short or lengthy, simple or complex. In any case, it is the principle requirement in project development as it gives the owner an early assessment of the viability of the project and the degree of risk involved. The outcome of the feasibility study helps select a defined project that meets the stated project objectives, together with a broad plan of implementation. Figure 4.15 illustrates various stages for a feasibility study.

Technical studies are performed to analyze that the proposed facility is suitable for intended use by the owner/user. The proposed project is economically feasible if the total value of the benefits that result from the project exceeds the cost that results from the project. Economic feasibility depends on technical feasibility because the facility must be suitable for intended use. Financial feasibility may or may not be related to economic feasibility.

The project study is usually performed by the owner through his or her own team or by engaging individuals/organizations involved in preparation of economic and financial studies. Once the project definition has been ascertained, the owner selects other team members of the project and finalizes the contract delivery system for the project. If the feasibility study shows that the objectives of the owner are best met through the ideas generated, then the owner will select and engage a project team based on the project delivery system to develop his or her notional ideas into a more workable form.

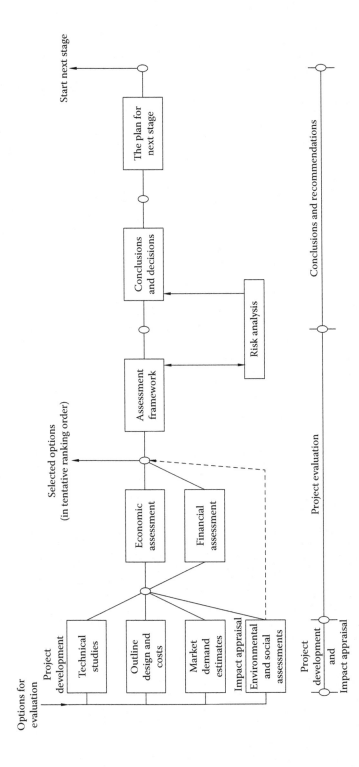

FIGURE 4.15

Project feasibility stage (modified). (*Source:* Corrie, R.K. (1991), *Engineering Management Project Evaluation*, Reprinted with permission from Thomas Telford Publishing, UK.)

4.2.3 Identification of Alternatives

Table 4.4 shows a quantitative comparison of functional alternatives to cost alternatives that may be analyzed to select the preferred alternative.

Each alternative is based on a predetermined set of performance measures to meet the owner's requirements. In case of construction projects, it is mainly the extensive review of development options that are discussed between the owner and the designer/consultant. The consultant engineer provides engineering advice to the owner to enable him or her to assess its feasibility and the relative merits of various alternative schemes to meet his or her requirements. Social, economical, and environmental impacts, functional capability, safety, and reliability should be taken into account while considering the development of alternatives. Each alternative is evaluated based on the predetermined set of performance measures to meet the owner's requirements. Figure 4.16 summarizes the general steps in the systematic process of studying project alternatives and evaluating associated impacts.

Quantitative comparison and evaluation of conceptual alternatives are carried out by considering the advantages and disadvantages of each item systematically. The designer makes a brief presentation to the owner, and the project is selected based on preferred conceptual alternatives. Various possibilities are considered during this stage, and the technological and economical feasibility is assessed and compared to select the best possible alternative.

TABLE 4.4

Conceptual Alternatives

Functional Alternatives	Cost Alternatives
• Materials handling methods	• Design cost
• Traffic flow arrangements (patterns in air, water, land, people, or products)	• Capital cost of construction
• Types of travel modes (vehicle type, size, style)	• Operation and maintenance costs
• Method to provide fish passage at barriers in waterways	• Life-expectancy or design-life periods
• Space allocations	• Return on investment
• Clear-span requirements in buildings	• Project phasing (initial opening or operating segments)
• Public/private (joint development) options	• Extra cost for aesthetics
• Methods to avoid or minimize impacts to the natural environment	• Cost/benefit ratios

Source: Quality in the Constructed Project (2000) by ASCE. Reprinted with permission from ASCE.

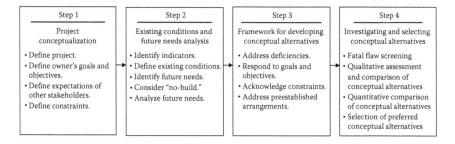

Step 1	Step 2	Step 3	Step 4
Project conceptualization	Existing conditions and future needs analysis	Framework for developing conceptual alternatives	Investigating and selecting conceptual alternatives
• Define project. • Define owner's goals and objectives. • Define expectations of other stakeholders. • Define constraints.	• Identify indicators. • Define existing conditions. • Identify future needs. • Consider "no-build." • Analyze future needs.	• Address deficiencies. • Respond to goals and objectives. • Acknowledge constraints. • Address preestablished arrangements.	• Fatal flaw screening • Qualitative assessment and comparison of conceptual alternatives • Quantitative comparison of conceptual alternatives • Selection of preferred conceptual alternatives

FIGURE 4.16
Alternative study and impact analysis process. (*Source*: Quality in the Constructed Project (2000) by ASCE. Reprinted with permission from ASCE.)

4.2.4 Identification of Project Team

Most construction projects involve three major groups or parties:

1. *Owner*: A person or an organization that articulated the need for the facility and responsible for arranging the financial resources for the creation of the facility.

2. *Designer/consultant*: This consists of architects, engineers, or consultants. They are the owner's appointed entity accountable for converting the owner's conception and need into a specific facility with detailed directions through drawings and specifications, within the economic objectives and schedule. They are responsible for the design process and assist the owner in preparation of tender and contract documents. The owner may engage the designer to supervise construction.

3. *Contractor*: A construction firm engaged by the owner to complete the specified facility by providing the necessary staff, workforce, materials, equipment, tools, and other accessories to complete the project to the satisfaction of the owner/end user in compliance with the contract documents.

The owner is the first member of the project team. The owner's relationship with other team members and his or her responsibilities depend on the type of deliverable system the owner would prefer to go with. There are many types of contract delivery systems; however, design/bid/build is the most predominantly used contracting system preferred by the owners.

For the design/bid/build type of contract system, the first thing the owner has to do is select design professionals/consultants. Generally, the owner selects a designer/consultant with whom he or she has worked before with satisfactory results. The owner can use his or her preferred designer/consultant or select one by obtaining proposals from several design professionals/consultants.

The owner may contract a designer or architect and engineer (A&E) to provide site supervision during the construction process. Thus, the designer or A&E firm acts as the prime professional to design the project and also supervises the construction.

Once the project delivery system is finalized and the designer/consultant is selected and contracted by the owner to proceed with the project design, a terms of reference (TOR) is issued to the designer/consultant to prepare a design proposal and contract documents. A TOR is a document that describes the purpose and structure of a project. It gives the project team a clear understanding of the development of the project.

Table 4.5 illustrates the typical requirements of project team members. Table 4.6 illustrates the typical responsibilities of project team members. Figure 4.17 illustrates the responsibilities of different parties.

The TOR generally requires the designer (consultant) to perform the following:

Predevelopment studies, which includes data collection and analysis related to the project

Development of conceptual alternatives

Evaluation of conceptual alternatives and selection of preferred alternatives in consultation with the owner

TABLE 4.5

Typical Requirements of Project Team Members

Owner/Project Manager	Design Professional/A/E	Constructor
• Adequate function and appearance of the new facility	• An adequate project scope definition	• A well-defined set of contract documents
	• An adequate budget	• A reasonable schedule
• Project completion on time and within budget	• A reasonable schedule	• Timely decisions from the owner and design professional
	• Timely decisions from the owner	
• Desirable balance of life cycle and initial capital costs	• Realistic and fair sharing of project risks	• Realistic and fair sharing of project risks
• Addressing of environmental, health, permitting, safety, user impacts, and sustainable development considerations	• Adequate communication with the owner regarding performance	• Adequate communication with the owner regarding performance
	• A fair and reasonable process for resolving disputes	• A fair and reasonable process for resolving disputes
• A fair and reasonable process for resolving disputes	• Timely payment and a reasonable profit	• Timely payment and a reasonable profit

Source: Quality in the Constructed Project (2000) by ASCE. Reprinted with permission from ASCE.

TABLE 4.6

Typical Responsibilities of Project Team Members

Owner/Project Manager	Design Professional/A/E	Constructor
• Fulfillment of contractual obligations to other team members, including furnishing site and related information, and timely payment • Compliance with applicable laws, regulations, codes, standards, and practices	• Fulfillment of contractual obligations to other team members • Compliance with applicable laws, regulations, codes, standards, and practices	• Fulfillment of contractual obligations to other team members • Compliance with applicable laws, regulations, codes, standards, and practices
• Provision of adequate funding • Provision of necessary real estate or right(s) of way • Provision of project goals and objectives • Fulfillment of insurance and legal requirements • Assignment of site safety responsibility	• Fulfillment of professional standards • Development and drafting of well-defined contract documents • Responsiveness to project schedule, budget, and program • Provision of construction-phase design services	• Interpretation of plans and specifications • Construction of facility as described in contract documents • Management of construction site activities and safety program • Management, quality control, and payment of subcontractors and vendors
• Acceptance of completed facility		

Source: Quality in the Constructed Project (2000) by ASCE. Reprinted with permission from ASCE.

Preparation of preliminary design, budget, and schedule and obtaining authorities' approvals

Preparation of detailed design and contract documents for tendering purpose

Table 4.7 illustrates the contribution of various participants during all the phases of the construction project life cycle for the design/bid/build type of contracting system.

4.2.5 Development of Concept Design

The selected preferred alternative is the base for development of the concept design. The designer can use techniques such as quality function deployment to translate the owner's need into technical specifications. Figure 4.18 illustrates the house of quality concept for an office building project based on certain specific requirements by the customer.

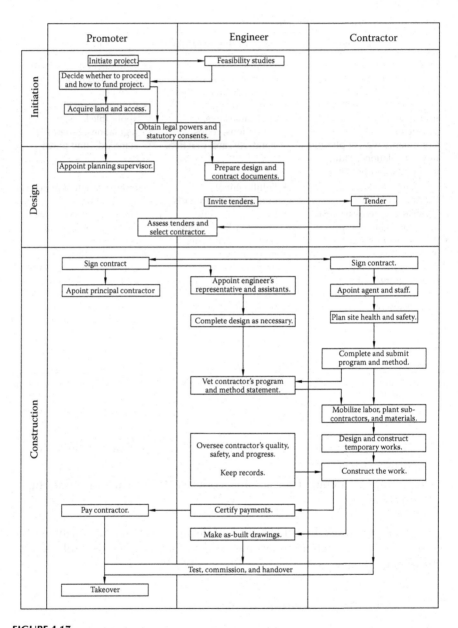

FIGURE 4.17
Division of responsibility between promoter, engineer, and contractor. (*Source*: Corrie, R.K. (1991), *Engineering Management Project Evaluation*, Reprinted with permission from ICE, UK.)

TABLE 4.7

Contribution of Various Participants (Design–Bid–Build Type of Contracts)

Phase	Owner	Example of Contribution	
		Designer	Contractor
Conceptual design	• Identification of need • Selection of alternative • Selection of team members • Approval of time schedule • Approval of budget • TOR	• Feasibility • Development of alternatives • Cost estimates • Schedule • Development of concept design	• Collection of tender documents • Preparation of proposal • Submission of bid
Preliminary design	• Approval of preliminary (schematic) design	• Develop general layout/scope of facility/project • Regulatory approval • Budget • Schedule • Contract terms and conditions	
Detail design	• Approval of budget • Approval of time schedule • Approval of design • Contract negotiation • Signing of contract	• Development of detail design • Authorities' approval • Detail plan • Budget • Schedule • BOQ • Tender documents • Evaluation of bids	

(Continued)

TABLE 4.7 (*Continued*)

Contribution of Various Participants (Design–Bid–Build Type of Contracts)

Phase	Owner	Example of Contribution	
		Designer	Contractor
Construction	• Approve subcontractor(s) • Approve contractor's core staff • Legal/regulatory clearance • SWI • VO • Payments	• Supervision • Approve plan • Monitor work progress • Approve shop drawings • Approve material • Recommend payment	• Execution of work • Contract management • Selection of subcontractor(s) • Planning • Resources • Procurement • Quality • Safety
Testing commissioning and handover	• Training • Acceptance of project • Substantial completion certificate • Payments	• Witness tests • Check closeout requirements • Recommend take over • Recommend issuance of substantial completion certificate	• Testing • Commissioning • Authorities' approvals • Documents • Training • Handover

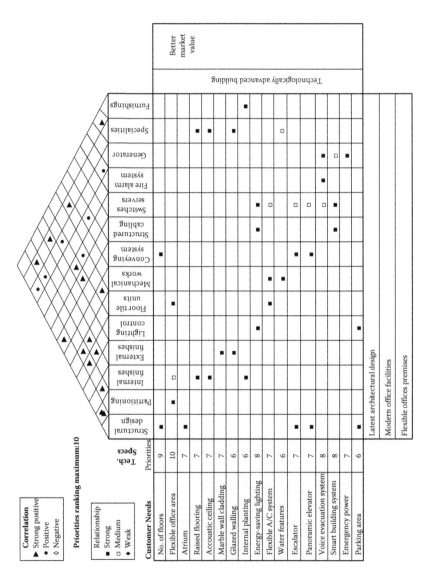

FIGURE 4.18
House of quality for office building project.

While developing the concept design, the designer must consider the following:

1. Project goals
2. Usage
3. Technical and functional capability
4. Aesthetics
5. Constructability
6. Sustainability (environmental, social, and economical)
7. Availability of resources
8. Health and safety
9. Reliability
10. Environmental compatibility
11. Fire protection measures
12. Supportability during maintenance/maintainability
13. Cost-effectiveness over the entire life cycle (economy)

It is the designer's responsibility to pay greater attention to improving the environment and to achieve sustainable development. Numerous UN meetings (such as the first United Nations Conference on Human Development held in Stockholm in 1972; the 1992 Earth Summit in Rio de Janeiro; the 2002 Earth Summit in Johannesburg; the 2005 World Summit; and the Brundtland Commission on Environment and Development in 1987) have emphasized "sustainability," whether it be sustainable environment, sustainable economic development, sustainable agricultural and rural development, and so on. Accordingly, the designer has to address environmental and social issues and comply with local environmental protection codes. A number of tools and rating systems have been created by LEED (the United States), BREEAM (the United Kingdom), and HQE (France) in order to assess and compare the environmental performance of the buildings. These initiatives have a great impact on how buildings are designed, constructed, and maintained. Therefore, during implementation of building projects, the following need to be considered:

1. Accretion with the environment by using natural resources such as sunlight, solar energy, and appropriate ventilation configuration
2. Energy conservation by energy-efficient measures to diminish energy consumption
3. Environmental protection to reduce environmental impact
4. Use of materials harmonizing with the environment
5. Aesthetic match between a structure and its surrounding natural and built environment

6. Good air quality

7. Comfortable temperature

8. Comfortable lighting

9. Comfortable sound

10. Clean water

11. Less water consumption

12. Integration with social and cultural environment

During the design stage, the designer must work jointly with the owner to develop details regarding the owner's needs and give due consideration to each part of the requirements. The owner on his or her part should ensure that the project objectives are

Specific

Realistic

Measurable

Agreed upon by all the team members

Possible to complete within the stipulated time

Within the budget

The following are the requirements for a building construction project, normally mentioned in the TOR, to be prepared by the designer during the conceptual phase for submission to the owner:

1. Site plan
   a. Civil
   b. Services
   c. Landscaping
   d. Irrigation
2. Architectural design
3. Building and engineering systems
   a. Structural
   b. Mechanical (heating, ventilation, and air-conditioning [HVAC])
   c. Public health
   d. Fire suppression systems
   e. Electrical
   f. Low-voltage systems
   g. Others
4. Cost estimates
5. Schedules

The designer is required to submit all the above mentioned submissions in the form of

- Report
- Drawings
- Models
- Presentation

4.2.6 Estimate Time Schedule

The duration of a construction project is finite and has a definite beginning and a definite end; therefore, during the conceptual phase the expected time schedule for the completion of the project/facility is worked out. The expected time schedule is important from both financial and acquisition of the facility by the owner/end user. It is the owner's goal and objective that the facility is completed in time. Figure 4.19 illustrates a time schedule for a typical construction project.

4.2.7 Estimate Project Cost

The next step is to refine cost estimates for the conceptual alternatives as this is required by the owner to determine the capital cost of construction, so that he or she can arrange the finances. It is the owner's responsibility to provide an approved maximum finance to complete the facility. It is required that the owner formulate his or her thoughts on project financing, as the financial

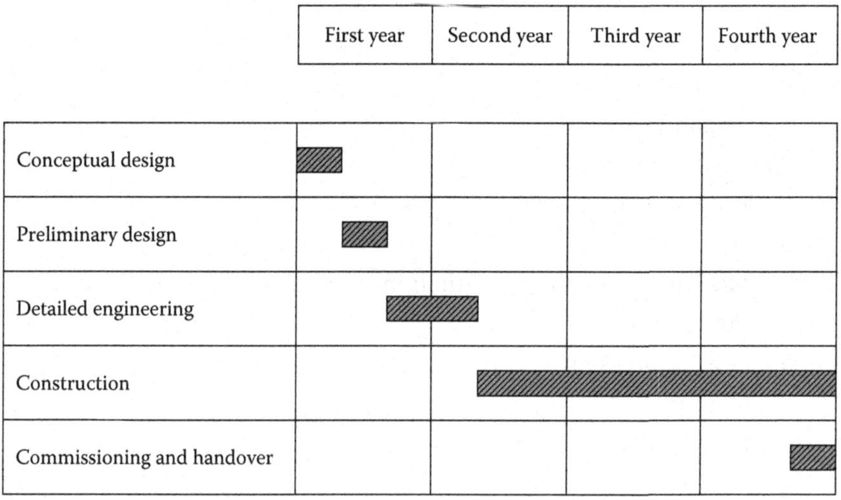

	First year	Second year	Third year	Fourth year
Conceptual design	▨			
Preliminary design	▨			
Detailed engineering	▨▨			
Construction		▨▨▨▨▨▨▨▨▨▨	
Commissioning and handover				▨

FIGURE 4.19
Typical time schedule of construction project.

conditions will affect the possible options from the beginning. Normally, the following points should be considered:

1. What are sources of funding?
2. What criteria or rules apply?
3. How could the project best respond to those rules?

In case any funding agency is involved in financing the project, it may impose certain conditions that affect the project feasibility and implementation. It is likely that such funding agencies may also insist on the adoption of a particular contract strategy.

4.2.8 Manage Quality

During this phase, the designer has to plan and establish quality criteria for the project. This includes mainly the following:

- Owner's requirements
- Quality standards and codes to be complied
- Regulatory requirements
- Conformance to owner's requirements
- Conformance to requirements listed under TOR
- Design review procedure
- Drawings review procedure
- Document review procedure
- Quality management during all the phases of project life cycle

4.2.9 Estimate Resources

Designer has to estimate the resources required to complete the project. This includes the estimation of manpower required during construction phase, testing, commissioning, and handover phase. The designer has to also estimate the manpower required during design, and tendering stages of the construction project.

4.2.10 Identify/Manage Risk

The designer has to identify the risks that will affect the successful completion of the project. The following are typical risks that normally occur during the conceptual design phase:

- Lack of input from owner about the project goals and objectives.
- Project objectives not defined clearly.

- Feasibility study not done properly.
- Alternative selection is not suitable for further development.
- The related project data and information collected are incomplete.
- The related project data and information collected are likely to be incorrect and wrongly estimated.
- Environmental consideration.
- Regulatory requirements.
- Errors in estimating the project schedule.
- Errors in cost estimation.

The designer has to take into account the mentioned risk factors while developing the concept design.

Further, the designer has to consider the following risks while planning the duration for completion of the conceptual phase:

- Impractical conceptual design preparation schedule
- Delay to obtain authorities' approval
- Delay in environmental approval
- Delay in data collection
- Delay in deciding project delivery system

4.2.11 Finalize Concept Design

Final designs for submission to the owner/client are prepared incorporating the comments, if any, found during analysis and review of the drawings and documents taking into consideration risks.

4.3 Preliminary Design

Preliminary design is mainly a refinement of the elements in the conceptual design phase. Preliminary design is also known as *schematic design*. During this phase, the project is planned to a level where sufficient details are available for the initial cost and schedule. This phase also includes the initial preparation of all documents necessary to implement the facility/construction project. The central activity of preliminary design is the architect's design concept of the owner's objective, which can help make the detailed engineering and design for the required facility. Preliminary design is a subjective process transforming ideas and information into plans, drawings, and specifications of the facility to be built. Component/equipment configurations, material specifications, and functional performance are decided during this stage. At this stage, the owner can alter the scope and consider alternatives. The owner seeks to optimize certain facility features within the constraints of other factors such as cost, schedule, vendor capabilities, and so on.

Design is a complex process. Before design is begun, the scope must adequately define deliverables, that is, what will be furnished. These deliverables are design drawings, contract specifications, types of contracts, construction inspection record drawings, and reimbursable expenses.

Preliminary design is the basic responsibility of the architect (designer/consultant or A&E). In the case of building construction projects, a preliminary design determines

1. General layout of the facility/building/project
2. Required number of buildings/number of floors in each building/area of each floor
3. Different types of functional facilities required such as offices, stores, workshops, recreation, training centers, parking, etc.
4. Type of construction such as reinforcement concrete or steel structure, precast, or cast in situ
5. Type of electromechanical services required
6. Type of infrastructure facilities inside the facilities area
7. Type of landscape

The designer has to consider the following points while preparing the preliminary design:

1. Concept design deliverables
2. Calculations to support the design
3. System schematics for electromechanical system
4. Coordination with other members of the project team
5. Authorities' requirements
6. Availability of resources
7. Constructability
8. Health and safety
9. Reliability
10. Energy conservation issues
11. Environmental issues
12. Selection of systems and products that support the functional goals of the entire facility
13. Sustainability
14. Requirements of all stakeholders
15. Optimized life cycle cost (value engineering [VE])

4.3.1 Identification of Preliminary Design Requirements

In order to identify requirements to develop schematic design, the designer has to gather comments made by the owner/project manager on the submitted concept design and collect TOR requirements, regulatory requirements, and other related data to ensure that the developed design is error-free and with minimum omissions.

4.3.2 Identification of Project Team

During this phase, most project team members such as owner's representative, project manager (design), and other design personnel are selected and identified.

Design team members are selected based on the organizational structure and suitable skills required to perform the job. Normally, the design team consists of

1. Project manager
2. Design managers (one for each trade)
3. Quality manager
4. Team leader (principal engineer) for each trade
5. Team members (engineers, CAD technicians for each discipline)
6. Quantity surveyor (cost engineer)

4.3.3 Develop Preliminary Design

The purpose of a general scope of works is to provide sufficient information to identify the works to be performed and to allow detail design to proceed without significant changes that may adversely affect the project budget and schedule.

At the preliminary design stage, the scope must define deliverables, that is, what will be furnished. It should include a schedule of dates for delivering drawings, specifications, calculations and other information, forecasts, estimates, contracts, materials, and construction. The designer develops a design concept with the plan, elevation, and other related information that meet the owner's requirements. The designer also develops a concept of how various systems, such as heating and cooling systems, communication systems, etc., will fit into the system.

Bennet (2003) has given a list of preliminary design drawings required for preliminary approval quoting one building agency of a U.S. state. These are

- The basic design approach drawn at an agreed-upon scale
- Site location in relationship to the existing environment
- Relation to master plans
- Circulation

- Organization of building functions
- Functional/aesthetic aspects of the design concepts under study
- Graphic description of critical details
- Visual and functional relationship
- Compatibility with the surrounding environment (p. 55)

Bennet (2003) further states that the same agency requires the design professional to prepare a narrative description of the following building systems upon completion of schematic design:

- Structure
- Floor grade and systems
- Roof
- Exterior/interior walls and partitions
- Interior finishes
- Sight lines
- Stairs and elevators
- Specialty items
- Electrical systems
- Mechanical systems
- Built-in equipment
- Site construction (p. 55)

4.3.4 Regulatory Approval

Once the preliminary design is approved, it should be submitted to regulatory bodies for their review and approval for compliance with the regulations, codes, and licensing procedure.

4.3.5 Contract Terms and Conditions

Normally, it is the consultant/designer team that is responsible for developing a set of contract documents that meet the owner's needs and specify the required level of quality, budget, and schedule. At this stage, a contract exists between the consultant and the client for the development of the project, and any good management test will demand that the contract be clearly understood by all parties associated with it. There are numerous combinations of contract arrangements for handling the construction projects; however, design/bid/build is predominantly used in most construction project contracts. This delivery system has been chosen by owners for many centuries and is called the traditional contracting system. In the traditional contracting system, the detailed design for the project is completed before tenders for construction are invited. The detailed engineering is carried out by the consultant/design professional to make the project qualitative and economical.

Based on the type of contracting arrangements with which the owner would like to handle the project, necessary documents are prepared by establishing a framework for execution of the project. Generally, FIDIC's model conditions for international civil engineering contracts are used as a guide to prepare these contract documents. Preliminary specifications and documents are prepared in line with model contract documents.

4.3.6 Schedule

After the preliminary scope of works, the preliminary design and budget for the facility/project are finalized; the logic of the construction program is set. On the basis of logic, a critical path method (CPM) schedule (bar chart) is prepared to determine the critical path and set the contract milestones.

4.3.7 Budget

Based on the preliminary design, the budget is prepared by estimating the cost of activities and resources. The preparation of the budget is an important activity that results in a timed phased plan summarizing the expected expenses toward the contract and also the income or the generation of funds necessary to achieve the milestone. The budget for a construction project is the maximum amount the owner is willing to spend for design and construction of the facility that meets the owner's need. The budget is determined by estimating the cost of activities and resources and is related to the schedule of the project. If the cash flow or resulting budget is not acceptable, the project schedule should be modified. It is required that while preparing the budget, the risk assessment of the project is also performed.

4.3.8 Manage Quality

In order to minimize design errors, minimize design omissions, and reduce rework during schematic design, the designer has to plan quality (planning of design work), perform quality assurance, and control quality for preparing schematic design. This will mainly consist of the following:

1. Plan quality
    - Establish owner's requirements
    - Determine number of drawings to be produced
    - Establish scope of work
    - Identify quality standards and codes to be complied
    - Establish design criteria
    - Identify regulatory requirements
    - Identify requirements listed under TOR
    - Establish quality organization with responsibility matrix

- Develop design (drawings and documents) review procedure
- Establish submittal plan
- Establish design review procedure

2. Quality assurance
- Collect data
- Investigate site conditions
- Prepare preliminary drawings
- Prepare outline specifications
- Ensure functional and technical compatibility
- Coordinate with all disciplines
- Select material to meet the owner's objectives

3. Control quality
- Check design drawings
- Check specifications/contract documents
- Check for regulatory compliance
- Check preliminary schedule
- Check cost of project (preliminary cost)

Building Information Modeling (BIM) tool can be used to control quality of the project.

4.3.9 Estimate Resources

The designer has to estimate the resources required to complete the project. At this stage of the project, more details about the activities and works to be performed during the construction and testing, commissioning phase are available, the designer has to update the earlier estimated resources and prepare manpower histogram. Also, the designer can estimate the total number of design team members to develop design and construction documents.

4.3.10 Identify/Manage Risk

The following are typical risks that normally occur during the schematic design phase:

- Concept design deliverables and review comments are not taken into consideration while preparing the schematic design.
- Regulatory authorities' requirements are not taken into consideration.
- Schematic design scope of work is incomplete.
- The related project data and information collected are incomplete.

- The related project data and information collected are likely to be incorrect and wrongly estimated.
- Site investigations for existing conditions are not carried out.
- Fire and safety considerations.
- Environmental consideration.
- Incomplete design.
- Prediction of possible changes in design during construction phase.
- Inadequate and ambiguous specifications.
- Wrong selection of materials and systems.
- Undersize HVAC equipment selection.
- Incorrect water supply requirements.
- Estimated total electrical load is much lower than expected actual consumption.
- Errors in calculating traffic study for conveying system.
- Errors in estimating the project schedule.
- Errors in cost estimation.
- Number of drawings not as per TOR requirements.

The designer has to take into account the mentioned risk factors while developing the schematic design.

Further, the designer has to consider the following risks while planning the duration for completion of the schematic phase:

- Impractical schematic design preparation schedule
- Delay to obtain authorities' approval
- Delay in site investigations
- Delay in data collection

4.3.11 Perform Value Engineering Study

VE studies can be conducted at various phases of a construction project; however, the studies conducted in the early stage of a project tend to provide the greatest benefit. In most projects, VE studies are performed during the schematic phase of the project. At this stage, the design professionals have considerable flexibility to implement the recommendations made by the VE team, without significant impacts on the project's schedule or design budget. In certain countries for a project over US $5 million, a VE study must be conducted as part of the schematic design process. The team members who perform the VE study depend on the client's/owner's requirement. It is advisable that a SAVE international registered certified value specialist be assigned to lead this study.

4.3.12 Finalize Preliminary Design

Final schematic design for submission to the owner/client is prepared incorporating the comments, if any, found during analysis and review of the drawings and documents.

4.4 Design Development

Detailed design is the third phase of the construction project life cycle. It follows the preliminary design phase and takes into consideration the configuration and the allocated baseline derived during the preliminary phase. **Design development** is also known as **detailed design/detailed engineering**. During this phase, all suggested changes are reevaluated to ensure that the changes will not detract from meeting the project design goals/objectives. Detailed design involves the process of successively breaking down, analyzing, and designing the structure and its components so that it complies with the recognized codes and standards of safety, and performance while rendering the design in the form of drawings and specifications that will tell the contractors exactly how to build the facility to meet the owner's need. During this phase, detail design of the work, contract documents, detail plan, budget, estimated cash flow, regulatory approval, and tender/bidding documents are prepared. Depending on the type of contract the owner would like to have for completing the facility, the designer (consultant) can start preparing the detailed design. The success of a project is highly correlated with the quality and depth of the engineering plans prepared during this phase.

Figure 4.20 illustrates major activities in the detailed design phase.

4.4.1 Identification of Design Development Requirements

In order to identify requirements to develop detail design, the designer has to gather comments made by the owner/project manager on the submitted schematic design and collect TOR requirements, regulatory requirements, and other related data to ensure that the developed design is accurate to the possible extent, free of errors, and with minimum omissions. Detail design activities are similar, although more in-depth than the design activities in the preliminary design stage. The size, shape, levels, performance characteristics, and technical details and requirements of all the individual components are established and integrated into the design. Design engineers of different trades have to take into consideration all these as a minimum while preparing the scope of works. The range of design work is determined by the nature of the construction project.

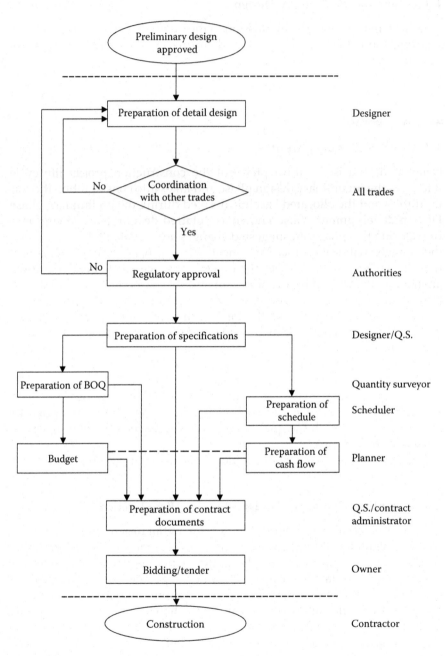

FIGURE 4.20
Major activities in the detailed design phase.
– – – – Functional relationship.

For development of detail design, the designer has to consider the following points:

1. Review of comments on the preliminary design by the client (project manager)
2. Review of comments on the preliminary design by the regulatory authorities
3. Preparation of detail design for all the works
4. Interdisciplinary coordination to resolve conflict
5. Obtain regulatory approval
6. Prepare project schedule
7. Prepare project budget
8. Prepare BOQ
9. Preparing specifications

4.4.2 Identification of Project Team

Generally, the project team members selected to develop schematic design continue during the detail design phase. Additional design personnel, if required, are added to meet the work load to develop the design.

4.4.2.1 Identify Design Team Members

Figure 4.21 illustrates the design management team and their major responsibilities.

Each of the managers has many other team members. These members are selected based on the organizational structure and suitable skills required to perform the job. These include

1. Team leader (principal engineer) for each trade
2. Team members (engineers and CAD technicians for each discipline)
3. Quantity surveyor (cost engineer)
4. Owner's representative
5. End user

4.4.3 Develop Detail Design of the Works

The detail design process starts once the preliminary design is approved by the owner. Detail design is enhancement of work carried out during the preliminary stage. During this phase, a comprehensive design of the works

FIGURE 4.21
Design management team. (*Source*: Rumane, A.R. (2013), *Quality Tools for Managing Construction Projects*, CRC Press, Boca Raton, FL. Reprinted with permission from Taylor & Francis Group.)

with a detailed work breakdown structure (WBS) and work packages is prepared. In general, specific and detailed scopes of works lead to better-quality projects. The detail design phase is the traditional realm of design professionals, including architects, interior designers, landscape architects, and several other disciplines such as civil, electrical, mechanical, and other engineers as needed.

Accuracy in the project design is a key consideration of the life cycle of the project; therefore, it is required that the designer/consultant be not only an expert in the technical field but also should have a broad understanding of engineering principles, construction methods, and VE. The designer must know the availability of the latest products in the market and to use proven technology, methods, and materials to meet the owner's objectives. He or she must refrain from using a monopolistic product, unless its use is important or critical for proper functioning of the system. He or she must ensure that at least two or three sources are available in the market producing the same type of product that complies with all its required features and intent of use. This will help the owner get competitive bidding during the tender stage.

The authors of *Quality in the Constructed Project* (ASCE 2000) have listed the general functions and responsibilities of the design professional as follows:

- Being fully qualified and licensed to offer and provide the services contractually undertaken and provided
- Applying appropriate skills to the design
- Being proactive and clear in communication
- Being responsive to the established budget schedule and program
- Making timely interpretation, evaluation, and decisions
- Disclosing fully related external interests
- Avoiding conflicts of interest
- Complying with applicable codes, regulations, and laws
- Interpreting contract documents impartially
- Representing the owner's interests as required by contract
- Performing project-specific duties outlined in the contract between the design professional and owner (p. 30)

The authors of *Quality in the Constructed Project* (ASCE 2000) further describe that the design professional (or in-house design team) can help ensure project quality through several activities, including

- Developing a scope of services that meet the owner's requirements and the project goals and objectives
- Developing a design activity plan for the project
- Defining project design guidelines
- Estimating accurately the hours of effort and costs involved in achieving a quality design
- Building flexibility into the design activity plan to allow for changes and future project development, as well as associated budget and schedule revisions
- Developing a realistic schedule with appropriate milestones to confirm progress
- Monitoring design progress constantly (p. 80)

Detail design activities are similar, although more in-depth than the design activities in the preliminary design stage. The size, shape, levels, performance characteristics, technical details, and requirements of all the individual components are established and integrated into the design. Design engineers of different trades have to take into consideration all these at a minimum while preparing the scope of works. The range of design work is determined by the nature of the construction project.

The following are the aspects of works to be considered by design professionals while preparing the detail design. These can be considered as a base for development of design to meet customer requirements and will help achieve the qualitative project.

4.4.3.1 Architectural Design

- Intent/use of building/facility
- Property limits
- Aesthetic look of the building
- Environmental conditions
- Elevations
- Plans
- Axis, grids, levels
- Room size to suit the occupancy and purpose
- Zoning as per usage/authorities' requirements
- Identification of zones, areas, and rooms
- Modules to match with structural layout/plan
- Number of floors
- Ventilation
- Thermal insulation details
- Stairs, elevators (horizontal and vertical transportation)
- Fire exits
- Ceiling height and details
- Reflected ceiling plan
- Internal finishes
- Internal cladding
- Partition details
- Masonry details
- Joinery details
- Schedule of doors and windows
- Utility services
- Toilet details
- Required electromechanical services
- External finishes
- External cladding
- Glazing details
- Finishes schedule
- Door schedule
- Windows schedule
- Hardware schedule

- Special equipment
- Fabrication of items, such as space frame, steel construction, retaining wall, having special importance for appearance/finishes
- Special material/product to be considered, if any
- Any new material/product to be introduced
- Conveying system core details
- Maintenance access for equipment/services requirements
- Ramp details
- Hard and soft landscape
- Parking areas
- Provision for future expansion (if required)

4.4.3.2 Concrete Structure

- Property limits/surrounding areas
- Type of foundation
- Energy-efficient foundation
- Design of foundation based on field and laboratory test of soil investigation that gives the following information:
  a. Subsurface profiles, subsurface conditions, and subsurface drainage
  b. Allowable bearing pressure and immediate and long-term settlement of footing
  c. Coefficient of sliding on foundation soil
  d. Degree of difficulty for excavation
  e. Required depth of stripping and wasting
  f. Methods of protecting below grade concrete members against impact of soil and groundwater (water and moisture problems, termite control, and radon where appropriate)
  g. Geotechnical design parameters, such as angle of shear resistance, cohesion, soil density, modulus of deformation, modulus of subgrade reaction, and predominant soil type
  h. Design loads, such as dead load, live load, wind load, and seismic load
- Footings
- Grade and type of concrete
- Size of bars for reinforcement and the characteristic strength of bars

- Clear cover for reinforcement for
    a. Raft foundation
    b. Underground structure
    c. Exposed to weather structures such as columns, beams, slabs, walls, and joists
    d. Not exposed to weather columns, beams, slabs, walls, and joists
- Reinforcement bar schedule, stirrup spacing
- Expansion joints
- Concrete tanks (water)
- Insulation
- Services requirements (shafts, pits)
- Shafts and pits for conveying system
- Location of columns in coordination with architectural requirements
- Number of floors
- Height of each floor
- Beam size and height of beam
- Openings for services
- Substructure
    a. Columns
    b. Retaining walls
    c. Walls
    d. Stairs
    e. Beams
    f. Slab
- Superstructure
    g. Columns
    h. Stairs
    i. Walls
    j. Beams
    k. Slabs
- Consideration of water proofing requirements for roof slab against water leakage
- Deflection that may cause fatigue of structural elements; crack or failure of fixtures, fittings, or partitions; or discomfort of occupants
- Movement and forces due to temperature
- Equipment vibration criteria

- Load sensors to measure deflection
- Reinforcement bar schedule, stirrup spacing
- Building services to fit in the building
- Environmental compatibility
- Parapet wall
- Excavation
- Dewatering
- Shoring
- Backfilling

4.4.3.3 Elevator Works

- Type of elevator
- Loading capacity
- Speed
- Number of stops
- Travel height
- Cabin, cabin accessories, cabin finishes, and car operating system
- Door, door finishes, and door system
- Safety features
- Drive, size, and type of motor
- Floor indicators, call button
- Control system
- Cab overhead dimensions
- Pit depth
- Hoist way
- Machine room
- Operating system

4.4.3.4 Fire Protection System

The fire protection system provides protection against fire to life and property. The system is designed taking into consideration the local fire code and NFPA standards. The system includes the following:

- Sprinkler system for fire suppression in all the areas of the building
- Hydrants (landing valve) for professional fighting
- Hose reel for public use throughout the building

- Gaseous fire protection system for communication rooms
- Fire protection system for diesel generator room
- Size of fire pumps and controls
- Water storage facility
- Interface with other related systems

4.4.3.5 Plumbing Works

- Maximum working pressure to have adequate pressure and flow of water supply
- Maximum design velocity
- Maximum probable demand
- Demand weight of fixture in fixture units for public uses
- Friction loss calculation
- Maximum hot water temperature at fixture outlet
- Water heater outlet hot temperature
- Providing isolating valves to ensure that the system is easily maintainable
- Hot water system
- Central water storage capacities
- Size of pumps and controls
- Location of storage tank
- Schematic diagram for water distribution system

4.4.3.6 Drainage System

While designing the drainage system, the schedule of foul drainage demand units and frequency factors for the following items should be considered for sizing the piping system, number of manholes, and capacity of sump pump and sump pit:

- Washbasins
- Showers
- Urinals
- Restrooms
- Kitchen sinks
- Other equipment such as dishwashers and washing machines

4.4.3.7 HVAC Works

- Environmental conditions
  - Outdoor design conditions
  - Indoor design conditions
- Air-conditioning calculations
  - Cooling load calculations
  - Heating load calculations
  - Space temperature and humidity at required set point
  - Occupancy load
  - Lighting load
- Room pressurizing and leakage calculations
- Energy consumption calculations
- Air-conditioning calculations for IT equipment room(s) based on heat emission of equipment
- Air distribution system calculations
- Smokes extract ventilation calculations
- Exhaust ventilation calculations
- Ductwork sizing calculations
- Selection of the ductwork components such as balancing dampers, constant volume boxes, variable air volume boxes, attenuators, grilles and diffusers, fire dampers, pressure relief dampers, etc.
- Pipework sizing calculations
- Selection of the inline pipework components, for example, valves, strainers, air vents, commissioning sets, flexible connections, and sensors
- Selection of boilers, pressurization units, and air-conditioning calculations
- Pipework and ductwork insulation selection
- Details of grilles and diffusers, control valves, etc.
- Selection of the ductwork systems plant and equipment, for example, air handling units, fan coil units, filters, coils, fans, humidifiers, and duct heaters
- Selection of chillers, cooling towers
- Selection of pumps
- Selection of fans
- Equipment system calculations
- Space requirements for chillers, cooling towers, pumps, and other equipment (plant room)

- Mechanical room location and access
- Preparation of the plan and section layouts and plant room drawings
- Electrical load calculations
- Comparison of electrical consumption with electrical conservation code
- Preparation of equipment schedules
- HVAC-related electrical works
- Control details
- Starter panels, Motor Control Center (MCC) panels, and schematic diagram of MCC
- Selection of program equipment
- Preparation of point schedule for building management system (BMS)
- Schematic diagram for BMS

4.4.3.8 Electrical System

- Lighting calculations for different areas based on illumination level recommended by CIE/CEN/CIBSE and Isolux diagrams
- Selection of light fittings, type of lamps
- Selection of control gear for light fixture
- Environmental consideration for selection of light fixture and control gear
- Exit/emergency lighting system
- Circuiting references, normal as well emergency
- Sizing of conduits
- Power for wiring devices
- Power supply for equipment (HVAC, PH&FF, conveying system, and others)
- Sizing of cable tray
- Sizing of cable trunking
- Selection (type and size) of wires and cable
- Voltage drop calculations for wires and cables
- Selection of upstream and downstream breakers
- Derating factor
- Sensitiveness of breakers (degree of protection)
- Selection of isolators
- IP ratings (degree of ingress protection) of panels, boards, and isolators

- Schedule of distribution boards, switch boards, and main low-tension boards
- Cable entry details
- Location of distribution boards, switch boards, and low-tension panels
- Short-circuit calculations
- Sizing of diesel generator set for emergency power supply
- Sizing of automatic transfer switch
- Generator room layout
- Sizing of capacitor bank
- Provision for solar system integration
- Schematic diagrams
- Sizing of transformers
- Substation layout
- Calculations for grounding (earthing) system
- Grounding system layout
- Calculations for lightning protection system
- Lightning protection system layout

4.4.3.9 Fire Alarm System

A fire alarm system is designed taking into consideration the local fire code and NFPA standards. The system includes the following:

- Conduiting and raceways
- Type of system: analog/digital/addressable
- Types of detectors based on the area and spacing between the detectors and the walls
- Break glass/pull station
- Types of horns/bells
- Voice evacuation system, if required
- Types of wires and cables
- Mimic panel, if required
- Repeater panel, if required
- Main control panel
- Interface with other systems such as HVAC, elevator
- Riser diagram

4.4.3.10 Telephone/Communication System

- Structured cabling considering type and size of cable: copper, fiber optic
- Type and size of the cables
- Racks
- Wiring accessories/devices
- Access/distribution switches
- Internet switches
- Core switch
- Access gateway
- Router
- Network management system
- Servers
- Telephone handsets

4.4.3.11 Public Address System

- Conduiting and raceways
- Type of system: analog/digital/IP based
- Types of wires and cables
- Types of speakers
- Distribution of speakers
- Required noise level in different areas
- Calculations for sound pressure level
- Zoning of system, if required
- Size and type of premixer
- Size and type of amplifier
- Microphones
- Paging system
- Message recorder/player
- Interface with other systems

4.4.3.12 Audiovisual System

- Conduiting and raceways
- Type of system: analog/digital/IP based
- Types of wires and cables

- Racks
- Type, size, and brightness of projectors
- Type and size of speakers and sound pressure level
- Type and size of screens
- Microphones
- Cameras (visualizers)
- CD/DVD players and recorders
- Control processors
- Video switch matrix
- Mounting details of equipment

4.4.3.13 Security System/CCTV

- Type of system: digital/IP based
- Conduiting and raceways
- Wires and cabling network
- Level of security required
- Type and size of cameras
- Types of monitors/screens
- Video/event recording
- Video servers
- Database server
- System software
- Schematic diagram
- System console

4.4.3.14 Security System Access Control

- Conduiting and raceways
- Wires and cabling network
- Proximity RFID reader
- Fingerprint and proximity combine reader
- Magnetic lock
- Release button
- Door contact
- RFID card
- Reader control panel

- Server
- Multiplexer
- Monitors
- Workstation
- Metal detector

4.4.3.15 Landscape

As a landscape architect, the following points are to be considered while designing the landscape system:

- Property boundaries
- Size and shape of the plot
- Shape and type of dwelling
- Integration with surrounding areas
- Orientation to the sun and wind
- Climatic/environmental conditions
- Ecological constraints (soil, vegetation, etc.)
- Location of pedestrian paths and walkways
- Pavement
- Garage and driveway
- Vehicular circulation
- Location of sidewalk
- Play areas and other social/community requirements
- Outdoor seating
- Location of services, positions of both under- and aboveground utilities and their levels
- Location of existing plants, rocks, or other features
- Site clearance requirements
- Foundation for paving, including front drive
- Top soiling, or top soil replacement
- Soil for planting
- Planting of trees, shrubs, and ground covers
- Grass area
- Sowing grass or turfing
- Lighting poles/bollards
- Special features, if required

- Signage, if required
- Surveillance, if required
- Installation of irrigation system
- Marking out the borders
- Storage for landscape maintenance material

4.4.3.16 External Works (Infrastructure and Road)

External works are part of the contract requirements of a project that involves construction of a service road and other infrastructure facilities to be connected to the building and also includes care of existing services passing through the project boundary line. The designer has to consider the following while designing external works:

- Grading material
- Asphalt paving for road or street
- Pavement
- Pavement marking
- Precast concrete curbs
- Curbstones
- External lighting
- Cable routes
- Piping routes for water, drainage, storm water system
- Sump pump(s) for drainage, storm water
- Trenches or tunnels
- Bollards
- Manholes and hand holes
- Traffic marking
- Traffic signals
- Boundary wall/retaining wall, if required

4.4.3.17 Bridges

The designer should use relevant authorities' design manual and standards and consider the following points while designing bridges:

- Soil stability
- Alignment with road width, property lines

- Speed
- Intersections/interchanges
- Number of lanes, width
- Right-of-way lines
- Exits, approaches, and access
- Elevation datum
- Superelevation
- Clearance with respect to railroad, roadway, navigation (if applicable)
- High and low levels of water (if applicable)
- Utilities passing through the bridge length
- Slopes
- Number and length of span
- Live loads, bearing capacity
- Water load, wind load, earthquake effect (seismic effect)
- Bridge rails, protecting screening, guard rails, barriers
- Shoulder width
- Footings, columns, and piles
- Abutment
- Beams
- Substructure
- Superstructure, deck slab
- Girders
- Slab thickness
- Reinforcement
- Supporting components, deck hanger, tied arch
- Expansion and fixed joints
- Retaining walls, crash wall
- Drainage
- Lighting
- Aesthetic
- Sidewalk, pedestrian, and bike facilities
- Signage, signals
- Durability
- Sustainability

4.4.3.18 Highways

The designer should use relevant authorities' design manual and standards and consider the following points while designing highways:

- Type of highway
- Soil stability
- Speed
- Number of lanes, width
- Shoulder width
- Gradation
- Type of pavement and thickness
- Right-of-way lines
- Exits, approaches, access, and ramp
- Superelevation
- Slopes, curvature, turning
- Median, barriers, curb
- Sidewalks, driveways
- Pedestrian accommodation
- Bridge roadway width
- Drainage
- Gutter
- Special conditions, such as snow and rain
- Pump(s) for drainage, rain water
- Lighting
- Signage, signals
- Durability
- Sustainability

4.4.3.19 Furnishings/Furniture (Loose)

In building construction projects, loose furnishings/furniture is tendered as a special package and is normally not a part of the main contract. In order to give sufficient information about the product, the descriptive features and specifications of the furnishing/furniture products are accompanied by a pictorial view/cutout sheet/photo of the product and the furniture layout. Figures 4.22 through 4.27 illustrate the detailed specifications for the furnishing of the director's or manager's room of an office building.

Director and office manager and advisor room layouts

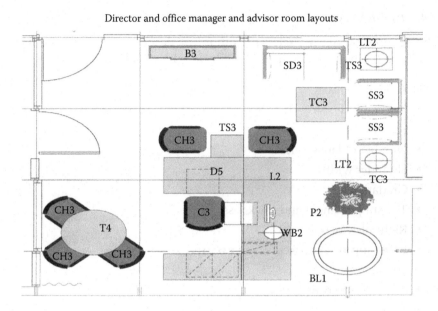

FIGURE 4.22
Room layout.

It is unlikely that the design of a construction project will be right in every detail the first time. Effective management and design professionals who are experienced and knowledgeable in the assigned task will greatly reduce the chances of error and oversight. However, so many aspects must be considered, especially for designs involving multiple disciplines and enfaces, and changes will be inevitable. The design should be reviewed taking into consideration requirements of all the disciplines before release of design drawings for a construction contract. Engineering design has significant importance for the construction projects and must meet the customer's requirement at the start of project implementation. Engineering weakness can adversely impact the quality of design to such an extent that marginal changes can easily increase costs beyond the budget, which may affect schedule. Some areas are deemed critical to the proper design of a product; therefore, explicit design, material specification, and grades of the material specified in documentation have great importance. Most of the products used in construction projects are produced by other construction-related industries/manufacturers; therefore, the designer, while specifying the products, must specify related codes, standards, and technical compliance of these products.

CII Publication 10-1 (1989) has summarized that deviation costs averaged 12.4% of the total installed project cost; design deviation averaged 79% of the total deviation costs and 9.5% of the total project cost. Furthermore, design

FIGURE 4.23
Furniture index.

Director's room finish index

Wooden furnishings
Stained cherry wood
Walnut, antique finishes (MW1)

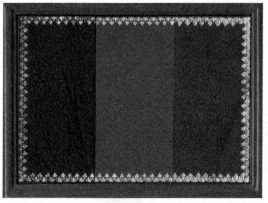

Chair upholestry
Full aniline leather
Sage (ML2)

Single- and double-seat
sofa upholestry
Full aniline leather
Sage (ML2)

FIGURE 4.24
Finish index.

DESCRIPTION: Desk units			CODE: D5 & D6		QUANTITY:
LOCATION	RECOMMENDED MANUFACTURERS	OPTION ONE	OPTION TWO	OPTION THREE	OR APPROVED EQUAL
		HAWORTH	STEELCASE	KIMBALL	

DIMENSIONS: (as shown on the drawing)
D5: H 760 mm × D 914 mm × W 2030 mm, Credenza H760 mm × D 508 mm × W 2030 mm and Bridge: H 760 mm × D 610 mm × L 1200 mm ;
D6: H 760 mm × D 914 mm × W 1830 mm, Credenza H760 mm × D 508 mm × W 1830 mm and Bridge: H 760 mm × D 610 mm × L 1200 mm ;

SPECIFICATIONS / CONSTRUCTION: The contractor to check the number of right-handed and left-handed desk units prior to order. High grade executive desk with full front modesty panel made of kiln dried cherry wood with walnut antique finish as specified as per engineer's control sample with solid traditional moulding polished in walnut finish. The table top is constructed of 22 mm high density marine grade MDF with stained walnut with three pencil drawers as shown in figure. The pattern along the top edge should match the wall paneling pattern and is to be approved prior to order and manufacture. The desk with bridge and credenza's top to have tanned aniline dyed leather inlay of color same as desk chair.

Two full fixed pedestal to be with box/ box/ box and the credenza to have four full fixed pedestals with box/ file fixed below the pencil drawer and should also have back finished panel. The central drawer to have pencil tray. For detailed specifications, refer appendix C.

FINISHES:
Top to have leather inlay and all hardware and other accessories to be in chrome finish (brass finish for level 8 only). The contractor to submit different ranges of the specified color to get the approval and match engineer's control sample and prior to order. All exposed wood machine sanded; Pieces pre-stained to check for scratches; Pre-stain completely sanded off; lacquered on spray - catalyzed urethane or catalyzed conversion varnish topcoat. Wooden parts must be stained walnut, antique finish; to match engineer's sample prior to order.

REMARKS: This specification work formats is just design intent, must be read in conjunction with other schedules and all relevant drawings, all dimensions must be verified on site prior to order or manufacture of materials or production of shop drawings. The Contractor must fully coordinate with all other engineering works, especially the audio & visual works; to ensure proper installation. All necessary accessories, optional features, hardwares including locks with keys etc., all to be included in order to make each item complete fully functional with maximum utility. The finishes required must be verified with respect to architectural interiors: wall paneling and coordinate its approval with respect to the engineer's control sample prior to order or manufacture.

FIGURE 4.25
Specification for desk units.

DESCRIPTION: Desk Lamp

CODE:

QUANTITY:

LOCATION	RECOMMENDED MANUFACTURERS	OPTION ONE	OPTION TWO	OPTION THREE	OR APPROVED EQUAL
		Louis Poulsen	W & D	PORTA ROMANA	

DIMENSIONS: (shape and size to be verified as per architectural interiors' drawings and prepared shop drawings at site)
H 472 mm x W 290mm

SPECIFICATIONS/CONSTRUCTION:
The shade is of blown glass with a top plate and base of flow formed high polished chrome. The stem is high luster steel plated brass. The desk lamp is fitted with a 2.5m black plastic cable with plug and toggle switch in the base plate. It should be of IP20 protection code and electric shock protection Class II. Light source 150 W .

FINISHES:
Transparent blown glass

All hardwares and other accessories to be in polished chrome finish (from level 3-7) & brass finish (for level 8).

The contractor to submit different ranges of colors to get the approval and match engineer's control sample and prior to order.

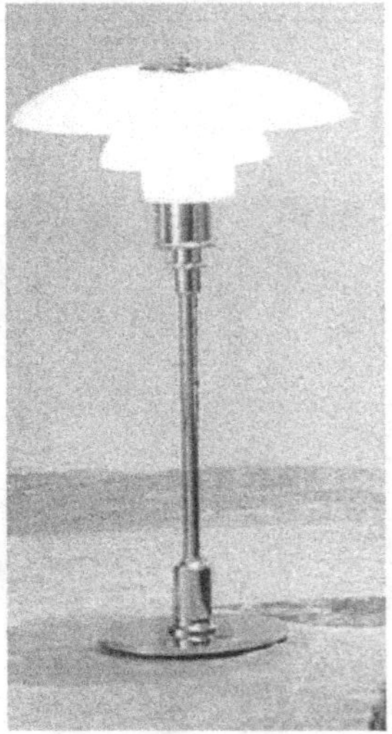

REMARKS: This specification work formats is just design intent, must be read in conjunction with other schedules and all relevant drawings, all dimensions must be verified on site prior to order or manufacture of materials or production of shop drawings. The Contractor must fully coordinate with all other engineering works, especially the audio & visual works; to ensure proper installation. All necessary accessories, optional features, hardwares including locks with keys etc., all to be included in order to make each item complete fully functional with maximum utility. The finishes required must be verified with respect to architectural interiors: wall paneling and coordinate its approval with respect to the engineer's control sample prior to order or manufacture.

FIGURE 4.26
Specification for desk lamp.

DESCRIPTION: Single- and Two-Seater Sofa		CODE:		QUANTITY:
LOCATION	**RECOMMENDED MANUFACTURERS**	**OPTION ONE**	**OPTION TWO**	**OPTION THREE**
		David Edward		OR APPROVED EQUAL

GRAPHICAL REPRESENTATION

DIMENSIONS:
SD3: W 1350 × D 750 × H 810 MM
SS3: W 720 × D 750 × H 810 MM

SPECIFICATIONS/CONSTRUCTION: Executive hand craved family sofas fully upholstered seat and back, with closed wooden topped arms upholstered in full aniline leather. The frame is constructed of kiln dried hard cherry; Mortise/tenon on main joinery; in solid parts but stained in walnut, antique finish. The chair with carved legs. The seat and back are upholstered with leather approved by the engineer and are trim nailed with brass pins. Ultraflex seat construction stretched an additional 20% for firm seat; Back: high strength nylon weave, 2430 Propex, stretched & stapled to frame; Polyurethane foam; All construction materials to meet CAL 117; Pre-upholstered muslin or bonded Dacron. Eight-way tied seat springs, pocket sprung back with frames all glued, screwed and dowelled together. For additional specifications for wood, refer appendix C and for upholstery specification, refer appendix G. It consists of spring down seat cushion that is made of; an insulating box of high quality density polyurethane foam filled with coil springs individually wrapped and held in place by fabric pockets. The unit is inserted into a down-proof envelope of fabric ticking that is channeled into sections on the top and bottom. Each ticking section is filled with a special blend of down, feathers and polyester fiber to achieve luxurious softness and resilience. The down content is 25% and 75% feathers.

FINISHES: Seat and both sides of back are upholstered in fully in full aniline leather, The contractor to submit different ranges of colors to get the approval and match engineer's control sample and prior to order. All exposed wood machine sanded; Pieces pre-stained to check for scratches; Pre-stain completely sanded off; lacquered on spray -catalyzed urethane or catalyzed conversion varnish top coat. Wooden parts must be stained walnut, antique finish; to match engineer's sample prior to order.

REMARKS: This specification work formats is just design intent, must be read in conjunction with other schedules and all relevant drawings, all dimensions must be verified on site prior to order or manufacture of materials or production of shop drawings. The Contractor must fully coordinate with all other engineering works, especially the audio & visual works; to ensure proper installation. All necessary accessories, optional features, hardwares including locks with keys etc., all to be included in order to make each item complete fully functional with maximum utility. The finishes required must be verified with respect to architectural interiors: wall paneling and coordinate its approval with respect to the engineer's control sample prior to order or manufacture.

FIGURE 4.27
Specification for sofa.

changes accounted for two-thirds of the design deviations. It has also given construction deviation averages, which are

17% of the total deviation costs

2.5% of the total installed project cost

It further states that design deviation related to construction projects are results of design errors and design omissions. Design errors are the result of mistakes or errors made in the project design. Design omissions result when a necessary item or component is omitted from the design. Design changes occur when changes are made in the project design or requirements.

Table 4.8 shows the major causes of rework.

In order to reduce the rework resulting from quality deviation in design, CII Publication 10-1 (1989) has made the following recommendations:

1. Reduce the number of design changes by
   - Establishing definitive project scope
   - Performing periodic reviews with participation of all parties
   - Establishing procedures to limit scope modifications
2. Implement a quality management program that has total commitment at all levels of the firm
3. Adopt the standard set of quality-related terminology
4. Develop and implement a system that incorporates a database to identify deviation costs and quality problem areas
5. Implement a quality performance management system to identify costs associated with both quality management and correcting deviation costs (p. 11)

It is, therefore, necessary to have quality control personnel from the project team review and check the design for quality assurance using thorough itemized review checklists to ensure that design drawings fully meet the

TABLE 4.8

Major Causes of Rework, by Phase

Primary Cause	When Detected (Phase)			
Party and Type	Design	Procurement	Construction	Startup
Owner change	×	×	×	×
Designer error/omission	×	×	×	×
Designer change	×	×	×	×
Vendor error/omission	×	×	×	×
Vendor change	×	×	×	×
Constructor error/omission			×	
Constructor change			×	
Transporter error		×	×	×

Source: CII Publication 10-3. Reprinted with permission of CII, University of Texas.

owner's objectives/goals. Appendix A contains illustrative design review checklists for architectural work, structural work, HVAC work, plumbing and fire protection systems, electrical systems, and landscape and infrastructure works. It is also necessary to review the design with the owner prior to initiation of work to ensure a mutual understanding of the build process. The design drawings should be fully coordinated with all the trades. The installation specification details are comprehensively and correctly described, and the installation quality requirements for systems are specified in detail.

Figure 4.28 illustrates the design data review cycle, which can be applied to review construction project design drawings. This process can be termed as continuous improvement of design.

4.4.4 Regulatory/Authorities' Approval

Government agency regulatory requirements have a considerable impact on precontract planning. Some agencies require that the design drawings be submitted for their preliminary review and approval to ensure that the designs are compatible with local codes and regulations. These include submission of drawings to electrical authorities showing the anticipated electrical load required for the facility, approval of fire alarm and firefighting system drawings, and approval of drawings for water supply and drainage system. Technical details of the conveying system are also required to be submitted for approval from the concerned authorities.

4.4.5 Prepare Contract Documents and Specifications

Preparation of detailed documents and specifications as per master format is one of the activities performed during this phase of the construction project. The contract documents must specify the scope of works, location, quality, and duration for completion of the facility. As regards the technical specifications of the construction project, master format specifications are included in the contract documents. The master format is a master list of section titles and numbers for organizing information about construction requirements, products, and activities into a standard sequence. It is a uniform system for organizing information in project manuals, for organizing cost data, for filling product information and other technical data, for identifying drawing objects, and for presenting construction market data. MasterFormat® (1995 edition) consisted of 16 divisions; however, MasterFormat® (2004 edition) consists of 48 divisions (49 is reserved). MasterFormat® contract documents produced jointly by the Construction Specifications Institute (CSI) and Construction Specifications Canada (CSC) are widely accepted as standard practice for preparation of contract documents.

Table 4.9 lists division numbers and titles of MasterFormat 2016 published by the CSI and CSC.

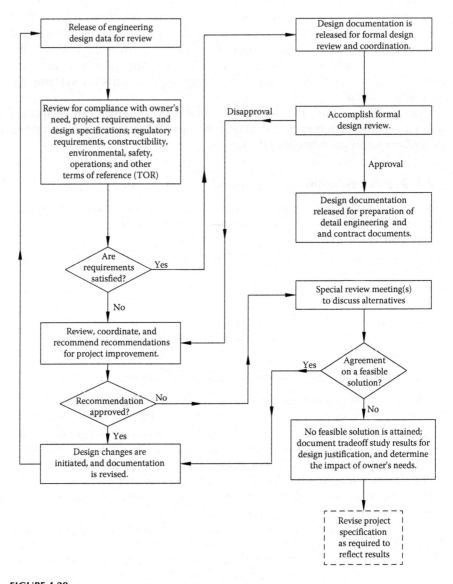

FIGURE 4.28
Design data review cycle. (*Source*: Balanchard, B.S. and Fabrycky, W.J. (1998). *Systems Engineering and Analysis*, Reprinted with permission from Pearson Education.)

TABLE 4.9

MasterFormat® 2016 Division Numbers and Titles

Procurement and Contracting Requirements Group
Division 00 Procurement and Contracting Requirements

Specifications Group

General Requirements Subgroup

Division 01   General Requirements

Facility Construction Subgroup

Division 02   Existing Conditions

Division 03   Concrete

Division 04   Masonry

Division 05   Metals

Division 06   Wood, Plastics, and Composites

Division 07   Thermal and Moisture Protection

Division 08   Openings

Division 09   Finishes

Division 10   Specialties

Division 11   Equipment

Division 12   Furnishings

Division 13   Special Construction

Division 14   Conveying Equipment

Division 15   Reserved

Division 16   Reserved

Division 17   Reserved

Division 18   Reserved

Division 19   Reserved

Facility Services Subgroup

Division 20   Reserved

Division 21   Fire Suppression

Division 22   Plumbing

Division 23   HVAC

Division 24   Reserved

Division 25   Integrated Automation

Site and Infrastructure Subgroup

Division 30   Reserved

Division 31   Earthwork

Division 32   Exterior Improvements

Division 33   Utilities

Division 34   Transportation

Division 35   Waterway and Marine Construction

Division 36   Reserved

Division 37   Reserved

Division 38   Reserved

Division 39   Reserved

Process Equipment Subgroup

Division 40   Process Interconnections

Division 41   Material Processing and Handling Equipment

Division 42   Process Heating Cooling, and Drying Equipment

Division 43   Process Gas and Liquid Handling, Purification, and Storage Equipment

Division 44   Pollution and Waste Control Equipment

Division 45   Industry-Specific Manufacturing Equipment

Division 46   Water and Wastewater Equipment

Division 47   Reserved

Division 48   Electric Power Generation

Division 49   Reserved

(Continued)

TABLE 4.9 (*Continued*)

MasterFormat® 2016 Division Numbers and Titles

Division 26	Electrical
Division 27	Communications
Division 28	Electronic Safety and Security
Division 29	Reserved

Source: The Construction Specifications Institute and Construction Specifications Canada. Reprinted with Permission from CSI.

Particular specifications consist of many sections related to a specific topic. Detailed requirements are written in these sections to enable the contractor understand the product or system to be installed in the construction project. The designer has to interact with the project team members and owner while preparing the contract documents.

Typical sections are as follows:

Section No.

Title

Part 1—General

> 1.01—General reference/related sections
>
> 1.02—Description of work
>
> 1.03—Related work specified elsewhere in other sections
>
> 1.04—Submittals
>
> 1.05—Delivery, handling, and storage
>
> 1.06—Spare parts
>
> 1.07—Warranties

In addition to the foregoing, a reference is made for items such as preparation of mock-up, quality control plan (QCP), and any other specific requirements related to the product or system specified herein.

Part 2—Product

> 2.01—Materials
>
> 2.02—List of recommended manufacturers

Part 3—Execution

> 3.01—Installation
>
> 3.02—Site quality control

Shop Drawing and Materials Submittals

The detailed procedure for submitting shop drawings, materials, and samples is specified under the section titled "Submittal" of contract specifications. The contractor has to submit the specifications to the owner/consultant for review and approval. The following are the details of preparation of shop drawings and materials.

A—Shop Drawings

The contractor is required to prepare shop drawings taking into account the following partial list of considerations:

1. Reference to contract drawings. This helps the A&E (consultant) to compare and review the shop drawing with the contract drawing.
2. Detail plans and information based on the contract drawings.
3. Notes of changes or alterations from the contract documents.
4. Detailed information about fabrication or installation of works.
5. Verification of all dimensions at the job site.
6. Identification of product.
7. Installation information about the materials to be used.
8. Types of finishes, color, and textures.
9. Installation details relating to the axis or grid of the project.
10. Roughing in and setting diagram.
11. Coordination certification from all other related trades (subcontractors).

Shop drawings are to be drawn accurately to scale and shall have project-specific information in it. They should not be reproductions of contract drawings.

Immediately after approval of individual trade shop drawings, the contractor has to submit builder's workshop drawings and composite/coordinated shop drawings taking into consideration the following at a minimum.

A1—Builder's Workshop Drawings

Builder's workshop drawings indicate the openings required in the civil or architectural work for services and other trades. These drawings indicate the size of openings, sleeves, and level references with the help of detailed elevation and plans.

A2—Composite/Coordination Shop Drawings

The composite drawings indicate the relationship of components shown on the related shop drawings and indicate the required installation sequence. Composite drawings should show the interrelationship of all services with one another and with the surrounding civil and architectural work. Composite drawings should also show the detailed coordinated cross sections, elevations, reflected plans, etc., resolving all conflicts in levels, alignment, access, space, etc. These drawings are to be prepared taking into consideration the actual physical dimensions required for installation within the available space.

B—Materials

Similarly, the contractor has to submit the following, at a minimum, to the owners/consultants to get their review and approval of materials, products, equipment, and systems. The contractor cannot use these items unless they are approved for use in the project.

B1—Product Data

The contractor has to submit the following details:

Manufacturer's technical specifications related to the proposed product

Installation methods recommended by the manufacturer

Relevant sheets of manufacturer's catalogs

Confirmation of compliance with recognized international quality standards

Mill reports (if applicable)

Performance characteristic and curves (if applicable)

Manufacturer's standard schematic drawings and diagrams to supplement standard information related to project requirements and configuration of the same to indicate product application for the specified works (if applicable)

Compatibility certificate (if applicable)

Single-source liability (this is normally required for systems approval when different manufacturers' items are used)

B2—Compliance Statement

The contractor has to submit a specification comparison statement along with the material transmittal.

The consultant reviews the transmittals and action as follows:

a. Approved

b. Approved as noted

c. Approved as noted, resubmit

d. Not approved

In certain projects, the owner is involved in approval of materials.

In case of any deviation from specifications, the contractor has to submit a schedule of such deviations listing all the points not conforming to the specification.

B3—Samples

The contractor has to submit (if required) the samples from the approved material to be used for the work. The samples are mainly required to

Verify color, texture, and pattern

Verify that the product is physically identical to the proposed and approved material

Compare with products and materials used in the works

At times, it may be specified to install the samples in such a manner as to facilitate review of qualities indicated in the specifications.

4.4.6 Detail Plan

According to PMBOK, a project plan is used to

- Guide project execution
- Document project planning assumptions
- Document project planning decisions regarding alternatives chosen
- Define key management views regarding content, scope, and timing
- Provide a baseline for progress measurement and project control

A project plan is a formal, approved document used to manage project execution. It is the evaluation of time and effort to complete the project. Based on the detailed engineering and design drawings and contract documents, the design team (consultant) prepares a detail plan for construction. The plan is based on the following:

Assessment of owner's capabilities and final estimated cost (budget)

Scheduling information

Resource management, which includes availability of financial resources, expected cash flow statement, supplies, and human resources

A typical preliminary work program prepared based on the contracted construction documents is illustrated in Figure 4.29.

4.4.7 Budget

The budget for a project is the maximum amount of money the owner is willing to spend for its design and construction. The preparation of a budget is an important activity that results in a time-phased plan summarizing expected expenditure, income, and milestones. Normally, project budgeting starts with the identification of need; however, the detailed cost estimate is done during the engineering phase. On the basis of work packages, the consultant/designer starts computing the project budget. A bill of material or BOQ is prepared based on the approved design drawings. The BOQ is considered as a base for computing the budget. If the budget exceeds the

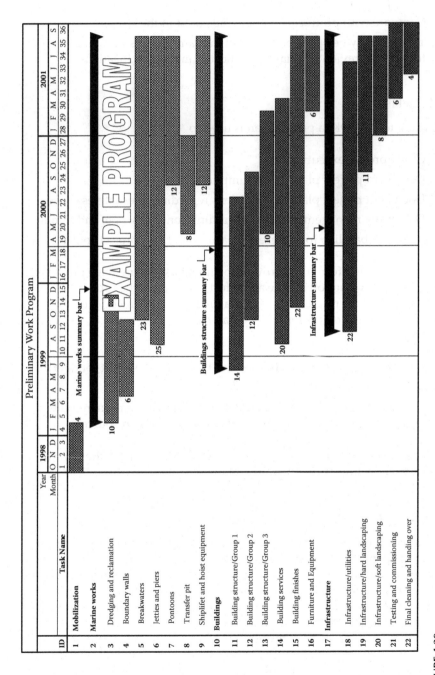

FIGURE 4.29
Preliminary work program.

owner's capability of financing the project, then the designs are reviewed to ensure that it meets the owner's estimated cost to build the facility.

Figure 4.30 illustrates a project S-curve for a building construction project.

4.4.7.1 Cash Flow

The estimate of cash flow requirement for the project is prepared from the preliminary estimate and preliminary work program. An accurate cash flow projection helps owners plan the payments on time as per the schedule for the project. A simple cash flow projection based on prior planning helps owners make available all the resources required from their side. Cash flow is used as part of the control package during construction.

4.4.8 Manage Quality

In order to reduce errors and omission, it is necessary to review and check the design for quality assurance by the quality control personnel from the project team through itemized review checklists to ensure that design drawings fully meet owner's objectives/goals. It is also required to review the design with the owner to ensure a mutual understanding of the build process. The designer has to ensure that the installation/execution specification details are comprehensively and correctly described and also the installation quality requirements for systems are specified in detail.

The designer has to plan quality (planning of design work), perform quality assurance, and control quality for preparing detail design. This will mainly consist of the following.

4.4.8.1 Plan Quality

- Review comments on schematic design
- Determine number of drawings to be produced
- Establish scope of work for preparation of detail design
- Identify requirements listed under TOR
- Identify quality standards and codes to be complied
- Establish design criteria
- Identify regulatory requirements
- Identify environmental requirements
- Establish quality organization with responsibility matrix
- Develop design (drawings and documents) review procedure
- Establish submittal plan
- Establish design review procedure

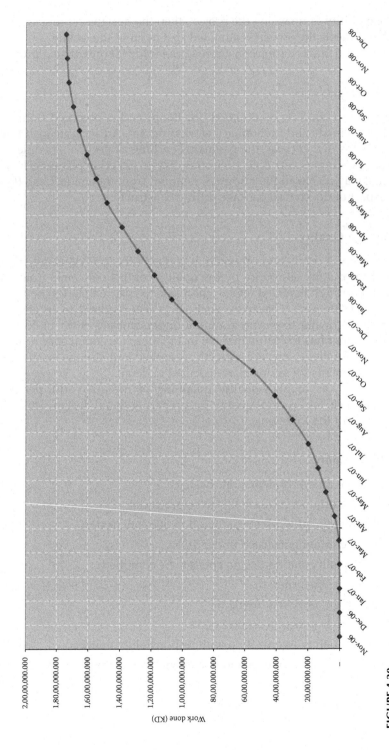

FIGURE 4.30
Project S-curve budgeted.

4.4.8.2 *Quality Assurance*

- Collect data
- Investigate site conditions
- Prepare design drawings
- Prepare detailed specifications
- Prepare contract documents
- Prepare BOQ
- Ensure functional and technical compatibility
- Ensure the design is constructible
- Ensure operational objectives are met
- Ensure drawings are fully coordinated with all disciplines
- Ensure the design is cost-effective
- Ensure selected/recommended material meets the owner's objectives
- Ensure that design fully meets the owner's objectives/goals

4.4.8.3 *Control Quality*

- Check quality of design drawings
- Check accuracy and correctness of design
- Verify BOQ
- Check specifications
- Check contract documents
- Check for regulatory compliance
- Check project schedule
- Check project cost
- Check interdisciplinary requirements
- Check required number of drawings prepared drawing

Table 4.10 lists items to be verified and checked internally by the designer before submission to the owner/project manager. Table 4.11 illustrates a mistake-proofing chart to eliminate design errors.

4.4.9 Estimate Resources

The designer has to estimate the resources required to complete the project. During the design development phase, detailed information is available to estimate manpower resources during the construction phase.

TABLE 4.10

Checklist for Design Drawings

Serial Number	Items to Be Checked
1	Whether design meets owner requirements and complete scope of work (TOR)
2	Whether designs were prepared using authenticated and approved software
3	Whether design calculation sheets are included in the set of documents
4	Whether design is fully coordinated for conflict between different trades
5	Whether design has taken into consideration relevant collected data requirements
6	Whether reviewer's comments responded
7	Whether regulatory approval is obtained and comments, if any, incorporated and all review comments responded
8	Whether design has environmental compatibility
9	Whether energy efficiency measures are considered
10	Whether design constructability is considered
11	Whether design matches with property limits
12	Whether legends match with layout
13	Whether design drawings are properly numbered
14	Whether design drawings have owner logo, designer logo as per standard format
15	Whether the design format of different trades has uniformity
16	Whether project name and contract reference are shown on the drawing

Source: Rumane, A.R. (2013), *Quality Tools for Managing Construction Projects*, CRC Press, Boca Raton, FL. Reprinted with permission from Taylor & Francis Group.

4.4.10 Identify/Manage Risk

The following are typical risks that normally occur during the design development phase:

- Schematic design deliverables and review comments are not taken into consideration while preparing the detail design.
- Regulatory authorities' requirements are not taken into consideration.
- Detail design scope of work is not properly established and is incomplete.
- The related project data and information collected are incomplete.
- The related project data and information collected are likely to be incorrect and wrongly estimated.
- Site investigations for existing conditions are not verified.
- Fire and safety considerations recommended by the authorities are not incorporated in the design.
- Environmental consideration.

TABLE 4.11

Mistake Proofing for Eliminating Design Errors

Serial Number	Items	Points to Be Considered to Avoid Mistakes
1	Information	1. Terms of reference (TOR)
		2. Client's preferred requirements matrix
		3. Data collection
		4. Regulatory requirements
		5. Codes and standards
		6. Historical data
		7. Organizational requirements
2	Mismanagement	1. Compare production with actual requirements
		2. Interdisciplinary coordination
		3. Application of different codes and standards
		4. Drawing size of different trades/specialist consultants
3	Omission	1. Review and check design with TOR
		2. Review and check design with client requirements
		3. Review and check design with regulatory requirements
		4. Review and check design with codes and standards
		5. Check for all required documents
4	Selection	1. Qualified team members
		2. Available material
		3. Installation methods

Source: Rumane, A.R. (2013), *Quality Tools for Managing Construction Projects*, CRC Press, Boca Raton, FL. Reprinted with permission from Taylor & Francis Group.

- Incomplete design drawings and related information.
- Inappropriate construction method.
- Conflict with different trades.
- Interdisciplinary coordination not done.
- Wrong selection of materials and systems.
- Undersize HVAC equipment selection.
- Incorrect water supply requirements.
- Estimated total electrical load is much lower than expected actual consumption.
- Traffic study for conveying system is not verified taking into consideration final load.
- Prediction of possible changes in design during construction phase.
- Inadequate and ambiguous specifications.
- Project schedule not updated as per detailed data and project assumptions.
- Errors in detail cost estimation.
- Number of drawings not as per TOR requirements.

The designer has to take into account the mentioned risk factors while developing the detail design.

Further, the designer has to consider the following risks while planning the duration for completion of the design development phase:

- Impractical design development preparation schedule
- Duration to obtain authorities' approval

4.4.11 Finalize Detail Design

Final detail design is prepared incorporating the comments, if any, found during analysis, review, and interdisciplinary coordination of the drawings and documents for submission to the owner/client.

4.5 Construction Documents

The construction development phase is the fourth phase of construction project life cycle. During this phase, the drawings and specifications prepared during the design development phase are further developed into the working drawings. All the drawings, specifications, documents, and other related elements necessary for construction of the project are assembled and subsequently released for bidding and tendering.

4.5.1 Identification of Construction Documents Requirements

In order to identify requirements to assemble contract documents, the designer has to gather the comments on the submitted detail design by the owner/project manager, collect TOR requirements, and regulatory requirements, identify owner requirements, and all other related information to ensure that nothing is missed.

4.5.2 Identification of Project Team

The following project team members have direct involvement in the construction document phase:

- Owner
- Consultant
- Designer
- Project/construction manager (if the owner decided to engage them during this phase, depending on the type of project delivery system)

During this phase, the quantity surveyor/contract administrator has great responsibilities. His/her team under the leadership of project manager (design) is responsible to coordinate and assemble all the required documents.

4.5.3 Development of Construction Documents

The following items are mainly developed during the construction document phase:

1. Working drawings
2. Technical specifications
3. Documents

4.5.4 Prepare Project Schedule

The project schedule is developed using bottom-up planning details using key activities/events.

4.5.5 Estimate Cost/Budget

The cost estimate during this phase is based on detail costing methodology. During this phase, all the project activities are known and detail BOQ is available for costing purpose.

4.5.6 Quality Management

In order to reduce errors and omission, it is necessary to review and check the design for quality assurance by the quality control personnel from the project team through itemized review checklists to ensure that working drawings are suitable for construction. The designer has to ensure that the installation/execution specification details are comprehensively and correctly described and coordinated with working drawings and also the installation quality requirements for systems are specified in detail.

The designer has to plan quality, perform quality assurance, and control quality for preparing contract documents. This will mainly consist of the following.

4.5.6.1 Plan Quality

- Review comments on design development package
- Determine the number of drawings to be produced
- Establish scope of work for preparation of construction documents
- Identify requirements listed under TOR

- Identify quality standards and codes to be complied
- Identify regulatory requirements
- Identify environmental requirements
- Establish quality organization with responsibility matrix
- Develop review procedure for the produced working drawings
- Develop review procedure for the specifications and contract documents
- Establish submittal plan for construction documents

4.5.6.2 Quality Assurance

- Prepare working drawings
- Prepare detailed specifications
- Prepare contract documents
- Prepare BOQ and schedule of rates
- Ensure functional and technical compatibility
- Ensure the design is constructible
- Ensure operational objectives are met
- Ensure drawings are fully coordinated with all disciplines
- Ensure the design is cost-effective
- Ensure selected/recommended material meets the owner's objectives
- Ensure that design fully meets the owner's objectives/goals
- Ensure that construction documents match with approved project delivery system
- Ensure type of contracting/pricing as per adopted methodology

4.5.6.3 Control Quality

- Check quality of design drawings
- Check accuracy and correctness of design
- Verify BOQ for correctness as per working drawings
- Check complete specifications are prepared and coordinated to match working drawings and BOQ
- Check contract documents as per project delivery system
- Check for regulatory compliance
- Check project schedule
- Check project cost

- Check calculations
- Review studies and reports
- Check accuracy of design
- Check interdisciplinary requirements
- Check required number of prepared drawings comply with requirements

4.5.7 Estimate Resources

At this stage, the designer can estimate resources having accuracy as more details are available to estimate exact resources.

4.5.8 Risk Management

The following are typical risks that normally occur during the construction document phase:

- Design development deliverables and review comments are not taken into consideration while preparing the construction documents.
- Scope of work to produce construction documents is not properly established and is incomplete.
- Documents not matching as per project delivery system.
- Documents not as per type of contract/pricing methodology.
- Regulatory authorities' requirements are not taken into consideration.
- Latest environmental consideration not considered.
- Conflict with different trades.
- Conflict between working drawings and specifications.
- Prediction of possible changes in design during the construction phase.
- Inadequate and ambiguous specifications.
- Project schedule not updated as per detailed data and project assumptions.
- Errors in definitive cost estimation.
- Number of drawings not as per TOR requirements.
- It is likely that owner-supplied items, if any, are not included in the documents.

The designer has to take into account the mentioned risk factors while preparing the construction documents.

Further, the designer has to consider the following risk while planning the duration for completion of the construction document phase:

- Impractical construction document preparation schedule

4.5.9 Finalize Construction Documents

Final construction documents package is prepared taking into consideration review comments and identified risk by the designer and comments from the owner/project manager.

4.6 Bidding and Tendering

Most of the cost of the construction project is expended during the construction phase. In most cases, the contractor is responsible for procurement of all the material, providing construction equipment and tools, and supplying the manpower to complete the project in compliance with the contract documents.

Table 4.12 lists contract documents consisting of tendering procedures, contract conditions, and technical conditions of major construction projects in Kuwait.

In many countries, it is a legal requirement that government-funded projects employ the competitive bidding method. This requirement gives an

TABLE 4.12

Contract Documents

Document No. (I) Tendering Procedures Consisting of the following	
I.1	Tendering Invitation
I.2	Instructions for Tenderers
I.3	Form of Tender and Appendix
I.4	Initial Bond (Form of Bank Guarantee)
I.5	Performance Bond (Form of Bank Guarantee)
I.6	Form of Agreement
I.7	List of Tender Documents
I.8	Declaration No.(1)
Document No. (II) Contract Conditions Consisting of the following	
II.1	General Conditions (Legal Clauses and Conditions 1971 (May 1985 edition) and Amendments until Closing Date of Tender)
II.2	Particular Conditions
II.3	Kuwait Tender Law (Currently Valid)
Document No. (III) Technical Conditions and Amendments Consisting of the following	
III.1	General Specifications for Building & Engineering works, Specific to Ministry of Public Works. 1990 Edition and All Amendments
III.2	Particular Specifications
III.3	Drawings
III.4	Bills of Quantities
III.5	Price Analysis Schedule
III.6	Addenda (if any)
III.7	Technical Requirements (if any), and Any Other Instructions Issued by the Employer

opportunity to all qualified contractors to participate in the tender, and normally the contract is awarded to the lowest bidder. Private-funded projects have more flexibility in evaluating the tender proposal. Private owners may adopt the competitive bidding system, or the owner may select a specific contractor and negotiate the contract terms. Negotiated contract systems have flexibility of pricing arrangement as well as the selection of the contractor based on his expertise or the owner's past experience with the contractor successfully completing one of his or her projects. The negotiated contract systems are based on the following forms of payment:

1. *Cost plus contracts*: It is a type of contract in which the contractor agrees to do the work for the cost of time and material, plus an agreed-upon amount of profit. The following are the different types of cost plus contracts:
   a. Cost plus percentage fee contract
   b. Cost plus fixed fee contract
   c. Cost plus incentive fee contract
2. *Reimbursement contracts*: It is a type of contract in which the contractor agrees to do the work for the cost per schedule of rates, or BOQ, or bill of material.
3. *Fixed price contracts*: With this type of contract, the contractor agrees to work with a fixed price (it is also called *lump sum*) for the specified and contracted work. Any extra work is executed only upon receipt of instruction from the owner. Fixed price contracts are generally inappropriate for work involving major uncertainties, such as work involving new technologies.
4. *Target cost contracts*: A target cost contract is based on the concept of a top-down approach, which provides a fixed price for an agreed range of out-turn costs around the target. In this type of contract, overrun or underspend are shared by the owner and the contractor at predetermined agreed-upon percentages.
5. *Guaranteed maximum price contracts* (cost plus guaranteed maximum price): With this type of contract, the owner and contractor agree to a project cost guaranteed by the contractor as maximum.

It is the owner's desire that his or her facility be of good quality and the price reasonable. In order to achieve this, the owner has to share risks and/or provide incentives and safeguards to enhance the quality of construction. The risks involved in various types of contracts based on forms of payment are as follows:

1. Cost plus—high risk
2. Reimbursement—intermediate
3. Fixed price—low risk

In order to maintain a climate of mutual cooperation during construction, the owner has to develop an understanding with the contractor. The contract needs to be adapted through mutual agreement with the contractor. The contract strategy needs to provide incentives and safeguards to deal with the risks.

Turner (2003) has suggested a twofold methodology in contract selection. The aim is

1. To develop a cooperative project organization
2. To appropriately allocate resources

Turner (2003) has further described that, as per Oliver Williamson (1995, 1996), there are two schemas or vectors to describe the ability of contracts to provide ex ante incentivization and flexible, farsighted ex post governance. The first schema has three parameters:

1. The reward it provides the contractor to share the owner's objectives and perform
2. The associated risk
3. The safeguard provided by the owner through contract to shield the contractor from the risk

Figure 4.31 illustrates a sample contractual schema for ex ante incentivization.

Risk	Present	High incentive	Medium incentive
	Absent	Low incentive	No safeguard or incentive necessary
		Absent	Present
		Safeguard	

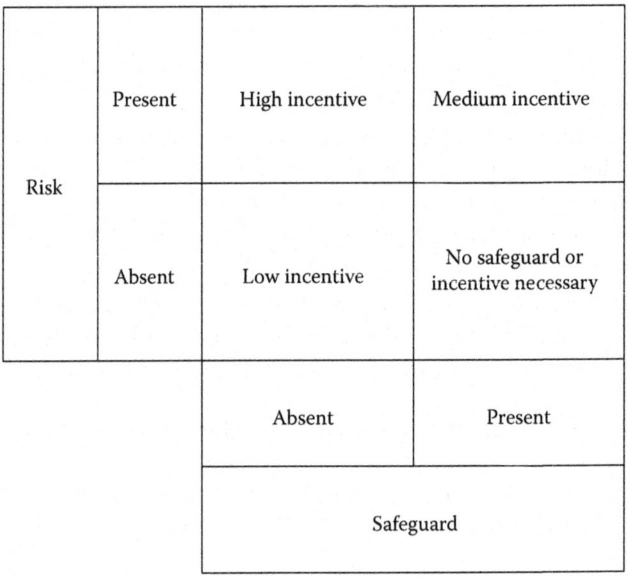

FIGURE 4.31
A simple contract schema for ex ante incentivization. (*Source*: Turner, J.R. (2003), *Contracting for Project Management*, Reprinted with permission from Taylor & Francis Group.)

Although the schema in Figure 4.31 assumes a safeguard risk, it can deal with only a risk that is foreseen. Williamson has further proposed four parameters to describe the ability of a contract form to provide flexible, far-sighted ex post governance:

1. The incentive intensity
2. The ease of making uncontested, bilateral adaptations to contract
3. The reliance on monitoring and related administrative controls (transaction costs)
4. The reliance on court ordering

The incentive profiles of the contract types are summarized in Table 4.13, and the governance profiles in Table 4.14.

In the case of a competitive bidding system, it is necessary that the detailed design and specifications for the project be prepared by the designer for

TABLE 4.13

Contract Forms and Ex Ante Incentivization

Contract Form	Reward	Risk	Safeguard
Cost plus			
Cost + %fee	High but misaligned	High	High
Cost + fixed fee	Medium but misaligned	High	High
Cost + incentive fee	Medium	High	High
Alliance	Medium	High	Medium
Remeasurement			
r-sor	Low and misaligned	Low	High
r-boq	Low	Medium	Medium
r-bom	Low	Medium	Low
Fixed price			
Build only	Low	Low	Low
Specification a	Low	Low	Low
Specification b	Medium	Medium	Low
Cardinal points	High	High	Low (insurance)
Other			
Target price	Medium	Medium	Medium
Time and materials	High	High	Low
Budget or gmp			
Routine contracts			
Market	High	Low	Low
Hierarchy	Low	High	High

Source: Turner, J.R. (2003), *Contracting for Project Management*, Reprinted with permission from Taylor & Francis Group.

r, Remeasurement based; sor, schedule of rates; boq, bill of quantities; bom, bill of material.

TABLE 4.14

Contract Forms and Flexible, Farsighted, Ex Post Governance

Contract Form	Incentive Intensity	Adaptiveness	Transaction Cost	Safeguard
Cost plus				
Cost + %fee	Misaligned	High	High	Low
Cost + fixed fee	Low	High	High	Low
Cost + incentive fee	Medium	High	High	Low
Alliance	High	High	High	Low
Remeasurement				
r-sor	Misaligned	Medium	High	Low
r-boq	Low	Medium	Medium	Low
r-bom	Low	Medium	Low	Low
Fixed price				
Build only	Low	Low	High	High
Specification a	Low	Low	High	High
Specification b	Medium	Medium	Medium	Medium
Cardinal points	High	High	Low	Low
Other				
Target price	Medium	Low	High	Medium
Time and materials	Low	Low	High	High
Budget or gmp				
Routine contracts				
Market	High	Low	Low	High
Hierarchy	Low	High	High	Low

Source: Turner, J.R. (2003), *Contracting for Project Management*, Reprinted with permission from Taylor & Francis Group.
r, Remeasurement based; sor, schedule of rates; boq, bill of quantities; bom, bill of material.

bidding purposes. Under the competitive bidding system, normally there are four stages in tendering of a construction project:

1. Selection of tenderer (prequalification)
2. Invitation to bid
3. Tender preparation and submission
4. Appraisal of tenders, negotiation, and decision

For most construction projects, selection of a tenderer is based on the lowest tender price. Tenders received are opened and evaluated by the owner/owner's representative. Normally, tender results are declared in the official gazette or by some sort of notifications. The successful tenderer is informed of the acceptance of the proposal and is invited to sign the contract. The tenderer has to submit the performance bond before the formal contract agreement is signed. If a successful tenderer fails to submit the performance bond within the specified period or withdraws his tender, then the contractor loses the initial bond and may be subjected to other regulatory applicable conditions.

The signing of contract agreement between the owner/owner's representative and the contractor binds both parties to fulfill their contractual obligations.

4.6.1 Organize Tender Documents

The owner hands over the approved construction documents/tender documents to the tender committee for further action. The bid documents are prepared as per the procurement method and contract strategy adopted during early stage of the project. Tendering procedure documents submitted by the designer are updated, and necessary owner-related information is inserted in the tender documents. The bid advertisement material is prepared and upon approval from the owner the bid notification is announced through different media as per the organization's/agency's policy.

4.6.2 Identification of Project Team

The following project team members have direct involvement in the bidding and tendering phase:

- Owner
- Tender committee
- Designer (consultant)
- Project/construction manager (if the owner decided to engage them during this phase, depending on the type of project delivery system)
- Bidders

4.6.3 Identification of Bidder

Short listing of bidders is done with prequalification questionnaires and their responses.

4.6.4 Manage Tender Documents

Normally, tender documents are distributed to eligible bidders against payment of fee announced in the bid notification, which is nonrefundable.

The owner conducts prebid meeting to provide an opportunity for the contractors bidding the project to review and discuss the construction documents and to discuss

- General scope of the project
- Any particular requirements of bidders that may have been difficult to specify
- Explain details of complex matters
- Engagement of subcontractor, specialist subcontractors
- Particular risks
- Any other matters that will contribute to the efficient delivery of project

The meeting is attended by the designer (consultant), bidders (contractors), project/construction manager, and tender committee member. Queries from the contractors pertaining to construction documents are noted, and the designer (consultant) provides written response to these queries by clarifying all the points. The bidders have to consider the clarification points and incorporate the requirements while calculating the bid price. The responses recorded in the meeting become a part of contract documents (part of addendum), which is signed by the owner and successful bidder. Figure 4.32 illustrates bid clarification form that becomes part of contract documents.

Bids are received in accordance with the Instructions to Bidders section of the tender documents. The bid should be accompanied with initial bond in

			Project Name		
			Project Number		
			Bid Clarification Form		
Serial Number	Name of Contractor	Item No. and Clause Reference	Queries	Owner/Consultant's Clarification	Remarks
			SAMPLE FORM		
			SAMPLE FORM		

FIGURE 4.32

Bid clarification. (*Source*: Rumane, A.R., (2016), *Handbook of Construction Management: Scope, Schedule, and Cost Control*, CRC Press, Boca Raton, FL. Reprinted with permission from Taylor & Francis Group.)

favor of owner/tender committee to be valid for a period mentioned in tendering procedures. All the bids received are documented and notified. The tender, which is submitted as sealed document, is opened as mentioned in tendering procedures.

4.6.5 Identify/Manage Risk

The following are the typical risks likely to occur during this phase:

- Not all the qualified bidders take part in bidding for the project
- Bidders noticing errors and omissions in construction documents resulting in delay in submission of bids
- BOQ not matching with design drawings
- Amendment to construction documents
- Addendum
- Delay in submission of bids than the notified one
- Bid value exceeding the estimated definitive cost
- Successful bidder fails to submit performance bond

The owner/designer has to consider these risks and plan the phase duration accordingly.

Table 4.15 lists the risks the contractor has to manage during the bidding and tendering phase.

TABLE 4.15

Major Risk Factors Affecting Contractor

Serial Number		Risk Factor
1		**Bidding/Tendering**
	1.1	Low bid
	1.2	Poor definition of scope of work
	1.3	Overall understanding of project
	1.4	Review of contract specs with bill of quantities
	1.5	Errors in resource estimation
	1.6	Errors in resource productivity
	1.7	Errors in resource availability
	1.8	Errors in material price
	1.9	Improper schedule
	1.10	Quality standards
	1.11	Exchange rate
	1.12	Review of contract document requirements with regulatory requirements
	1.13	Unenforceable conditions or contract clauses

4.6.6 Select Contractor

The contractor is selected based on the procurement strategy adopted by the owner.

4.6.7 Award Contract

Figure 4.33 illustrates the contract award process.

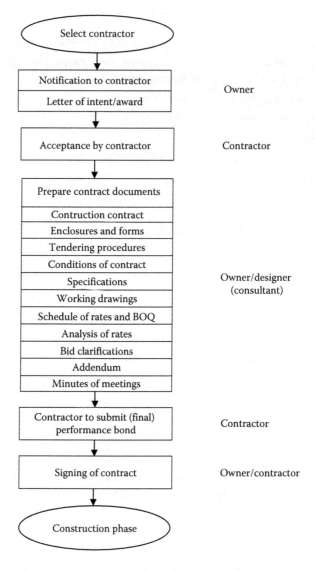

FIGURE 4.33

Contract award process. (*Source*: Rumane, A.R., (2016), *Handbook of Construction Management: Scope, Schedule, and Cost Control*, CRC Press, Boca Raton, FL. Reprinted with permission from Taylor & Francis Group.)

4.7 Construction

Construction is the translation of the owner's goals and objectives into a facility built by the contractor as stipulated in the contract documents, plans, and specifications within budget and on schedule. The construction phase is an important phase in construction projects. A majority of total project budget and schedule is expended during construction. Similar to costs, the time required for the construction phase of the project is much higher than the time required for the preceding phases. Construction usually requires a large workforce and a variety of activities. Construction activities involve erection, installation, or construction of any part of the project. Construction activities are actually carried out by the contractor's own workforce or by subcontractors. Construction therefore requires more detailed attention of its planning, organizations, monitoring and control of project schedule, budget, quality, safety, and environment concerns.

Oberlender (2000) has described the importance of construction in the following words:

> The construction phase is important because the quality of the completed project is highly dependent on the workmanship and management of construction. The quality of construction depends on the completeness and quality of the contract documents that are prepared by the designer and three other factors: laborers who have the skills necessary to produce the work, field supervisors who have the ability to coordinate the numerous activities that are required to construct the project in the field, and the quality of materials that are used for construction of project. Skilled laborers and effective management of the skilled laborers are both required to achieve a quality project. (p. 258)

It is usual to invite contractors to compete for a contract for construction work, in the expectation that they will plan to do the work efficiently and therefore at a minimum cost. Once the contract is awarded to the successful bidder (contractor), it is the responsibility of the contractor to respond to the needs of the client (owner) by building the facility as specified in the contract documents, drawings, and specifications within the budget and on time.

The owner also appoints an engineer to supervise the construction process. It is a normal practice for the designer/consultant of the project to be contracted by the owner to supervise the construction process. The engineer is responsible for achieving project quality goals and is also responsible for implementing the procedures specified in the contract documents. Table 4.16 lists the responsibilities the owner delegates to the engineer (consultant).

Sometimes, the owner engages a construction manager or project manager during the construction process to act as the owner's representative

TABLE 4.16

Responsibilities of Supervision Consultant

Serial Number	Description
1	Achieving the quality goal as specified
2	Review contract drawings and resolve technical discrepancies/errors in the contract documents
3	Review construction methodology
4	Approval of contractor's construction schedule
5	Regular inspection and checking of executed works
6	Review and approval of construction materials
7	Review and approval of shop drawings
8	Inspection of construction material
9	Monitoring and controlling construction expenditure
10	Monitoring and controlling construction time
11	Maintaining project record
12	Conduct progress and technical coordination meetings
13	Coordination of owner's requirements and comments related to site activities
14	Project-related communication with contractor
15	Coordination with regulatory authorities
16	Processing of site work instruction for owner's action
17	Evaluation and processing of variation order/change order
18	Recommendation of contractor's payment to owner
19	Evaluating and making decisions related to unforeseen conditions
20	Monitor safety at site
21	Supervise testing, commissioning, and handover of the project
22	Issue substantial completion certificate

and delegates the following activities, thus leaving the engineer to perform project quality–related work only:

1. Review of contract documents
2. Approval of contractor's construction schedule
3. Cost control
4. Time control
5. Project methodology

The engineer appoints an engineer's representative to supervise the project construction process. The engineer's representative is supported by a supervision team consisting of professionals having experience and expertise in supervision and administration of similar construction projects. The engineer's representative is also called the resident engineer. Depending on the type and size of the project, the supervision team usually consists of the following personnel:

1. Resident engineer
2. Contract administrator/quantity surveyor
3. Planning/scheduling engineer
4. Engineers from different trades such as architectural, structural, mechanical, HVAC, electrical, low-voltage system, landscape, and infrastructure
5. Inspectors from different trades
6. Interior designer
7. Document controller
8. Office secretary

The construction phase consists of various activities such as mobilization, execution of work, planning and scheduling, control and monitoring, management of resources/procurement, quality, and inspection. Figure 4.34 illustrates major activities to be performed during the construction phase.

These activities are performed by various parties having contractual responsibilities to complete the specified work. Coordination among these parties is essential to ensure that the constructed facility meets the owner's objectives.

4.7.1 Mobilization

The contractor is given a few weeks to start the construction work after the signing of the contract. The activities to be performed during the mobilization period are defined in the contract documents. During this period, the contractor is required to perform many of the activities before the beginning of actual construction work at the site. Necessary permits are obtained from the relevant authorities to start the construction work. After being granted access to the construction site by the owner, the contractor starts mobilization work, which consists of preparation of site offices/field offices for the owner, supervision team (consultant), and for the contractor himself. This includes all the necessary on-site facilities and services necessary to carry out specific tasks. Mobilization activities usually occur at the beginning of a project but can occur anytime during a project when specific on-site facilities are required.

Examples of mobilization activities include

- Set up site offices and storage
- Construct temporary access roads, lay down areas and perimeter fences

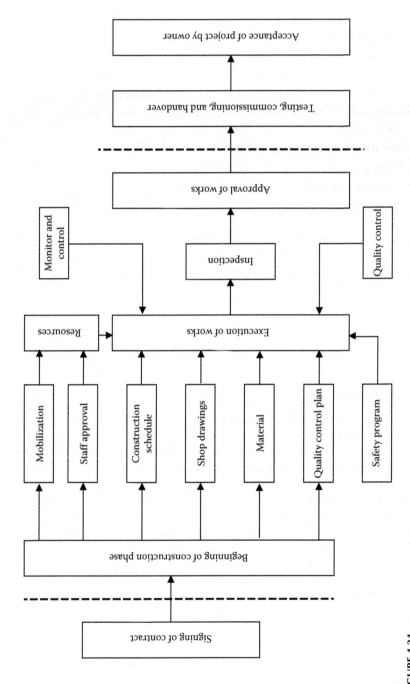

FIGURE 4.34
Major activities during construction phase.

- Install the necessary utilities for construction
- Set up a temporary firefighting system
- Perform site survey and testing
- Satisfy health and site safety requirements
- Submit preliminary construction program
- Selection of core staff as mentioned in the contract documents
- Insurance policies
- Selection of subcontractor (this may be an ongoing activity as per the approved schedule)

In anticipation of the award of contract, the contractor begins the following activities much in advance, but these are part of contract documents, and the contractor's action is required immediately after signing of the contract in order to start construction:

- Mobilization of construction equipment and tools
- Workforce to execute the project

For a smooth flow of construction process activities, proper communication and submittal procedure need to be established among all concerned parties at the beginning of the construction activities. Table 4.17 illustrates an example matrix for site administration of a building construction project.

Proper adherence to these duties helps smooth implementation of the project. Correspondence between consultant and contractor is normally through letters or job site instructions. Figure 4.35 is a job site instruction form used by the consultant to communicate with the contractor.

4.7.2 Identification of Project Team

The following project team members have involvement in the project during the construction phase:

1. Owner
   - Owner's representative/project manager
2. Construction supervisor
   - Construction manager (agency)
   - Consultant (designer)
   - Specialist contractor

TABLE 4.17

Matrix for Site Administration and Communication

Serial Number	Description of Activities	Contractor	Consultant	Owner
1	*Communication*			
	1.1 General correspondence	P	P	P
	1.2 Job site instruction	D	P	C
	1.3 Site work instruction	D	P/B	A
	1.4 Request for information	P	A	C
	1.5 Request for modification	P	B	A
2	*Submittals*			
	2.1 Subcontractor	P	B/R	A
	2.2 Materials	P	A	C
	2.3 Shop drawings	P	A	C
	2.4 Staff approval	P	B	A
	2.5 Premeeting submittals	P	D	C
3	*Plans and Programs*			
	3.1 Construction schedule	P	R	C
	3.2 Submittal logs	P	R	C
	3.3 Procurement logs	P	R	C
	3.4 Schedule update	P	R	C
4	*Monitor and Control*			
	4.1 Progress	D	P	C
	4.2 Time	D	P	C
	4.3 Payments	P	R/B	A
	4.4 Variations	P	R/B	A
	4.5 Claims	P	R/B	A
5	*Quality*			
	5.1 Quality control plan	P	R	C
	5.2 Checklists	P	D	C
	5.3 Method statements	P	A	C
	5.4 Mock-up	P	A	B
	5.5 Samples	P	A	B
	5.6 Remedial notes	D	P	C
	5.7 Nonconformance report	D	P	C
	5.8 Inspections	P	D	C
	5.9 Testing	P	A	B
6	*Site Safety*			
	6.1 Safety program	P	A	C
	6.2 Accident report	P	R	C
7	*Meetings*			
	7.1 Progress	E	P	E
	7.2 Coordination	E	P	C

(*Continued*)

TABLE 4.17 (*Continued*)

Matrix for Site Administration and Communication

Serial Number	Description of Activities	Contractor	Consultant	Owner
	7.3 Technical	E	P	C
	7.4 Quality	P	C	C
	7.5 Safety	P	C	C
	7.6 Closeout		P	
8	*Reports*			
	8.1 Daily report	P	R	C
	8.2 Monthly report	P	R	C
	8.3 Progress report		P	A
	8.4 Progress photographs		P	A
9	*Closeout*			
	9.1 Snag list	P	P	C
	9.2 Authorities' approvals	P	C	C
	9.3 As-built drawings	P	D/A	C
	9.4 Spare parts	P	A	C
	9.5 Manuals and documents	P	R/B	A
	9.6 Warranties	P	R/B	A
	9.7 Training	P	C	A
	9.8 Handover	P	B	A
	9.9 Substantial completion certificate	P	B/P	A

Note: P, Prepare/initiate; B, advise/assist; R, review/comment; A, approve; D, action; E, attend; C, information.

3. Contractor
   - Main contractor
   - Subcontractor
   - Supplier
4. Regulatory authorities
5. End user

4.7.3 Planning and Scheduling

Project planning is a logical process to ensure that the work of the project is carried out

- In an organized and structured manner
- Reducing uncertainties to a minimum
- Reducing risk to a minimum
- Establishing quality standards
- Achieving results within budget and scheduled time

Project Name
Consultant Name

JOB SITE INSTRUCTION (JSI)

CONTRACTOR: _____   JSI No. : _____

CONTRACT No.: _____   DATE : _____

The work shall be carried out in accordance with the Contract Documents without change in Contract Sum or Contract Time. Proceeding with the work in accordance with these instructions indicates your acknowledgement that there will be no change in the Contract Sum or Contract Time.

Subject: _____

_____

SAMPLE FORM

_____

_____

_____

_____

ATTACHMENTS: (List attached documents that support description.)

_____

_____

_____

_____

_____

Signed:	Received by Contractor :
Resident Engineer	Date:

Distribution: ☐ Owner   ☐ A/E   ☐ Contractor

FIGURE 4.35
Job site instruction.

Prior to the start of execution of a project or immediately after the actual project starts, the contractor prepares the project construction plans based on the contracted time schedule of the project. Detailed planning is needed at the start of construction to decide how to use resources such as laborers, plant, materials, finance, and subcontractors economically and safely to achieve the specified objectives. The plan shows the periods for all sections of the works and activities, indicating that everything can be completed by the date specified in the contract and ready for use or for installation of equipment by other contractors.

According to Oberlender (2000)

> Project Planning is the heart of good project management because it provides the central communication that coordinates the work of all parties. Planning also establishes the benchmark for the project control system to track the quantity, cost, and timing of work required to successfully complete the project. Although the most common desired result of planning is to finish the project on time, there are other benefits that can be derived from good project planning. (p. 140)

Effective project management requires planning, measuring, evaluating, forecasting, and controlling all aspects of project quality and quality of work, cost, and schedules. The purpose of the project plan is to successfully control the project to ensure completion within the budget and schedule constraints. Project planning is the evolution of the time and efforts to complete the project. Table 4.18 lists the benefits of project planning and scheduling.

Planning is a mechanism that conveys or communicates to project participants what activity is to be done, how, and in what order to meet the project objectives by scheduling the activities. Project planning is required to bring the project to completion on schedule, within budget, and in accordance with the owner's needs as specified in the contract. The planning process considers all the individual tasks, activities, or jobs

TABLE 4.18

Benefits of Project Planning and Scheduling

1.	Finish the project on time.
2.	Continuous (uninterrupted) flow of work (no delays).
3.	Reduced amount of rework (least amount of changes).
4.	Minimize confusion and misunderstandings.
5.	Increased knowledge of status of project by everyone.
6.	Meaningful and timely reports to management.
7.	You run the project instead of the project running you.
8.	Knowledge of scheduled times of key parts of the project.
9.	Knowledge of distribution of costs of the project.
10.	Accountability of people, defined responsibility/authority.
11.	Clear understanding of who does what, and how much.
12.	Integration of all work to ensure a quality project for the owner.

Source: Oberlender, G.D. (2000), *Project Management for Engineering.* Reprinted with permission of The McGraw-Hill Companies.

that make up the project and must be performed. It takes into account all the resources available, such as human resources, finances, materials, plant, and equipment. It also considers the works to be executed by the subcontractors.

The following is the list of activities of construction projects normally included in the construction program/plan:

A. General activities
   1. Mobilization
B. Engineering
   1. Subcontractor submittal and approval
   2. Materials submittal and approval
   3. Shop drawing submittal and approval
   4. Procurement
C. Site activities
   1. Site earthworks
   2. Dewatering and shoring
   3. Excavation and backfilling
   4. Raft works
   5. Retaining wall works
   6. Concrete foundation and grade beams
   7. Waterproofing
   8. Concrete columns and beams
   9. Casting of slabs
   10. Wall partitioning
   11. Interior finishes
   12. Furnishings
   13. External finishes
   14. Equipment
   15. Conveying system works
   16. Plumbing and public health works
   17. Firefighting works
   18. HVAC works
   19. Electrical works
   20. Fire alarm system works
   21. Communication system works
   22. Low-voltage system works

23. Landscape works
24. External works

D. Closeout
    1. Testing and commissioning
    2. Completion and handover

The contractor also submits the following:

1. Resources (equipment and manpower) schedule
2. Cost loading (schedule of item's pricing based on bill of quantities)

Planning and scheduling are often used synonymously for preparing a construction program because both are performed interactively. Planning is the process of identifying the activities necessary to complete the project, while scheduling is the process of determining the sequential order of the planned activities and the time required to complete the activity. Scheduling is the mechanical process of formalizing the planned functions, assigning the starting and completion dates to each part or activity of the work in such a manner that the whole work proceeds in a logical sequence and in an orderly and systematic manner.

The first step in preparation of a construction program is to establish the activities, and the next step is to establish the estimated time duration of each activity. The deadline for each activity is fixed, but it is often possible to reschedule by changing the sequence in which the tasks are performed, while retaining the original estimated time. Figure 4.36 illustrates the steps in project planning.

Construction projects are unique and nonrepetitive in nature. Construction projects consist of many activities aimed at the accomplishment of a desired objective. In order to achieve the quality objectives of the project, each activity has to be completed within the specified limit, using the specified product and approved method of installation. A construction project consists of a number of related activities that are dependent on other activities and cannot be started until others are completed, and some that can run in parallel. The most important point while starting the planning is to establish all the activities that constitute the project. Table 4.19 lists key principles for planning and scheduling.

Planning involves defining the objectives of the project; listing of tasks or jobs that must be performed; determining gross requirements for material, equipment, and manpower; and preparing costs and durations for the various jobs or activities needed for the satisfactory completion of the project. The techniques for planning vary depending on the project's size, complexity, duration, personnel, and owner's requirements. Techniques used during the construction phase of the project should make possible the evaluation of the project's progress against the plan. There are many

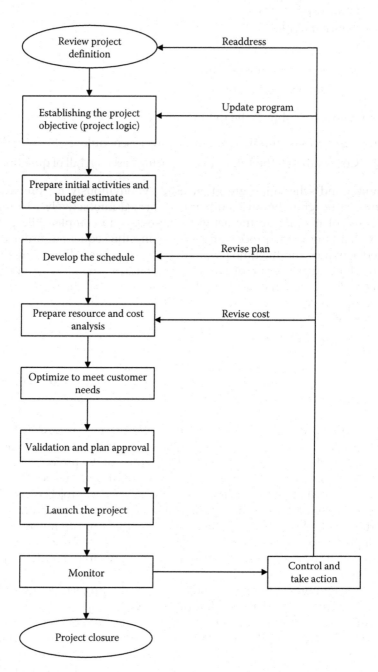

FIGURE 4.36
Project planning steps.

TABLE 4.19

Key Principles for Planning and Scheduling

1.	Begin planning before starting work, rather than after starting work.
2.	Involve people who will actually do the work in the planning and scheduling process.
3.	Include all aspects of the project: scope, budget, schedule, and quality.
4.	Build flexibility into the plan, include allowance for changes and time for reviews and approvals.
5.	Remember the schedule is the plan for doing the work, and it will never be precisely correct.
6.	Keep the plan simple; eliminate irrelevant details that prevent the plan from being readable.
7.	Communicate the plan to parties; any plan is worthless unless it is known.

Source: Oberlender, G.D. (2000), *Project Management for Engineering*. Reprinted with permission of The McGraw-Hill Companies.

different analytical and graphical techniques that are commonly used for planning of the project. These are

1. The bar chart
2. Critical path method
3. Progress curves
4. Matrix schedule

The most widely used forms of program are bar charts and network diagrams. The bar chart is the oldest planning method used in project management. It is a graphical representation of the estimated duration of each activity and the planned sequence of activities. The horizontal axis represents the time schedule, whereas the project activities are shown along the vertical axis.

Network diagrams such as program evaluation and review technique (PERT) and CPM are used for scheduling of complex projects. PERT/CPM diagrams consist of nodes and links and represent the entire project as a network of arrows (activities) and nodes (events). In order to draw a network diagram, work activities have to be identified, the relationships among the activities need to be specified, and a precedence relationship between the activities in a particular sequence needs to be established.

The most widely used scheduling technique is CPM. CPM analysis represents the set of sequence of predecessor/successor activities that will take the longest time to complete. The duration of the critical path is the sum of all the activity durations along the path. Thus, the critical path is the longest possible path of the project activities network. The duration of the critical path represents the minimum time required to complete the project.

There are many computer-based programs available for preparing the network and critical path of activities for construction projects. These programs can be used to analyze the use of resources, review project progress, and forecast the effects of changes in the schedule of works or other resources. Most computer programs automate preparation and presentation of various planning tools such as the bar chart, PERT, and CPM analysis. These programs are capable of storing enormous quantities of data and help process and update the program quickly. They manipulate data for multiple usages from the planning and scheduling perspectives.

In order to manage and control the project at different levels in the most effective manner, it is broken down into a group of smaller subprojects/subsystems and then into small, well-defined activities. This breakdown is necessary because of the size and complexity of construction projects and is referred to as work breakdown structure (WBS). To begin the preparation of the detailed construction program, the contractor prepares a WBS. Its purpose is to define various activities that must be executed to complete the project. WBS helps the construction project planner to

1. Plan and schedule the work
2. Estimate costs and budget
3. Control schedule, cost, and quality

Activities are those operations of the plan that take time to carry out and on which resources are expended. Depending on the size of the project, the project is divided into multiple zones, and relevant activities are considered for each zone to prepare the construction program. Appendix B illustrates an example guideline listing major activities performed during building construction projects and are commonly used by the contractor's scheduler/planner to prepare the contractor's construction schedule/program. Electromechanical activities are further divided into first fix, second fix, and final fix, depending on their relationship with civil and architectural works. The construction program is prepared by selecting appropriate activities relevant to a particular floor/zone. These activities are also considered for preparation of cost and resource loading schedule for the project. While preparing the program, the relationships between project activities and their dependency and precedence are considered by the planner. These activities are connected to their predecessor and successor activity based on the way the task is planned to be executed. There are four possible relationships that exist between various activities. These are the finish-to-start relationship, the start-to-start relationship, the finish-to-finish relationship, and the start-to-finish relationship.

In order to prepare a project plan, the logic is reviewed for correctness and ascertained that all activities are shown, the scope of the project has been interpreted correctly, and the resources that are required for performing each job are applied. Figure 4.37 illustrates a logic flowchart for firefighting works. Similarly, Figures 4.38 through 4.40 illustrate logic flowcharts for plumbing, HVAC, and electrical works, respectively.

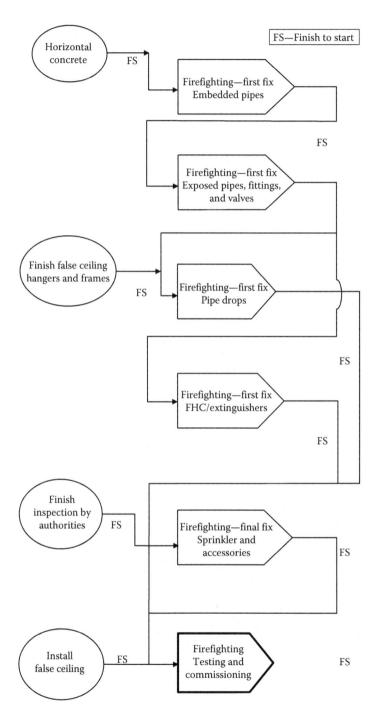

FIGURE 4.37
Logic flow diagram for firefighting works.

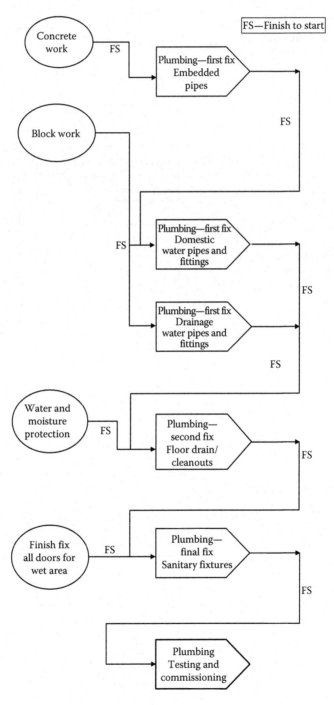

FIGURE 4.38
Logic flow diagram for plumbing works.

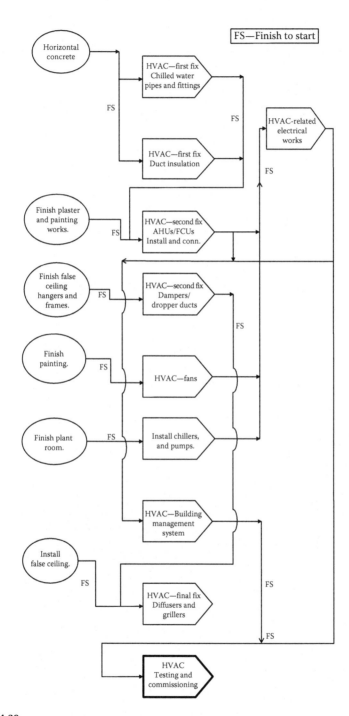

FIGURE 4.39
Logic flow diagram for HVAC works.

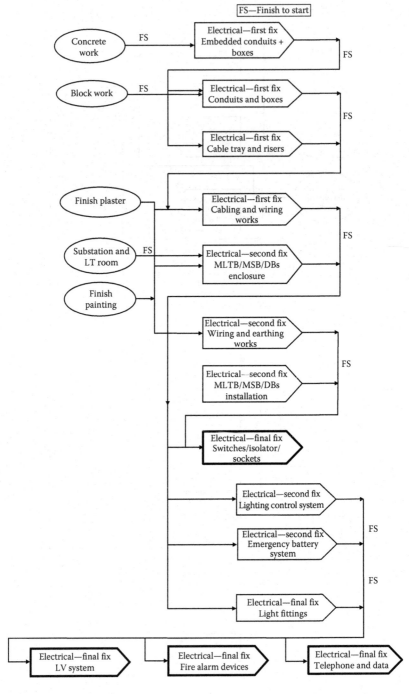

FIGURE 4.40
Logic flow diagram for electrical works.

Once all the activities are established by the planner and the estimated duration of each activity has been assigned, the planner prepares a detailed program fully coordinating all the construction activities.

CPM calculates the minimum completion time for a project along with the possible start and finish times for the project activities. The critical path is the longest in the network, whereas the other paths may be equal or shorter than that path. Therefore, there is a possibility that some of the events and activities can be completed before they are actually needed and, accordingly, it is possible to develop a number of activity schedules from the CPM analysis to delay the start of each activity as long as possible but still finish the project with minimum possible time without extending the completion date of the project. To develop such a schedule, it is required to find out when each activity needs to start and when it needs to be finished. There may be some activities in the project with some leeway for when they can start and finish. This is called slack time, or float, in an activity. For each activity in a project, there are four points in time: early start, early finish, late start, and late finish. Early start and early finish are the earliest times an activity can start and finish, respectively. Similarly, late start and late finish are the latest times an activity can start and finish, respectively. The difference between late start time and early start time is the slack time, or float.

With the advent of powerful computer programs such as Primavera® and Microsoft Project™, it is possible for the details of the work breakdown to be fed to these software programs. The software is capable of producing network diagrams and schedules and a limitless number of different reports, which also help in the efficient monitoring of the project schedule by comparing actual with planned progress. The software can be used to analyze the project for use of resources, forecasting the effects of changes in the schedule, and cost control. Figure 4.41 illustrates an example construction program summary of a smart building system project for an office building.

4.7.4 Execution of Works

According to the authors of *Civil Engineering Procedure* (Institution of Civil Engineers 1996):

> The contractor is responsible for construction and maintaining the works in accordance with the contract drawings, specifications and other documents and also further information and instruction issued in accordance with the contract. The contractor should be as free as possible to plan and execute the works in the way he wishes within the terms of his contract, so should the sub-contractors. Any requirements for part of a project to be finished before the rest and all limits of contractor's freedom should therefore have been stated in the tender document. (p. 77)

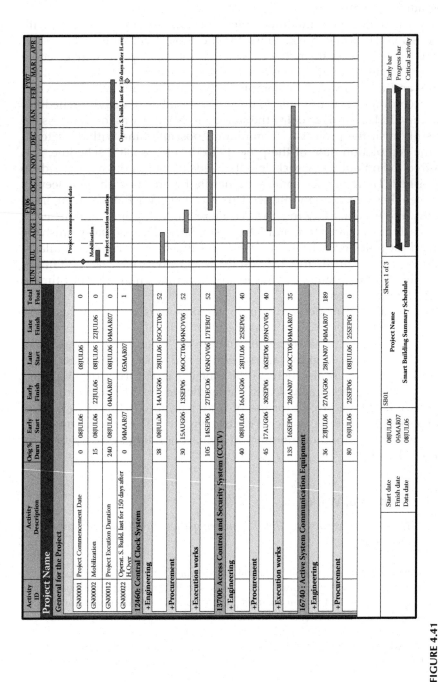

FIGURE 4.41
Summary construction program. (Copyright Primavera Systems, Inc.)

(Continued)

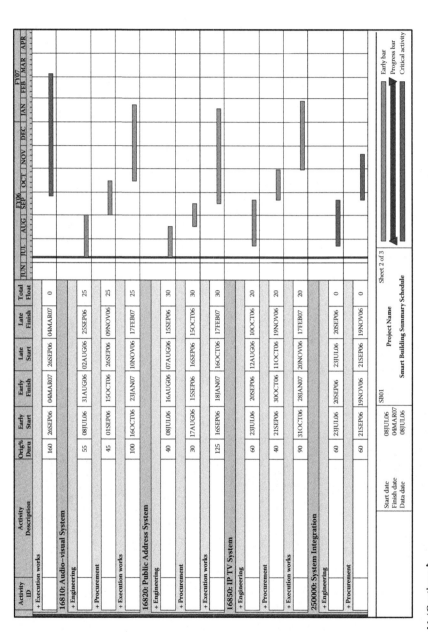

FIGURE 4.41 (*Continued*)
Summary construction program. (Copyright Primavera Systems, Inc.)

(*Continued*)

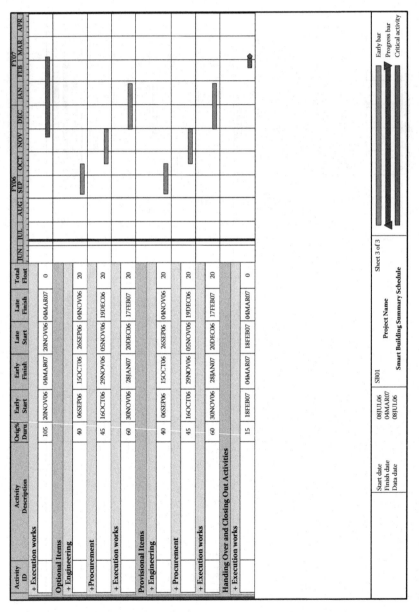

Activity ID	Activity Description	Orig% Durn	Early Start	Early Finish	Late Start	Late Finish	Total Float
	+ Execution works	105	20NOV06	04MAR07	20NOV06	04MAR07	0
	Optional Items						
	+ Engineering	40	06SEP06	15OCT06	26SEP06	04NOV06	20
	+ Procurement	45	16OCT06	29NOV06	05NOV06	19DEC06	20
	+ Execution works	60	30NOV06	28JAN07	20DEC06	17FEB07	20
	Provisional Items						
	+ Engineering	40	06SEP06	15OCT06	26SEP06	04NOV06	20
	+ Procurement	45	16OCT06	29NOV06	05NOV06	19DEC06	20
	+ Execution works	60	30NOV06	28JAN07	20DEC06	17FEB07	20
	Handing Over and Closing Out Activities						
	+ Execution works	15	18FEB07	04MAR07	18FEB07	04MAR07	0

Start date	08JUL06
Finish date	04MAR07
Data date	08JUL06

SB01	Project Name
Smart Building Summary Schedule	

Sheet 3 of 3

Early bar
Progress bar
Critical activity

FIGURE 4.41 (Continued)
Summary construction program. (Copyright Primavera Systems, Inc.)

Construction activities mainly consist of the following:

- Site work such as cleaning and excavation of project site
- Construction of foundations, including footings and grade beams
- Construction of columns and beams
- Forming, reinforcing, and placing of the floor slab
- Laying up masonry walls and partitions
- Installation of roofing system
- Finishes
- Furnishings
- Conveying system
- Installation of firefighting system
- Installation of water supply, plumbing, and public health system
- Installation of HVAC system
- Integrated automation system
- Installation of electrical lighting and power system
- Emergency power supply system
- Fire alarm system
- Communication system
- Electronic security and access control system
- Landscape works
- External works

4.7.5 Management of Resources/Procurement

In most construction projects, the contractor is responsible for engaging subcontractors, specialist installers, and suppliers, and arranging for materials, equipment, construction tools, and all types of human resources needed to complete the project as per contract documents and to the satisfaction of the owner or the owner's appointed supervision team. Workmanship is one of the most important factors to achieve quality in construction; therefore, it is required that the construction workforce be fully trained and have full knowledge of all the related activities to be performed during the construction process.

Once the contract is awarded, the contractor prepares a detailed plan for all the resources he or she needs to complete the project. The contractor also prepares a procurement log based on the project completion schedule.

Contract documents normally specify a list of the minimum number of core staff to be available on-site during the construction period. The absence of any of these staff may result in a penalty being imposed on the contractor by the owner.

The following is a typical list of a contractor's minimum core staff needed during the construction period for execution of the work of a major building construction project:

1. Project manager
2. Site senior engineer for civil works
3. Site senior engineer for architectural works
4. Site senior engineer for electrical works
5. Site senior engineer for mechanical works
6. Site senior engineer for HVAC works
7. Site senior engineer for infrastructure works
8. Planning engineer
9. Senior quantity surveyor/contract administrator
10. Civil works foreman
11. Architectural works foreman
12. Electrical works foreman
13. Mechanical works foreman
14. HVAC works foreman
15. Laboratory technician
16. Quality control engineer
17. Safety officer

The contractor has to submit the names of the staff for these positions for approval from the owner/consultant to work on the project. Figure 4.42 illustrates a staff approval request form used by the contractor to propose the staff to work for these positions and is submitted along with the qualification and experience certificates of the proposed staff.

The contractor's human resources mainly fall into two categories:

1. The contractor's own staff and workers
2. The subcontractor's staff and workers

The main contractor has to manage all these personnel by

1. Assigning daily activities
2. Observing their performance and work output
3. Daily attendance
4. Safety during the construction process

Project Name	
Consultant Name	
REQUEST FOR SITE STAFF APPROVAL	
CONTRACT NO. :	NO. :
CONTRACTOR :	DATE:

To : Owner   **SAMPLE FORM**

1.	Name	:	_____
2.	Profession	:	_____
3.	Position No. in Document-I	:	_____
4.	No. of years of Experience	:	_____
5.	Membership of Professional Body	:	Valid ☐ Not Valid ☐
6.	Requested Date of Commencement	:	_____
7.	Remarks	:	_____

_____
Signature Contractor's Project Manager

OWNER COMMENTS	APPROVED ☐	NOT APPROVED ☐

_____   _____
Owner Rep. Signature   Date

Distribution	OWNER	A/E	CONTRACTOR

FIGURE 4.42
Request for staff approval.

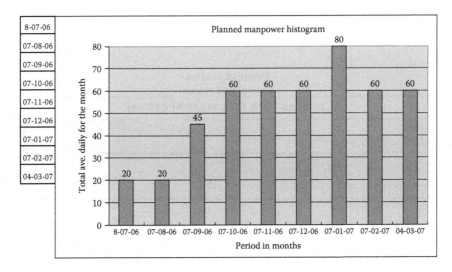

FIGURE 4.43
Manpower plan.

Figure 4.43 illustrates the contractor's manpower chart of a construction project.

Likewise, the contract documents specify that a minimum equipment set is to be available on-site during the construction process to ensure smooth operation of all the construction activities. These are

- Tower crane
- Mobile crane
- Normal mixture
- Concrete mixing plant
- Dump trucks
- Compressor
- Vibrators
- Water pumps
- Compactors
- Concrete pumps
- Trucks
- Concrete trucks
- Diesel generator sets

Figure 4.44 illustrates the equipment schedule, which lists the equipment the contractor has to make available for major building construction projects.

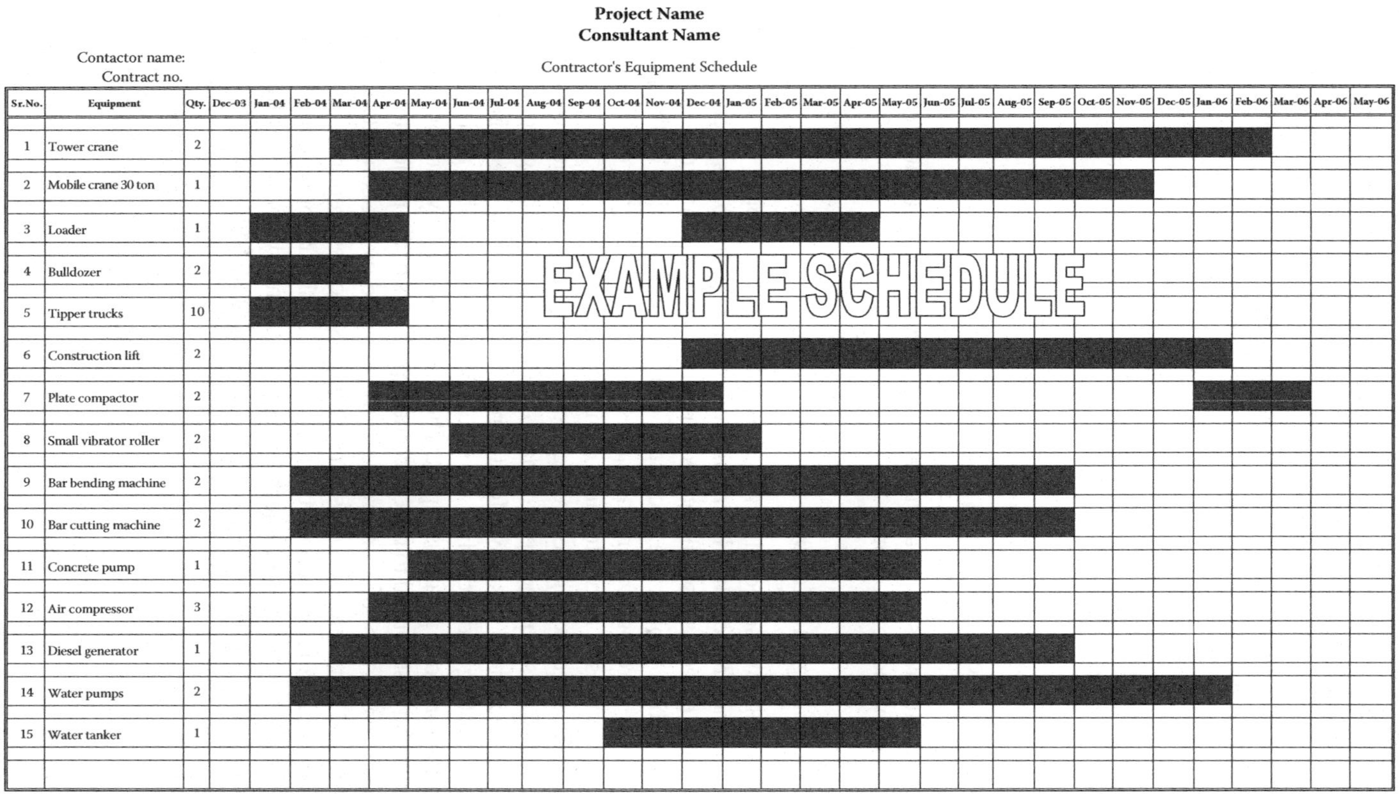

Project Name
Consultant Name

Contractor's Equipment Schedule

Contactor name:
Contract no.

Sr.No.	Equipment	Qty.
1	Tower crane	2
2	Mobile crane 30 ton	1
3	Loader	1
4	Bulldozer	2
5	Tipper trucks	10
6	Construction lift	2
7	Plate compactor	2
8	Small vibrator roller	2
9	Bar bending machine	2
10	Bar cutting machine	2
11	Concrete pump	1
12	Air compressor	3
13	Diesel generator	1
14	Water pumps	2
15	Water tanker	1

Columns (months): Dec-03, Jan-04, Feb-04, Mar-04, Apr-04, May-04, Jun-04, Jul-04, Aug-04, Sep-04, Oct-04, Nov-04, Dec-04, Jan-05, Feb-05, Mar-05, Apr-05, May-05, Jun-05, Jul-05, Aug-05, Sep-05, Oct-05, Nov-05, Dec-05, Jan-06, Feb-06, Mar-06, Apr-06, May-06

EXAMPLE SCHEDULE

FIGURE 4.44
Equipment schedule.

In most construction projects, the contractor is responsible for procurement of material, equipment, and systems to be installed on the project. Contractors have their own procurement strategies. While submitting the bid, the contractor obtains quotations from various suppliers/subcontractors. The contractor has to consider the following, at a minimum, while finalizing the procurement:

- Contractual commitment
- Specification compliance
- Statutory obligations
- Time
- Cost
- Performance

Figure 4.45 illustrates the material approval and procurement procedure to be followed by the contractor.

4.7.6 Monitoring and Control

Once planning and scheduling are complete and the project is under way, progress on the project must be monitored on an ongoing basis to ensure that the goals and objectives on the project are being met. Monitoring and control of a construction project are necessary during the execution of the project, and its aim is to recognize any obstacles encountered during execution and to apply measures to mitigate these difficulties.

Monitoring is collecting, recording, and reporting information concerning any and all aspects of project performance that the project manager or others in the organization need to know. Monitoring of construction projects is normally done by collecting and recording the status of various activities and compiling them in the form of a progress report. These are prepared by the consultant or contractor and distributed to the concerned members of the project team. Table 4.20 shows the content of a progress report prepared by the consultant's team and Table 4.21 illustrates contents of contractor's progress report.

Monitoring involves not only tracking time but also resources and budget. Monitoring in construction projects is normally done by compiling the status of various activities in the form of progress reports. These are prepared by the contractor, supervision team (consultant), and construction/project management team. The objectives of project monitoring and control are to

1. Report the necessary information in detail and in an appropriate form that can be interpreted by management and other concerned personnel to provide them with information about how the resources are being used to achieve project objectives

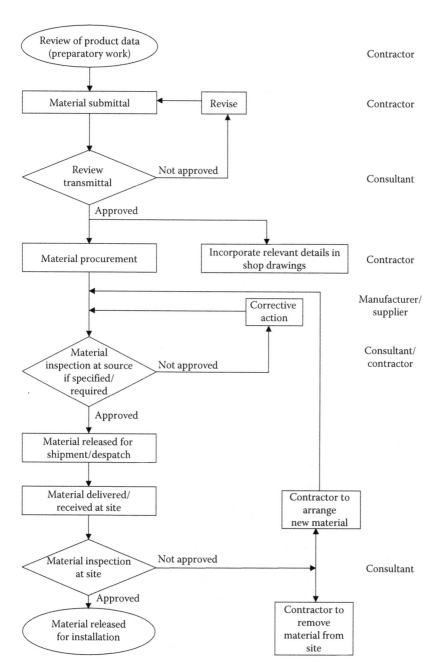

FIGURE 4.45
Material approval and procurement procedure.

TABLE 4.20

Contents of Progress Report

1.0	**Contract Particulars**	
	1.1	Project description
	1.2	Project data
2.0	**Construction Schedule**	
3.0	**Progress of Works**	
	3.1	Temporary facilities and mobilization
	3.2	Summary of construction progress
		3.2.1 Status
		3.2.2 On-shore progress
		3.2.3 Off-shore progress
4.0	**Time Control**	
	4.1	CPM schedule—level one (target vs. current)—summary by building/marine
	4.2	CPM Schedule—level two (target vs. current)—summary by building/division
	4.3	30 days look-ahead schedule
	4.4	Time control conclusion
5.0	**Cost Control**	
	5.1	Financial progress
	5.2	Cash flow curve and histogram
	5.3	Work-in-place S-curve and histogram
	5.4	Cost control conclusion
6.0	**Status of Contractor's Submittals**	
	6.1	Material status
	6.2	Shop drawing status
7.0	**Subcontractors**	
8.0	**Consultant's Staff**	
9.0	**Quality Control**	
10.0	**Meetings**	
11.0	**Site Work Instructions**	
12.0	**Variation Orders**	
13.0	**Construction Photographs**	
14.0	**Contractors Resources**	
15.0	**Other Matters**	
	15.1	Safety
	15.2	Weather conditions
	15.3	Important developments/proposals/submissions

TABLE 4.21

Monthly Progress Report

EXAMPLE CONTENTS

Contents of Monthly Report

Serial Number		Contents	Description
1		**Executive Summary—Tabular**	
	1.1	Summary status report	Brief description of the project status up to date, i.e., manpower, cash, activities
	1.2	NOC's report	No objection certificate report
	1.3	Project manager narratives	Narratives description of project status up to date
2		**Progress Layouts**	
	2.1	Updated milestone table	Comparison of planned vs. actual for contractual milestones per construction unit/design. You track the delays for major trades through the color theme
	2.2	Updated major of events table	Comparison of planned vs. actual for major trades per construction unit/design. Track the delays for major trades through the color theme
	2.3	Updated layouts	Same as (2.1) above but presentation per milestone phase (drawing)
3		**Updated Execution Program**	
	3.1	Updated milestone schedule—roll up "update versus latest target"	To indicate the status of the control and key milestones comparing the current status with the baseline
	3.2	Updated detailed schedule	All activities in detail
	3.3	One month look ahead program	Same as (3.2) but includes only the detailed activities for the coming month BUT ON EXCEL FORMAT
4		**Submittal Status Report (E1 Log)**	Updated status of submittals
	4.1	Submittal status report	Brief description of the project submittal status up to date
	4.2	Detailed E1 Log	Detailed description of the project submittal status up to date
5		**Procurement Status Report (E2 Log)**	Updated procurement status

(Continued)

TABLE 4.21 (*Continued*)

Monthly Progress Report

EXAMPLE CONTENTS

Contents of Monthly Report

Serial Number		Contents	Description
6		**Status of Information Requested**	
	6.1	RFI status report	Request for information (RFI)
	6.2	NCR report	Nonconformance request (NCR)
	6.3	PCO and NOV Log	Potential change order and notice of variation summary
7		**Updated Cost Loaded Program**	
	7.1	T-7 updated status report	Money progress = Physical * Budgeted cost for each running or completed activity
	7.2	Updated cost loaded schedule	Same as T-7 but money values not percentages
	7.3	Updated work in place (%) report	Histogram and cumulative curve
	7.4	Updated cash flow report	Histogram and cumulative curve
8		**Updated Manpower Histogram**	Histogram and cumulative curve
9		**Updated Schedule of Construction Equipment and Vehicles**	Tabular report
10		**Updated Critical Indicators**	
	10.1	Shop drawings status report	Histogram and cumulative curve
	10.2	Material status report	Histogram and cumulative curve
	10.3	Construction. Leading indicators	Each trade alone
	10.4	Line of balance diagram	It indicates all progress of the major trades as line cumulative chart
11		**Updated Progress Photographs**	—
12		**Updated Safety Inspection Checklist**	Tabular report
13		**Contractor Information**	Organization chart, tabular report

Source: Rumane, A.R. (2013), *Quality Tools for Managing Construction Projects*, CRC Press, Boca Raton, FL. Reprinted with permission from Taylor & Francis Group.

2. Provide an organized and efficient means of measuring, collecting, verifying, and quantifying data reflecting the progress and status of execution of project activities, with respect to schedule, cost, resources, procurement, and quality
3. Provide an organized, efficient, and accurate means of converting the data from the execution process into information
4. Identify and isolate the most important and critical information about the project activities to enable decision-making personnel to take corrective action for the benefit of the project
5. Forecast and predict the future progress of activities to be performed

Once the scheduled plan has been prepared and execution has commenced, control over progress of work has to be exercised to ensure completion of the work by the stipulated time. A project control mechanism can be used to determine deviations from the basic plan, the precise effect of these deviations on the plan, and to replan and reschedule to compensate for the deviations.

Based on the progress of work and the revised durations of unfinished activities due to delays, the network diagram is replanned and rescheduled; this process is known as updating. If the completion date is beyond the milestone dates, then the contractor is required to submit a recovery plan to compensate for the delay in execution of the project. Project control involves a regular comparison of performance against targets, a search for the cause of deviation, and a commitment to check adverse variance. It serves two major functions:

1. It ensures regular monitoring of performance.
2. It motivates project personnel to strive for achieving project objectives.

Construction projects have certain specific checkpoints or events where an evaluation is made to see that the project is on time and to ensure that project objectives are met. These significant events are termed *milestones*. A milestone is a point in time that marks the start or finish of a significant activity or group of activities, usually the completion of a major deliverable, for example, completion of concrete works, receipt of major equipment, completion of testing, etc. The milestone points are determined at the beginning of the project. Milestone evaluation is used to certify that all the work scheduled to be accomplished has been completed according to the specification requirements.

According to CII Source Document 61 (1990):

> Project control is frequently referenced in the literature as a highly problematic area of project management. It is concerned with scope control, timely decision making, control system integration, control techniques, and key estimating practices. One attribute of successful project control is a sound baseline (or plan) from which to compare actual performance. This baseline for the cost control component is the budget estimate. (p. 78)

A construction project control is exercised through knowing where to put in the main effort at a given time and maintaining good communication. There are mainly three areas where project control is required: (1) quality, (2) schedule, and (3) budget.

In most contracts, the cost and time required to complete the specified scope of works are defined in project documents. Control is thus a matter of reporting and regular progressing of project activities against fixed targets. Control of quality of materials and workmanship is achieved through a proper QCP and its implementation through a preset level of quality control, and inspection of various activities and materials.

Budget control is done through monitoring progress payments and variation costs.

The schedule is monitored by ensuring timely approval of materials, shop drawings, timely procurement of materials, and execution of works as planned.

Quality control is achieved through inspection of works during the construction process, ensuring the use of approved materials and workmanship.

There are different types of logs used in a construction project for monitoring purposes. The main logs used in a construction project are as follows.

4.7.6.1 Subcontractors Submittal and Approval Log

In most construction projects, the contractor engages special subcontractors to execute certain portions of the contracted project work. Areas of subcontracting are generally listed in the particular conditions of the contract document. Generally, the contractor has to submit subcontractors/specialist contractors to execute the following types of works:

1. Precast concrete works
2. Metal works
3. Space frame, roofing works
4. Wood works
5. Aluminum works
6. Internal finishes such as painting, false ceiling, tiling, cladding
7. Furnishings
8. Waterproofing and insulation works
9. Mechanical works
10. HVAC works
11. Electrical works
12. Low-voltage systems/smart building system
13. Landscape
14. External works
15. Any other specialized works

The contractor has to submit their names for approval to the owner prior to their engagement to perform any work at the site. The names of subcontractors for various subprojects have to be submitted for approval in a timely and orderly manner, as planned in the approved work program, so that work progresses in a smooth and efficient manner. This log helps remind the project manager to ask the main contractor to submit the subcontractors on time.

Figure 4.46 is a request for subcontractor approval to be submitted by the contractor for getting approval of any subcontractor proposed to work on the construction project.

The request includes all the related information to prove the subcontractor's capability of providing services that meet the required project quality, the resources available to meet the specified schedule, and past performance; also, if any quality system was implemented by the subcontractor, that fact is noted.

Sometimes, the owner/consultant nominates subcontractors to execute a portion of a contract; such a subcontractor is known as a nominated subcontractor.

4.7.6.2 Shop Drawings and Materials Logs—E1

The timely submission and approval of shop drawings and materials is a must for the smooth and efficient progress of any project. The project manager and the consultant team constantly monitor these logs on a biweekly basis. The bottlenecks are identified, and technical meetings are held to solve the problems.

Figure 4.47 illustrates a submittal status log form normally called Log E-1 that is submitted by the contractor to the construction manager/project manager/consultant.

Figure 4.48 illustrates a shop drawing status log.

These logs are based on the approved construction program.

4.7.6.3 Procurement Log—E2

Delivery of long-lead items has to be initiated at an early stage of the project and monitored closely. Late order placement for materials resulting in delayed delivery of material, which in turn affects the timely completion of the project, is a common scenario in construction projects. Hence, these logs have to be updated regularly, and prompt actions have to be taken to avoid delays. The contractor is required to provide twice a month, or at any time requested by the owner/consultant, full and complete details of the procurement data relating to all the approved products and systems that have been ordered and/or procured by the contractor for use in the construction project.

Figure 4.49 illustrates a procurement log.

	Project Name

Contract No.:
Contractor :

**CONTRACTOR REQUEST
FOR SUB-CONTRACTOR APPROVAL**

Serial No.: _ _ _ _

Kindly approve the following as a SUB-CONTRACTOR in the above mentioned Project.

Sub-Contract Works
Sub-Contractor
Address (Head Office)

Reference Letter

Attachments:

SAMPLE FORM

Commercial Register		Foundation Contract	
Experience		Resources	
Current Work		Financial Status	
Others (List, if any)			

Performance Bond (Yes/No)   No.

Signed by
Contractor's Representative   Date: _ _ _ _ _ _ _ _ _ _ _

OWNER'S NAME

Received by: _____ Signature : _____ Date: _ _ _ _ _ _ _ _ _ _

SITE SUPERVISION CONSULTANT

Received by: _____ Signature : _____ Date: _ _ _ _ _ _ _ _ _ _

Consultant's recommendation :

Signed by
Resident Engineer   Date: _ _ _ _ _ _ _ _ _ _ _

Distribution :   OWNER(Original)   Supervision Consultant (Copy)   Contractor (Copy)

Note :   Contractor shall submit Original to OWNER with copy to Site Supervision Consultant.

FIGURE 4.46
Request for subcontractor approval.

Project Name
Consultant Name

CONTRACTOR'S SUBMITTAL STATUS LOG (REPORT E-1)

CONTRACT NO. : ______________________ NO. : ________
CONTRACTOR : ______________________ DATE : ________

Document Specification or BOQ Page	Submittal Sequence No.	Rev	Description	Type of Submittal: Shop Drawing	Samples	Guarantee	Mf's Data	Mt's Data	Certificates	Test	Otr	Action Code : A = Approved as submitted B = Approved as noted C = Not Approved D = For Information						Notes
				SD	SM	GT	MF	MT	CT	TT	OT	Planned Sub. date	Submitted By Contrac.	Returned By A/E	Action	Approval Needed By	Required On Job	

SAMPLE FORM

FIGURE 4.47
Contractor's submittal status log.

Project Name
Consultant Name

CONTRACTOR'S SHOP DRAWINGS STATUS LOG

CONTRACT NO. : ______________________ NO. : ________
CONTRACTOR : ______________________ DATE : ________

Document Specification or BOQ Page	Item No.	Submittal Sequence No.						Drawing No.	Description	Submittal Type	Action Code : A = Approved as submitted B = Approved as noted BR = Approved as noted, Resubmit C = Not Approved D = For Information						Remarks
		Div	Sec	Typ	Sr.	Rev	Ser			SD	Planned Sub. date	Submitted By Contrac.	Returned By A/E	Action	Approval Needed By	Required On Job	

SAMPLE FORM

FIGURE 4.48
Contractor's shop drawing submittal log.

Project Name
Consultant Name

CONTRACTOR'S PROCUREMENT LOG (REPORT E-2)

CONTRACT NO. :
CONTRACTOR :

NO. :
DATE :

Activity No.	Description	R.O.Q./ Specification. No.	Estimated Quantity	Ordered Quantity	Reqd. Order Date	Date Pur. Ord. Issued	Purchase Order No.	Supplier	Method Of Shipping	Shipping Date In Kuwait	Required On Job Date	E.D.A. To Site	A.D.A. To Site	Lead Time	Remarks

SAMPLE FORM

FIGURE 4.49
Contractor's procurement log.

4.7.6.4 Equipment and Manpower Logs

The challenge for construction management is to bring together all the required resources (material, manpower, and equipment) in the correct quantity at the correct time. In the construction network programs, in addition to detailing how the project will be assembled, the resources required for each activity are also prepared. This process is called resource loading. It is necessary that all the construction resources be coordinated and brought together at the right time in order to complete the project on time and within the budget. All construction projects track equipment and manpower employed on-site by updating the respective logs.

Figure 4.50 illustrates a list of equipment available on-site during a particular month for a major construction project.

Figure 4.51 illustrates a histogram comparing manpower.

These logs provide necessary information about the status of materials, shop drawings, procurement, and availability of contractor's resources and help determine their effects on project schedule and project completion.

4.7.6.5 Project Payment/Progress Curve (S-Curve)

Monitoring and control of project payment is essential within the budgeted amount. This is done through monitoring cash flow with the help of S-curves and progress curves that give the exact status of payment and also identify if the budget is being exceeded. Uninterrupted cash flow is one of the most important elements in the overall success of the project. Conceptually, cash

Project Name
Consultant Name

Contractor name:
Contract no.

Contractor's Equipment Status

Sr. No.	Equipment	1	2	3	4	5	6	7	8	9	10	11	12	13	14	15	16	17	18	19	20	21	22	23	24	25	26	27	28	29	30	Total
1	Tower crane	2	2	2	2	2	2		2	2	2	2	2	2		2	2	2	2	2	2		2	2	2	2	2	2		2	2	52
2	Mobile crane 30 ton	1	1	1	1	1	1		1	1	1	1	1	1		1	1	1	1	1	1		1	1	1	1	1	1		1		26
3	Excavator								1	1	1	1	1	1													1	1				8
4	JCB/back hoe	1							1	1	1	1	1																			6
5	Loader	1							1	1	1	1	1	1																		7
6	Tractor	1		1	1	1			1	1	1	1	1	1		1					1							1		1	1	11
7	Water tanker	1	1	1	1	1	1		1	1	1	1	1	1		1	1	1	1	1	1		1	1	1	1	1	1		1	1	26
8	Tipper/trucks	5	4	3	6	2	5		6	5	7	5	4	3		4	3	2	5	5	2		4	3	2	5	3	4		2	5	104
9	Trailer	1	1	1	1	1	1		1	1	1	1	1	1		1	1	1	1	1	1		1	1	1	1	1	1		1	1	19
10	Truck crane	1	1	1	1	1	1		1	1	1	1	1	1		1	1	1	1	1	1		1	1	1	1	1	1		1	1	26
11	Construction lift	2	2	2	2	2	2		2	2	2	2	2	2		2	2	2	2	2	2		2	2	2	2	2	2		2	2	52
12	Concrete pump	2		2		2	2		2	3	2	1	1	1		1	2	2	1	1	1				2	2	1	1		1	2	33
13	Concrete vibrator	4				4	4		4	4	4	2	2	2		2	4	4	4	2	2				3	3	1	1		1	2	59
14	Concrete mixer	2				6	5		4	6	3	3	2	1		2	3	4	2	1	2		1		4	5	1	1			1	58
15	Bobcat	1	1	1	1	1	1		1	1	1	1	1	1		1	1	1	1	1	1		1	1	1	1	1	1		1	1	26
16	Fork lift	1	1	1	1	1	1		1	1	1	1	1	1		1	1	1	1	1	1		1	1	1	1	1	1		1	1	26
17	Bar bending machine	4	4	4	4	4	4		4	4	4	4	4	4		4	4	4	4	4	4		4	4	4	4	4	4		4	4	104
18	Bar cutting machine	4	4	4	4	4	4		4	4	4	4	4	4		4	4	4	4	4	4		4	4	4	4	4	4		4	4	104
19	Jackhammer	1	1	1	1	1	1		1	1	1	1	1	1		1	1	1	1	1	1		1	1	1	1	1	1		1	1	26
20	Anchor machine	1	1	1	1	1	1		1	1	1	1	1	1		1	1	1	1	1	1		1	1	1	1	1	1		1	1	26
21	Hammering machine	1	1	1	1	1	1		1	1	1	1	1	1		1	1	1	1	1	1		1	1	1	1	1	1		1	1	26
22	Air compressor	2	2	2	2	2	2		2	2	2	2	2	2		2	2	2	2	2	2		2	2	2	2	2	2		2	2	52
23	Diesel generator	2	2	2	2	2	2		2	2	2	2	2	2		2	2	2	2	2	2		2	2	2	2	2	2		2	2	52
24	Water pumps	1	2	2	2	2	2		2	2	2	2	2	2		2	2	2	2	2	2		2	2	2	2	2	2		2	2	52
25	Dewatering pumps	4	4	4	4	4	4		4	4	4	4	4	4		4	4	4	4	4	4		4	4	4	4	4	4		4	4	104
26	Half lorry	4	4	4	4	4	4		4	4	4	4	4	4		4	4	4	4	4	4		4	4	4	4	4	4		4	4	104
27	Jeep	2	2	2	2	2	2		2	2	2	2	2	2		2	2	2	2	2	2		2	2	2	2	2	2		2	2	52

EXAMPLE

FIGURE 4.50
Equipment status.

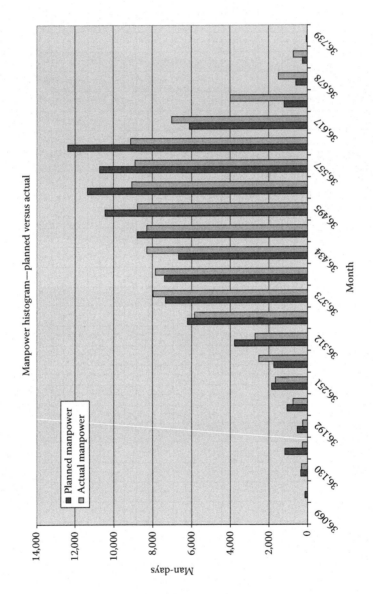

FIGURE 4.51
Manpower comparison histogram.

flow is a simple comparison of when revenue will be received and when the financial obligations must be paid. It is also an indication of the progress of work to be completed in a project. This is obtained by loading each activity in the approved schedule with the budgeted cost in the BOQ. The process of inputting the schedule of values is known as cost loading. The graphical representation of the foregoing is obtained as a curve and is known as an "S" curve. This also represents the planned progress of a project.

Figure 4.52 illustrates the planned work S-curve.

4.7.6.6 Time Control

Completion of a construction project within the defined schedule is most important. The time control status is prepared in different formats to monitor the project completion time. Figure 4.53 illustrates the project progress status of a building construction project. This chart presents the overall picture of the elapsed period of the project and the remaining period of the scheduled project duration, and actual progress versus planned progress.

4.7.6.7 Cash and Time Control

Figure 4.54 illustrates a budget control or work value control chart. This helps to monitor work progress and compare actual work done against planned work.

4.7.6.8 Progress Reports and Meetings

Apart from different types of logs and submittals, progress curves, and time control charts, the contractor's progress is monitored through various types of reports and meetings. The contractor's daily progress is monitored through a daily progress report submitted by the contractor on the morning of the working day following the day to which the report relates. It gives the status of all the resources available on-site for that particular day. It shows the details of the contractor's staff and manpower, the contractor's plant and equipment, and material received at site. Details of the subcontractor's work and resources are also included in the report. Figure 4.55 illustrates the daily progress report of a building construction project.

Along with the daily report, the contractor submits the work-in-progress report. Figure 4.56 shows this report.

Figure 4.57 shows the checklist status report, which is also to be submitted along with the daily report. This will help the contractor as well as the supervision consultant to monitor quality of work on a daily basis by knowing how many checklists are approved and how much of the work is not conforming to the specified requirements.

Progress meetings are conducted at an agreed-upon interval to review the progress of work and discuss the problems, if any, for smooth progress of

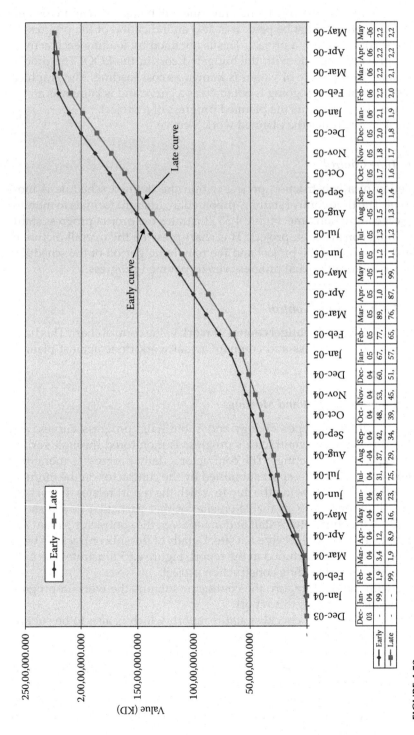

FIGURE 4.52
Planned S-curve.

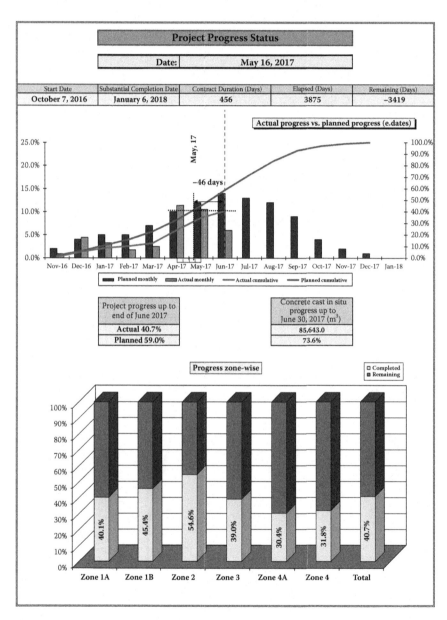

FIGURE 4.53
Project progress status.

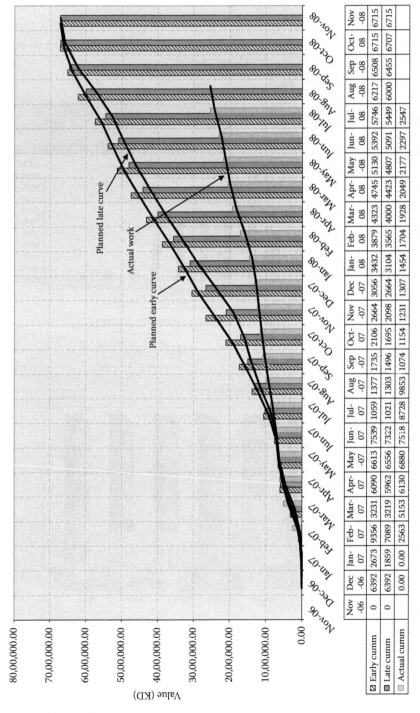

	Nov-06	Dec-06	Jan-07	Feb-07	Mar-07	Apr-07	May-07	Jun-07	Jul-07	Aug-07	Sep-07	Oct-07	Nov-07	Dec-07	Jan-08	Feb-08	Mar-08	Apr-08	May-08	Jun-08	Jul-08	Aug-08	Sep-08	Oct-08	Nov-08
▨ Early cumm	0	6392	2673	9356	3231	6090	6613	7539	1059	1377	1735	2106	2664	3056	3432	3879	4323	4745	5130	5392	5746	6217	6508	6715	6715
■ Late cumm	0	6392	1859	7089	3219	5962	6556	7322	1021	1303	1496	1695	2098	2664	3104	3565	4000	4423	4807	5091	5449	6000	6455	6707	6715
▨ Actual cumm		0.00	0.00	2563	5153	6130	6880	7518	8728	9853	1074	1154	1231	1307	1454	1704	1928	2049	2177	2297	2547				

FIGURE 4.54
Planned versus actual.

Project Name

Consultant Name

CONTRACTOR'S DAILY PROGRESS REPORT

Contract No.:-
Contractor :-

Contract Day No.
Date

SAMPLE FORM

This daily report is to be completed on both sides and submitted to the Resident Engineer
following the report date.

Contractor's Staff and Manpower Required			Contractor's Staff and Manpower Additional		
Job Description	No.	Actual	Job Description	Skilled	Un-skilled
Contractor's representative	1		Secretary		
Project manager	1		Store keeper		
Deputy project manager	1		Carpenter		
Planning manager	1		Steel bender		
Deputy planning manager	1		Concrete workers		
Quality control manager	1		Mason		
Quality control engineer	1		Plasterer		
Quantity surveyor	1		Tiler		
Assistant quantity surveyor	1		Marble		
Site engineer (architect)	2		Ceramic		
Site civil engineer	2		Stone		
Site engineer (water and sewerage)	2		Precast		
HVAC engineer	2		Safety officer		
Mechanical/firefighting engineer	1		Painter		
Electrical engineer	2		Plumber		
Communications engineer	1		HVAC		
Site engineer (marine)	1		Fire system		
Site engineer (roads and services)	1		Seaman (diver)		
Material engineer	1		Mechanical supervisor		
Assistant material engineer	1		Driver		
Landscape gardener	1		Operator		
Safety engineer	1		Welder		
Coordinator	1		Electrician		
Surveying engineer	1		Mech. and elec. workshop labour		
Surveyor	1		Labour		
Supervisor	4		Others		
Computer programmer	1				
Computer draftsman	2				
Draftsman	1				
Laboratory engineer	1				
Laboratory technician	1				
	40				

Distribution
Original : Resident Engineer
CC : Owner

Contractor...

(a)

FIGURE 4.55
(a–c) Daily progress report.

(Continued)

Project Name

Consultant Name

CONTRACTOR'S DAILY PROGRESS REPORT

Contract No.:-
Contractor :-

Contract Day No.
Date

SAMPLE FORM

This daily report is to be completed on both sides and submitted to the Resident Engineer
following the report date.

Contractor's Plant and Equipment Required			Contractor's Plant and Equipment Additional	
Description of Item	No.	Actual	Description of Item	No.
Tower crane	3		Loader	
Crane	4		Rock body truck	
Tipper truck	12		Boat with crew and radio	
Excavator	4		Radio communication system	
Grader	2		Bobcat	
Well point system with WP	4		Forklift	
Water tanker	4		Crane	
Compactor (plate)	8		Transit mixer	
Vibrator	8		Flat bed truck	
Conc. testing equipment	1		Floating crane 120 ton	
Soil testing equipment	1		Pile driving machine	
Compressor	5		Side crane	
Transit mixer	6		Tug	
Water pump	2		Tractor	
Vibrator compact roller	4		Truck with crane	
Automatic batching plant	1		Gantry crane	
Concrete pump	2		Buldozer	
Asphalt roller	4		Pickup	
Welding machine	4		Car	
Generator	4		Bus	
Bulldozer	2		Minibus	
Barge	1		Tug boat	
Split barges	1		Motor grader	
Crane pontoon	1			
Grab	2			
Diving equipment	4			
Automatic tide gauge	1			

These items are provided by supplier.

Distribution
Original : Resident Engineer   Contractor..
CC :   Owner

(b)

FIGURE 4.55 (*Continued*)
(a–c) Daily progress report.   (*Continued*)

Project Name

Consultant Name

CONTRACTOR'S DAILY PROGRESS REPORT

Contract No.:-   Contract Day No.
Contractor :-   SAMPLE FORM   Date

This daily report is to be completed on both sides and submitted to the Resident Engineer
following the report date.

Material Delivered to the Site		
Description of Material	Quantity	Unit

Distribution
Original : Resident Engineer   Contractor...
CC :   Owner
(c)

FIGURE 4.55 (*Continued*)
(a–c) Daily progress report.

Project Name
Consultant Name

WORK IN PROGRESS REPORT

Contract No.:   Contract Day No. :
Contractor: Date :

On-Site Activities										
During the Day					Expected Next Day					
No.	Description	Area	Unit	Qty.	No.	Description	Area	Unit	Qty.
Reasons for Delay, if any.										

SAMPLE FORM

Off-Site Work/Activities							
During the Day				Expected Next Day			
No.	Subcontractor Name	Work Description	Qty.	No.	Subcontractor Name	Work Description	Qty.
Reasons for Delay, if any.							

FIGURE 4.56
Work in progress.

construction activities. The contractor submits a premeeting submittal to the project manager/consultant, normally 2 days in advance of the scheduled meeting date. The submittal consists of

1. List of completed activities
2. List of current activities
3. Two weeks look ahead
4. Critical activities
5. Materials submittal log
6. Shop drawings submittal log
7. Procurement log

Project Name
Consultant Name

DAILY CHECKLIST STATUS

Contract No.:   Contract Day No. :
Contractor: Date :

Sr.No.	Check List No.	Description	Activity	Area/Location	Action	Remedial Action*	Remark
		SAMPLE FORM					

* For Not Approved Check List

FIGURE 4.57
Daily checklist status.

Apart from the issues related to progress of work and programs, site-safety- and quality-related matters are also discussed in these meetings. These meetings are normally attended by the owner's representative, designer/ consultant staff, contractor's representative, and subcontractor's responsible personnel.

Coordination meetings are also held from time to time to resolve coordina- tion matters among various trades. Safety meetings and quality meetings are also held to discuss related matters.

A monthly report giving details of all the site activities along with photo- graphs is submitted by the contractor to the consultant/owner.

4.7.6.9 Variation Orders

It is common that during the construction process, changes to the original contract will be made. Even under the most ideal circumstances, contract documents cannot provide complete information about every possible con- dition or circumstance that the construction team may encounter. These changes may occur due to the following:

1. Differences/errors in contract documents
2. Construction methodology
3. Nonavailability of specified material

4. Regulatory changes regarding use of certain types of material
5. Technological changes/introduction of new technology
6. Value engineering process
7. Additional work instructed
8. Omission of some work

These changes help build the facility in such a way as to achieve the project objective and are identified as the construction proceeds. Prompt identification of such requirements helps both the owner and contractor avoid unnecessary disruption of work and its impact on cost and time. Figure 4.58 illustrates a request for information (RFI) form that the contractor submits to the consultant to clarify differences/errors observed in the contract documents, changes in construction methodology, changes in the specified material, etc.

These queries are normally resolved by the concerned supervision engineer. However, it is likely that the matter has to be referred to the designer, as RFI has many other considerations to take care of that may be beyond the capacity of a supervision team member to resolve. Such queries may result in variation in the contract documents. Figure 4.59 shows a flow diagram for processing RFI. It is in the interest of both the owner and contractor to resolve RFI expeditiously to avoid its effect on construction schedule.

Figure 4.60 illustrates a site work instruction (SWI) form. It gives instructions to the contractor to proceed with the changes. All the necessary documents are sent along with the SWI to the contractor. The SWI is also used to instruct the contractor regarding owner-initiated changes.

Figure 4.61 shows a variation order proposal form that the contractor submits to the owner/consultant for approval of changes in the contract.

Similarly, if the contractor requires any modification to the specified method, then he or she submits a request for modification to the owner/consultant. Figure 4.62 shows the request for modification. Usually, these modifications are carried out by the contractor without any extra cost and time obligation toward the contract.

It is normal practice that, for the benefit of project, the engineer's representative assesses the cost and time related to SWI or request for change-over and obtains preliminary approval from the owner, and the contractor is asked to proceed with such changes. The cost and time implementation is negotiated and formalized simultaneously/later to issue the formal variation order. In all the circumstances where a change in contract is necessary, owner approval has to be obtained. Figure 4.63 shows the form used by the engineer's representative to obtain change order approval from the owner.

There are instances in the project where the variation is initiated by the owner. Figure 4.64 illustrates a flow diagram for processing of owner-initiated variation order.

Project Name

Consultant Name

REQUEST FOR INFORMATION (TECHNICAL)

CONTRACT NO. : _____   R.F.I. NO. _____

CONTRACTOR. : _____   DATE : _____

To: Resident Engineer _____

REF:

SUBJECT:

<u>**REQUEST FOR INFORMATION (Technical)**</u>

This form is used by the contractor to request information and is normally
sent to the A/E who responds on the same form.

SAMPLE FORM

CONTRACTOR: _____

DISTRIBUTION: Employer ☐   Engineer ☐   R.E. ☐

RESPONSE BY R.E.:

Signature of R. E.   _____   Date _____

RESPONSE RECEIVED:
FOR CONTRACTOR: _____   DATE: _____

DISTRIBUTION: Employer ☐   Engineer ☐   R.E. ☐

FIGURE 4.58
Request for information.

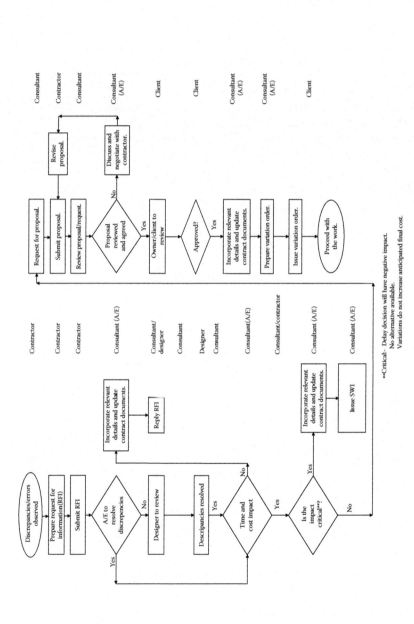

FIGURE 4.59

Flow diagram for processing RFI. (*Source:* Rumane, A.R. (2016), *Handbook of Construction Management: Scope, Schedule, and Cost Control*, CRC Press, Boca Raton, FL. Reprinted with permission from Taylor & Francis Group.)

Project Name
Consultant Name

SITE WORK INSTRUCTION

CONTRACT NO. : _____ NO. : _____
CONTRACTOR : _____ DATE : _____

SUBJECT:

SITE WORK INSTRUCTION (SWI)

SAMPLE FORM

S.W.I. involves an anticipated change in the work. All S.W.I. must be authorized
and signed by the Owner Representative OR Authorised Signatory.

A S.W.I. is an instruction to the contractor to proceed prior to the issuing of a Variation
Order (V.O.). Whenever time allows, a V.O. will be issued instead of S.W.I. .

Owner Rep. Signature   **Date**

THIS SITE WORKS INSTRUCTION (S.W.I.) IS A NOTICE TO PROCEED AND MAY INVOLVE CHANGE IN COST AND OR TIME.
YOU ARE REQUIRED TO ADVISE THE ENGINEER WITHIN 14 DAYS OF ANY ADDITIONAL COST AND
OR TIME REQUIRED TO COMPLY WITH THIS INSTRUCTION.

**RECEIVED FOR
CONTRACTOR:** _____ **DATE:** _____

DISTRIBUTION: Owner ☐   Engineer ☐   R.E. ☐

FIGURE 4.60
Site work instruction.

Project Name
Consultant Name
REQUEST FOR VARIATION

CONTRACT NO.	NO. : _____
CONTRACTOR:	DATE : _____
TO :	

SAMPLE FORM

PROPOSED VARIATION:

☐ PRODUCT   ☐ METHOD OF FABRICATION   ☐ METHOD OF INSTALLATION

SPECIFIED PRODUCT _____
PROPOSED PRODUCT _____
SPEC. SECTION # _____   PAGE # _____   ARTICLE # _____
DRG REF ____   DRG # _____   REV# _____
SPECIFIED MANUFACTURER _____
PROPOSED MANUFACTURER _____
BRIEF PRODUCT DESCRIPTION _____

REASON FOR PROPOSED VARIATION

CHANGE IN DESIGN	REQUIRED BY AUTHORITIES	SITE CONDITIONS	SWI

COST AND TIME EFFECT

COST   NO ☐   YES ☐   AMOUNT --------(ADDITION)
TIME   NO ☐   YES ☐   DAYS ----------

ATTACHMENTS:
  1 Schedule of additions/ommissions
  2 Bill Summary
  3 Rate Analysis
  4 Measurements
Technical and cost comparison sheets must be attached with this request, other wise it will not be reviewed. Contractor shall fill and submit two forms to the OWNER.
Front sheet only shall be returned to Contractor with OWNER action.

WE (THE MAIN CONTRACTOR) CERTIFIES AND UNDERTAKES THAT:

CONTRACTOR'S REP _____   DATE/TIME _____
RECEIVED BY A/E _____   DATE/TIME _____

REVIEW AND ACTION BY OWNER

☐ Approved   ☐ Not Approved   ☐ Approved as Noted   ☐ Incomplete Data Resubmit

COMMENTS

APPROVED SUBJECT TO COMPLIANCE WITH CONTRACT DOCUMENTS

Authorised Signature _____   DATE/TIME: _____

THE APPROVAL OF ANY VARIATION REQUEST SHALL BE SOLELY AT THE DIRECTION OF THE OWNER AND
SUCH APPROVAL SHALL IN NO WAY RELIEVE THE CONTRACTOR OF ANY OF HIS LIABILITIES AND OBLIGATIONS UNDER THE CONTRACT.

RECEIVED BY CONTRACTOR: _____   DATE/TIME: _____

cc: OWNER ☐   EMPLOYER ☐   R.E. ☐   ☐

FIGURE 4.61
Request for variation.

| **Project Name** |
| Consultant Name |

REQUEST FOR MODIFICATION
(AT NO EXTRA COST & OR TIME TO THE EMPLOYER)

CONTRACTOR : _____   DATE : _____
CONTRACT NO : _____   NO : _____

TO :   Owner Name   CC: A/E ☐

PROPOSED MODIFICATION TO:
☐ DESIGN DRAWING   ☐ METHOD OF FABRICATION   ☐ METHOD OF INSTALLATION
DESIGN DRAWING NO _____
SECTION #
CONTRACTOR'S PROPOSED DRAWING NO.   _____
REASON FOR MODIFICATION   _____

COST AND TIME SAVINGS   *At no extra cost and or time to the Employer*
COST   NO   ☐   YES   ☐   AMOUNT ------- (DEDUCTION) _____
TIME   NO   ☐   YES   ☐   DAYS
ATTACHMENTS:
Supplier confirmation letter

SAMPLE FORM

CONTRACTOR'S REP   _____   DATE/TIME   _____
RECEIVED BY A/E   _____   DATE/TIME   _____

REVIEW AND ACTION BY OWNER
☐ Approved   ☐ Not Approved   ☐ Approved as Noted   ☐ Incomplete Data Resubmit
COMMENTS

Authorised Signatories   _____   DATE/TIME: _____

RECEIVED BY CONTRACTOR: _____   DATE/TIME: _____
cc: Owner ☐   Engineer ☐   R.E. ☐   ☐

FIGURE 4.62
Request for modification.

Project Name
Consultant Name

VARIATION ORDER PROPOSAL (VOP)

CONTRACT NO: _____   VOP No. _____

CONTRACTOR: _____   Date: _____

The "Employer" decision is requested for approval / rejection of this Variation Order Proposal as described below.
Should the VOP be approved by the "Employer," an order to proceed shall be issued to the Contractor for his further action.

INITIATED / PROPOSED BY:

☐ ENGINEER   ☐ ENGINEER'S REP.   ☐ CONTRACTOR   ☐ OTHERS

REASONS:

     ☐ CHANGE IN DESIGN   ☐ SWI _____

     ☐ REQUIRED BY AUTHORITIES   _____

     ☐ REQUIRED BY SITE CONDITIONS

BRIEF DESCRIPTION & LOCATION:

SAMPLE FORM

BASIS OF V.O.EVALUATION:
PRO RATA BOQ PRICES: ☐

PROPOSAL from CONTRACTOR: ☐

APPROXIMATE COST IMPACT   _____   APPROXIMATE %   _____

APPROXIMATE TIME IMPACT   _____   ANY DELAY   _____

RELATED REFERENCES:

ENGINEER'S RECOMMENDATION:

RESIDENT ENGINEER'S SIGNATURE   _____   DATE: _____

Distribution: ☐ Employer   ☐ Engineer   ☐ Resident Engineer   ☐

FIGURE 4.63
Variation order proposal.

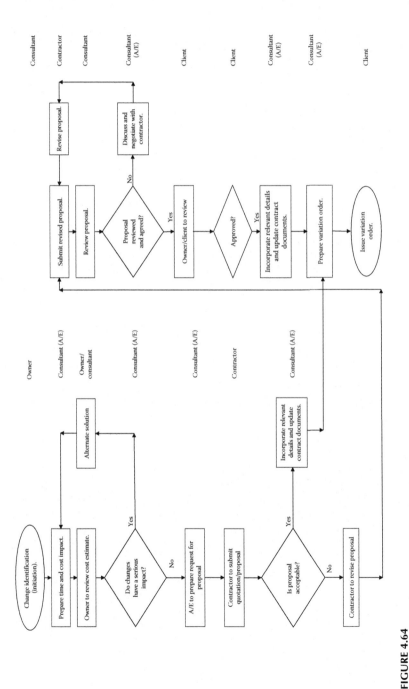

FIGURE 4.64

Flow diagram for processing variation order (owner initiated). (*Source:* Rumane, A.R. (2016), *Handbook of Construction Management: Scope, Schedule, and Cost Control*, CRC Press, Boca Raton, FL. Reprinted with permission from Taylor & Francis Group.)

Once cost and time implications are negotiated and finalized, and both the owner and contractor approve them, a variation order is issued to the contractor, and changes are adjusted with the contract sum and schedule. Figure 4.65a shows the variation order form issued to formalize the change order, and Figure 4.65b shows the attachment to the change order.

4.7.7 Quality Management

The construction project quality control process is the part of the contract documents that provides details about specific quality practices, resources, and activities relevant to the project. The purpose of quality control during construction is to ensure that the work is accomplished in accordance with the requirements specified in the contract. Inspection of construction works is carried out throughout the construction period either by the construction supervision team (consultant) or the appointed inspector agency. Quality is an important aspect of a construction project. The quality of a construction project must meet the requirements specified in the contract documents. Normally, the contractor provides on-site inspection and testing facilities at the construction site. On the construction site, inspection and testing are carried out in three stages during the construction period to ensure quality compliance:

1. During the construction process. This is carried out with the checklist request submitted by the contractor for testing ongoing work before proceeding to the next step.
2. Receipt of subcontractor or purchased material or services. The contractor submits a material inspection request to the consultant upon receipt of material.
3. Before final delivery or commissioning and handover.

Quality management in construction is a management function. In general, quality assurance and control programs are used to monitor design and construction conformance to established requirements as determined by the contract specifications. Instituting quality management programs reduces costs while producing the specified facility.

Regarding the importance of quality in construction, Chung (1999) states that

> In the construction industry, which is project-oriented, a contractor may be working at any one time for a number of clients on a variety of projects. Each project has its own characteristics and requirements. The contractor's quality system must be capable of being tailored to meet the specific needs of the project. The mechanism for doing so is the project quality plan, simply called the quality plan. Preparation of a quality plan ensures that specific requirements of the project are considered and planned for. It also demonstrates to the client how the requirements will be met. (p. 42)

FIGURE 4.65
(a) Variation order.   (*Continued*)

Project Name
Consultant Name

ATTACHMENT TO VARIATION ORDER

CONTRACTOR: _____ CONTRACT No.: _____

V. O. No. _____   Date: _____

1.   Site Works instructions incorporated into this Variation Order:

2.   Previous correspondence references (attached):

SAMPLE FORM

3.   Revised Drawings and Specifications (attached):

4.   Schedule of Omissions/Additions:

5.   Rate Analysis for New Items:

(b)

FIGURE 4.65 (*Continued*)
(b) VO attachment.

TABLE 4.22

Quality Assurance and Quality Control Typical Responsibilities

Responsibility ⇩	Owner	Design Professional	Constructor	Design Builder
Set overall approach to attitude and quality	●	◉	◉	◉
Develop and implement design phase QA/quality control (QC) program	◉	●	◉	●
Conduct design audits or review		●*	◉	●*
Incorporate QA/QC measures in construction contract bid documents	◉	●	◉	●
Evaluate construction bids	●	◉	O	◉
Develop and implement construction phase QA/QC program	◉	◉	●	●*
Set measurable goals or standards for construction quality	◉	◉	●*	●
Provide documentation of progress toward construction QA/QC goals	O	◉	●	●
Communicate regularly with team members regarding quality concerns	●	●	●	●

Source: Quality in the Constructed Project (2000) by ASCE. Reprinted with permission from ASCE.

●, Primary responsibility; ◉, Assist or advise; O, Review; ●*, May engage qualified consultant or other design staff not assigned to project for design reviews or audits.

Table 4.22 shows quality assurance and quality control typical responsibilities.

Construction project documents usually specify in detail the materials and workmanship required and also specify the type of inspection and tests to be performed to prove compliance. The contractor submits a QCP to achieve the required quality standards and objectives of the project.

With the commencement of the project, the activities related to mobilization such as setting up of site offices, stores, etc., start, followed by excavation and installation of dewatering and shoring systems.

Quality in construction is achieved through proper control at every stage of execution and installation of works. The contractor should have complete knowledge of the project he or she has been contracted to construct. Safe and reliable construction should be the objective of all the participants of the project. The contractor has to submit shop drawings, materials, products, equipment, and systems to the owner/consultant for their review and approval before they can be installed on the project. The procedure and requirements for submitting shop drawings, materials, and samples are specified in detail under the "Submittal" section of the contract specifications.

4.7.7.1 Shop Drawings Approval

The contract drawings and specifications prepared by the design professionals are indicative and are generally meant for determining the tender pricing and for planning the construction project. In many cases, they are not sufficient for installation or execution of works at various stages. More details are required during the construction phase to ensure the specified quality. These details are provided by the contractor on the shop drawings. Shop drawings are used by the contractor as reference documents to execute/install the works. A detailed shop drawing helps the contractor achieve zero defects in installation at the first stage, thus avoiding any rejection/rework. Based on the contract drawings, the contractor prepares shop drawings and submits them to the consultant for approval. All the work is executed as per the approved shop drawings. The contractor's shop drawings are prepared taking into consideration the following, at a minimum, to meet the design intent.

4.7.7.1.1 Architectural Works

The architectural works shop drawings mainly cover masonry, doors and windows, cladding, partitioning, reflected ceiling, stone flooring, toilet details, stairs details, and roofing. The elements of these shop drawings for these sections are as follows:

A. *Masonry*: The shop drawings for masonry works shall include at a minimum
   - Area layout
   - Guidelines
   - Height of masonry
   - Type and thickness of blocks used for masonry
   - Reinforcement details
   - Fixation details
   - Openings in the block works for other services
   - Sills
   - Lintels
   - Plastering details, if applicable
B. *Doors and windows*: The shop drawings for doors and windows shall include
   - Size of doors and windows
   - Type of material

- Thickness of frames
- Details of door leaves
- Details of glazing
- Fixing details
- Schedule of doors and windows

C. *Cladding*: The shop drawings for cladding shall include

- Type of cladding material
- Size of panels/tiles and thickness
- Elevations
- Fixation method, anchorage, and supports
- Openings for other services

D. *Partitioning*: The shop drawings for partitions shall include

- Type of partitioning material
- Size of panels
- Frame size and its installation details
- Partition support system
- Fixation details of panels

E. *Reflected ceiling*: The shop drawings for reflected ceiling shall include

- Type of ceiling material
- Size of ceiling panels, thickness of material
- Ceiling level
- Suspension system and framing
- Layout showing jointing layout, faceting, and boundaries between materials
- Location of light fittings, detectors, sprinklers, grills, etc.
- Location of access panel

F. *Stone flooring*: The shop drawings for flooring shall include

- Type of flooring material
- Size of flooring tiles
- Layout
- Screeding details
- Jointing details
- Flooring pattern
- Cut out for other services
- Control joints and jointing method
- Antislip inserts

- Separators
- Thresholds

G. *Toilet details*: The shop drawings for toilet details shall include
  - Layout
  - Plan
  - Sections
  - Installation details of sanitary ware and fixtures
  - Installation details of toilet accessories
  - Jointing method
  - Material finishes and surface textures
  - Installation details of toilet mirror

H. *Stairs details*: The shop drawings for stairs details shall include
  - Steps details, riser, treads, width, and height
  - Finishing details
  - Handrail details
  - Sections
  - Plan

I. *Roofing waterproofing*: The shop drawings for roofing details shall include
  - Insulation details
  - Installation details of insulation
  - Installation of waterproofing material
  - Control joints/expansion joints details
  - Tiling details
  - Flashings

4.7.7.1.2 Structural Works

The structural works shop drawings to be prepared according to ACI 315 will mainly cover reinforced concrete, formwork, precast concrete, and structural steel fabrication. The elements in these shop drawings are as follows:

A. *Reinforced concrete*: The reinforced concrete works shop drawings to be prepared according to ACI 315 should have the following details:
  - The size of reinforcement material
  - Bar schedule
  - Stirrup spacing
  - Bar bent diagram
  - Arrangement and support of concrete reinforcement

- Dimensional details
- Special reinforcement required for openings through concrete structure
- Type of cement and its strength

B. *Formwork*: Formwork has great importance from the safety point of view. The shop drawing of formwork shall include

- Details of individual panels
- Position, size, and spacing of adjustable props
- Position, size, and spacing of joints, solders, ties, and the like
- Details of formwork for columns, beams, parapet, slabs, and kickers
- Details of construction joints and expansion joints
- Details of retaining walls, core walls, and deep beams showing the position and size of ties, joints, solders, and sheeting, together with detailed information on erection and casting sequences and construction joints
- General assembly details, including propping, prop bearings, and through propping
- All penetrations through concrete
- Full design calculations

C. *Precast concrete*: The shop drawings for precast concrete works have great importance as the casting is carried out at a precast concrete factory. All the required details need to be shown on these drawings to ensure that precast elements are produced without any defects. The shop drawing should provide details of fabrication and installation of precast structural concrete units. It should indicate member location, plans, elevations, dimensions, shapes, cross sections, openings, and type of reinforcement, including special reinforcement, if any. The following information also has to be indicated in the shop drawings:

- Welded connections by American Welding Society's standards symbol
- Details of loose and cast-in hardware, insets, connections, joints, and all types of accessories
- Location and details of anchorage device to be embedded in other construction
- Comprehensive engineering analysis including fire-resistance analysis

D. *Structural steel fabrication*: The shop drawings for fabrication of structural steel shall include

- Details of cuts, connections, splices, camber, holes, and other pertinent data

- Welding standards symbols
- Size and length of bolts, indicating details of material and strength of bolts
- Details of steel bars, plates, angles, beams, channels
- Method of erection
- Details of anchorage details, templates, and installation details of bolts

4.7.7.1.3 Elevator Works

Overall elevation (vertical) for elevator area

- Floor levels
- Hoist way plan, sections, and anchoring details for installation of rails
- Machine room plan and installation details for drive and controller
- Equipment layout in machine room/hoist way
- Cuttings/openings, sleeves required in the concrete slab/walls
- Details and level of cab overhead
- Details and levels about the pit
- Location and level of hall button and hall position indicator
- Openings for landing door, entrance view, and finishes level
- Finishes of cab, door, indicators
- Power supply devices and equipment earthing
- Interface with elevator management system, if any

4.7.7.1.4 Mechanical Works (Firefighting and Plumbing)

The shop drawings for mechanical works should include, but not be limited to, the following:

- Sprinkler layout
- Hose reel details
- Hose reel cabinet size, location, and installation details
- Fire pump location and installation details
- Location of flow switches
- Interface with fire alarm and BMS
- Riser diagram for water supply system
- Size of piping

- Piping route and levels
- Size of isolating valves and their location
- Equipment plan layout
- Storage tank details
- Electrical power connection details
- Details of toilet accessories and connection details
- Riser diagram for rain water system
- Drainage system piping
- Riser diagram for storm water system
- Storm water system

4.7.7.1.5 HVAC Works

The shop drawing for HVAC works should include, but not be limited to, the following:

- Location of equipment and their configuration
- Piping size
- Piping route and levels
- Ducting size
- Ducting route and levels
- Insulation details
- Suspension/hanger details
- Equipment layout and plan
- Riser diagram for chilled water system
- Installation details of equipment
- Size of diffusers and grills
- Installation details of grills
- Exhaust and ventilation fans layout and details
- Riser diagram for exhaust air system
- Return air opening details
- Equipment schedule
- Electrical connection and power supply details
- Control details
- Sequence of operation
- Schematic diagram for HVAC system
- Schematic diagram for BMS by configuring all the equipment and components

4.7.7.1.6 Electrical Works

The shop drawings for electrical works should indicate, but not be limited to, the following:

- Size and type of conduits, raceways, and exact routing of the same
- Size of cable trays, cable trunking, installation methods, and exact route indicating the level
- Size of wires and cables
- Wiring accessories with circuit references
- Lighting layout with circuit references
- Installation details of light fixtures
- Small power layout
- Large power layout
- Installation details of distribution boards, switch boards, and panels
- Field installation wiring details for light, power, controls, and signals
- Panels, switch board layout in L.T. room
- Diesel generator installation details
- Schematic diagram showing the configuration of all the equipment
- Load schedules per actual connected loads
- Voltage drop calculations
- Earthing system
- Lightning protection system

4.7.7.1.7 HVAC Electrical Works

The shop drawings for HVAC electrical works should include

- Wiring diagram of individual components and accessories
- MCC panels, starter panels, and their installation details
- Schematic diagram including power and control wiring
- Type and size of conduit, and number and size of wires in the conduit
- Type and size of cable
- Details of protection and interlocks
- Details of instrumentation
- Description of sequence of operation of equipment

4.7.7.1.8 Low-Voltage Systems

The shop drawings for low-voltage systems such as BMS, fire alarm system, communication system, public address system, audiovisual system, CCTV/security system, and access control system should include

- Size of conduits and raceways, their routing and levels
- Location of components, wiring accessories
- Installation details of components, wiring accessories
- Wiring and cabling details
- Installation details of racks
- Installation details of equipment
- Schematic/riser diagram configuring all the components and equipment used in the system
- Interface with other systems

4.7.7.1.9 Landscape Works

The shop drawings for landscape works should include

- Boundary limit
- Excavation area
- Excavation level
- Type of soil in different areas
- Location of plants, trees, shrubs, etc.
- Foundation details for plants, trees, shrubs, etc.
- Location of sidewalk
- Location of driveways
- Foundation for paving
- Grass areas
- Location of services, ducts for utilities
- Location of light poles/bollards
- Foundation for light poles/bollards
- Details of special features
- Details of irrigation system

4.7.7.1.10 External Works (Infrastructure)

The shop drawing for external works (infrastructure) should include

- Width of the road
- Grading details

- Thicknesses of asphalt layers
- Pavement details
- Location of manholes and levels
- Services/utilities pipe routes
- Location of light poles
- Cable routes
- Trench details
- Road marking
- Traffic sign and signals

Figure 4.66 illustrates a site transmittal for shop drawings approval.

4.7.7.1.11 Builder's Workshop Drawings

The contractor submits builder's workshop drawings, which are prepared to show all the openings required in the civil or architectural work for services and other trades. These drawings are based on the approved shop drawings of all related trades. These drawings indicate the size of openings, sleeves, and level references with the help of detailed elevations and plans of the building, fully coordinating the requirements of all the trades. Detailed sections are also shown for complicated areas.

4.7.7.1.12 Composite/Coordination Shop Drawings

The contractor's composite drawings are prepared taking into consideration approved shop drawings, builder's workshop drawings, and material approval to show the relationship of components shown on the related shop drawings and to indicate required installation sequence and exact location. Composite drawings show the interrelationship of all services with each other and with the surrounding civil and architectural work. Composite drawings show detailed coordinated cross sections, elevations, reflected plans, etc., resolving all conflicts in levels, alignment, access, space, etc., and the actual physical dimensions required for installation of panels, equipment, plant, etc., within the available space.

Figure 4.67 illustrates a logic flow diagram for preparation of shop drawings/composite drawings.

4.7.7.2 Materials Approval

All the materials, equipment, and systems specified in the contract documents need prior approval from the owner/consultant. Figure 4.68 shows a site transmittal form for material approval.

Figure 4.69 shows specification comparison statement to be submitted along with the material transmittal.

Project Name
Consultant Name
SITE TRANSMITTAL
Request for Shop Drawings Approval

CONTRACT No.:   TRANSMITTAL NO. _____   REV. _____

CONTRACTOR   :

TO   :

WE REQUEST APPROVAL OF THE FOLLOWING ENCLOSED DRAWINGS

ITEM NO.	DWG., SPEC. OR BOQ. REF	DRAWING TITLE	DWG. NOS.	Rev.	SUBMITTAL CODE *	ACTION CODE **
		SAMPLE FORM				

N.B: We certify that the above items have been reviewed in detail & and are correct & in strict conformance with the Contract Drawings
& Specification except as otherwise stated.

CONTRACTOR'S REP. : _____   DATE : _____

RECEIVED BY CONSULTANT : _____   DATE : _____

cc: Owner Rep.

                    Resident Engineer to enter ACTION CODE and REMARKS

R.E.'s REMARKS :

_____

_____

_____

_____

                                           Initials _____   Date _____

Corrections or comments made relative to submittals during this review do not relieve Contractor from compliance with the requirements of the
Drawings and Specifications. This check is only for review of general conformance with the design concept of the project and general compliance
with the information given in the Contract Documents. Contractor is responsible for confirming and correlating all quantities and dimensions,
selecting fabrication process and techniques of construction; coordinating his work with that of other trades, and performing his work in a safe and.
satisfactory manner.

Resident Engineer : _____   DATE : _____

Received by Contractor : _____   DATE : _____

cc: Owner Rep.

cc: Project Manager	** ACTION CODE:			
1: Submitted for Approval	A:	Approved	C:	Not Approved
2: Submitted for your Information	B:	Approved as Noted	D:	For Information
3:				

FIGURE 4.66
Site transmittal for work shop drawings.

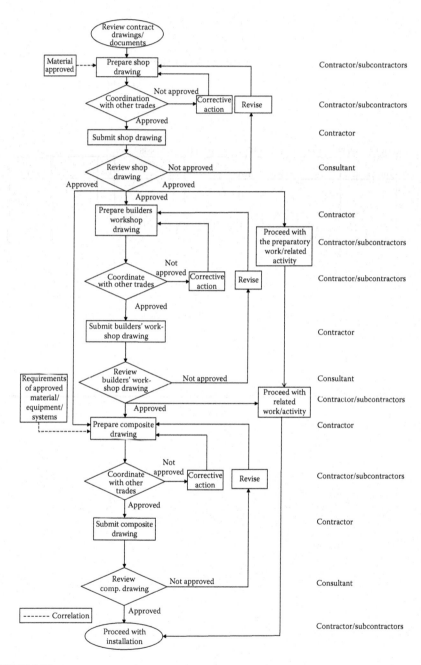

FIGURE 4.67
Logic flow diagram for shop drawings/composite drawings.

Project Name
Consultant Name
SITE TRANSMITTAL
Request for Material Approval

CONTRACT No. :   TRANSMITTAL NO. _____   REV. _____

CONTRACTOR   :

TO   :

WE REQUEST APPROVAL OF THE FOLLOWING MATERIALS/GOODS/PRODUCTS/EQUIPMENT

ITEM NO.	DWG., SPEC. OR BOQ. REF	DESCRIPTION	SUBMITTAL CODE *	ACTION CODE **

SAMPLE FORM

DETAILS OF INFORMATION, LITERATURE, CATALOG CUTS, AND THE LIKE ATTACHED ARE:

_____

SAMPLES:
Enclosed ☐   Submitted under separate cover ☐   Not applicable ☐

N.B: We certify that the above items have been reviewed in detail & and are correct & in strict conformance with the Contract
         Drawings & Specification except as otherwise stated.

CONTRACTOR'S REP. : _____   DATE : _____

RECEIVED BY CONSULTANT : _____   DATE : _____

cc: Owner Rep.

Resident Engineer to enter ACTION CODE and REMARKS

R.E.'s REMARKS :

_____
_____
_____
_____
                                             Initials _____   Date _____

Corrections or comments made relative to submittals during this review do not relieve Contractor from compliance with the requirements
of the Drawings and Specifications. This check is only for review of general conformance with the design concept of the project and
general compliance with the information given in the Contract Documents. Contractor is responsible for confirming and correlating all
quantities and dimensions, selecting fabrication process and techniques of construction; coordinating his work with that of other trades,
and performing his work in a safe and satisfactory manner.

Resident Engineer : _____   DATE : _____

Received by Contractor : _____   DATE : _____

cc: Owner Rep.

* SUBMITTAL CODE:	** ACTION CODE:		
1: Submitted for Approval	A: Approved	C:	Not Approved

FIGURE 4.68
Site transmittal for material.

	Project Name
	Consultant Name

SPECIFICATION COMPARISON STATEMENT (SCS)

Contractor: _____   Date : _____

Contract No. _____   A/S No.: _____

Submittal No. : _____	Revision: _____	Transmittal Ref.: _____
Submittal Title: _____		Specification Ref : _____

S No.	SPECIFICATION REQUIREMENTS	CONTRACTOR'S PROPOSAL	REMARKS
	SAMPLE FORM		

FIGURE 4.69
Specification comparison statement.

If the contractor is unable to obtain the specified product or is unable to find an approved equivalent product, then he or she may propose a substitute product for the approval of the consultant/owner for their review and approval. Figure 4.70a and b shows a request for substitution of material.

Upon receipt of material at site, the contractor submits the material inspection report.

Figure 4.71 shows a material inspection report.

The work at site is executed as per approved shop drawings with approved materials only.

Quality of construction has mainly three elements:

a. Defined scope
b. Schedule
c. Budget

The contractor has to comply with all the requirements specified under contract documents. The contractor executes the work as per the approved QCP. The contractor's QCP is the contractor's everyday tool to ensure quality.

During the construction process, the contractor has to submit the checklists to the consultant to inspect the work. Submission of checklists or requests

| Project Name |
| Consultant Name |

REQUEST FOR ALTERNATIVE OR SUBSTITUTION (A/S)

Contractor: _____   A/S No. _____

Contract No. _____

SUBJECT: _____   DATE: _____

"The Contractor shall comply with the requirements of Doc----- Section ------ prior to submission of this request."

DESCRIPTION OF CONTRACTED ITEM FOR WHICH ALTERNATIVE OR SUBSTITUTION IS PROPOSED:

SAMPLE FORM

Section: _____   Page: _____   Specification Ref: _____

REASON FOR PROPOSED ALTERNATIVE OR SUBSTITUTION:

PROPOSED ALTERNATIVE OR SUBSTITUTION:

☐ Product   ☐ Method of Fabrication   ☐ Method of Installation

Manufacturer: _____   Address: _____   Phone #: _____
Trade Name: _____   Model #: _____

☐ History   ☐ New Product   ☐ 2-5 Years Old   ☐ 5-10 Years Old ☐ More than 10 years old

Difference between proposed alternative OR substitution and specified:

☐ Point-by-point comparative data sheet is ATTACHED as per Ref: -------

Proposed alternative or substitution affects other parts of Work: ☐ No   ☐ Yes; explain

SIMILAR INSTALLATION OF PROPOSED ALTERNATIVE OR SUBSTITUTE:

Project: _____   Engineer: _____
Address: _____   Owner: _____
                                               Date Installed: _____

(a)

FIGURE 4.70
(a, b) Request for substitution.   (*Continued*)

Project Name
Consultant Name

REQUEST FOR ALTERNATIVE OR SUBSTITUTION (A/S)

Contractor: _____ A/S No. _____

COST & TIME:
COST SAVING: ☐ No   ☐ Yes   Amount: _____
TIME SAVING: ☐ No   ☐ Yes   Days: _____

SUPPORTING DATA ATTACHED: **SAMPLE FORM**

☐ Product Data   ☐ Drawings   ☐ Tests   ☐ Reports   ☐ Samples   ☐ Others: _____

THE CONTRACTOR CERTIFIES AND UNDERTAKES THAT:

- Same warranty will be furnished for proposed alternative or substitution as for specified product.
- Same maintenance service and source of replacement parts, as applicable is available.
- Cost data as stated above is complete. Claims for additional costs related to accepted alternative or substitution which may subsequently become apparent are to be waived. Other costs to other disciplines and/or contractors shall be borne by the Contractor.
- Proposed alternative or substitution does not affect dimensions and functional clearances.
- The cost for all changes to design, including architectural or engineering design, detailing, and construction cost including authorities approvals, caused by the requested alternative or substitution shall be fully the responsibility of the contractor at no cost to the employer.
- Coordination, installation, and changes in the Work as necessary for accepted alternative or substitution will be complete in all respects.

THE CONTRACTOR'S REP.: _____   DATE & TIME: _____

RECEIVED BY   : _____   DATE & TIME: _____

REVIEW AND ACTION BY THE ENGINEER'S REPRESENTATIVE:

☐   Alternative or Substitution approved – Make submittals in accordance with Specification Section -------
☐   Alternative or Substitution approved as Noted – Make submittals in accordance with Specification Section -------
☐   Alternative or Substitution rejected – Use specified Materials/Method.
☐   Alternative or Substitution request received too late – Use specified Materials / Method.
☐   Incomplete Data – Resubmit.

Additional Comments: _____

THE ENGINEER'S REP.: _____   DATE & TIME: _____

THE OWNER: _____   DATE & TIME: _____

The approval of any alternative or substitute product / method shall be solely at the direction of the Engineer and such approval shall in no way relieve the contractor of any of his liabilities and obligations under the contract. The approval does not entail any additional cost or time.

RECEIVED BY CONTRACTOR: _____   DATE & TIME: _____

Distribution: ☐ Owner   ☐ Resident Engineer   ☐ Contractor

(b)

FIGURE 4.70 (*Continued*)
(a, b) Request for substitution.

Project Name

Consultant Name

CONTRACT No.:

CONTRACTOR :

MATERIAL INSPECTION REPORT

Description of material for inspection: -		MIR No.	:	
		Date	:	
		Contract No.	:	
		Transmittal No.	:	
		Spec/Drg.ref.	:	

Qty. required	Qty. delivered	Total delivered	Attachments
	SAMPLE FORM		

Inspection Location :   Date of Material Delivery :

Contractor's Comments: -

Contractor's Signature :   ....   Date :

Inspection Comments :

Comply with Approved Transmittal: YES ☐   NO ☐

Signature of Inspection Engineer :   ....   Date : ....

Signature of R.E. :   ....   Date : ....

FIGURE 4.71
Material inspection report.

for inspection is an ongoing activity during the construction process to ensure proper quality control of construction. Concrete work is one of the most important components of building construction. The concrete work has to be inspected and checked at all the stages to avoid rejection or rework. Necessary care has to be taken right from the control of the design mix of the concrete until the casting is complete and cured. The contractor has to submit checklists at different stages of concrete work and has to ensure that tests specified in the contract are performed during casting of concrete. The following are the checklists that the contractor submits at different stages of the concreting process:

Figure 4.72 shows the checklist for quality control of formwork.

Figure 4.73 shows the notice for daily concrete casting.

Figure 4.74 shows the checklist for concrete casting.

Figure 4.75 shows the checklist for quality control of concreting.

Figure 4.76 shows the report on concrete casting.

Figure 4.77 shows the notice for testing at the laboratory.

Figure 4.78 shows the concrete quality control form.

In certain projects, the owner asks the contractor to involve an independent testing agency for quality control of the concrete. The agency is responsible for quality control of materials, performing tests, and submitting test reports to the owner.

The contractor has to submit checklists for all the installations and work executed as per the contract documents and the approved shop drawings and materials. Figure 4.79 shows the checklist for general works to be inspected by the consultant.

If the consultant finds that an item has been executed at site that is not in accordance with contract documents, specifications, or general code of practice, then a remedial note is issued by the consultant. Figure 4.80 shows a remedial note.

The contractor is required to reply on the same form after taking necessary action. Upon finalization of the issue, the withdrawal notice is issued by the consultant/supervision staff.

The consultant's supervisory staff always makes a routine inspection during the construction process. A nonconformance report is prepared and sent to the contractor to take corrective/preventive action toward this activity. Figure 4.81 shows a nonconformance report used in a building construction project.

In order to achieve quality in construction projects, all the work has to be executed as per approved shop drawings using approved material and fully coordinating with different trades. Figure 4.82 shows the sequence of execution of work.

Project Name

Consultant Name

QUALITY CONTROL OF FORMWORK / FALSEWORK / DIMENSIONS / LEVELS

CONTRACTOR: DATE : | / / |

CONTRACT NO:

Site Engineer   :

Inspected Element   :

(N.B.: This form is to be prepared by the Site Engineer and submitted to the R.E.)

(A=Acceptable, N=Needs Adjustment, U=Unsatisfactory)

1) Form Dimensions & Levels :

		A	N	U
1.1	Setting Out			
1.2	Top of Concrete Level Ready for Casting			
1.3	Dimensions			
1.4	Heights & Levels			
1.5	Chamfers			

2) Falsework :

		A	N	U
2.1	Supports			
2.2	Rigidity			
2.3	Bracing			
2.4	Screw Jacks			
2.5	Timber Straightness			
2.6	Splices of Vertical Members			

3) Formwork :

		A	N	U
3.1	Rigidity			
3.2	Water Tightness			
3.3	Steel Bolts / Rods / Ties			
3.4	Openings & Inserts			
3.5	Cleanliness			
3.6	Oiling			
3.7	Working Platforms and Walkways			

R.E's Comments :

Signature of Resident Engineer   Date :

FIGURE 4.72

Checklist for form work.

Project Name
Consultant Name

NOTICE OF DAILY CONCRETE CASTING

CONTRACTOR: DATE :
CONTRACT NO:

The following is the tentative schedule of our concrete casting for elements already inspected and accepted for casting this day :

NO.	LOCATION	ELEMENT	GRADE Mpa	QTY m^3	TENTATIVE CASTING TIME	
					START	FINISH
	SAMPLE FORM					
	TOTAL					

Your action for supervision of the above would be very much appreciated.

SITE ENGINEER	CONTRACTOR MANAGER

CONSULTANT RECEIVED: _____   Date: TIME: ____

R.E.'s COMMENTS:-
Resident Engineer   Date

FIGURE 4.73
Notice for daily concrete casting.

Following this sequence will help the contractor avoid rejection of works. Rejection of the checklist will result in rework, which will require time to redo the works and has cost implications to the contractor. Frequent rejection of works may delay the project, ultimately affecting the overall completion schedule. Table 4.23 summarizes the probable reasons for rejection of executed/installed works. Table 4.24 shows the responsibilities matrix for QC-related personnel.

PDCA is a well-known tool for continual process improvement. The contractor may use the PDCA cycle (Deming wheel) principle to improve the

Project Name
Consultant Name

CHECKLIST FOR CONCRETE CASTING

CONTRACTOR: _____

CONTRACT NO: _____ Date : [/ /]

Building / Structure	: _____
Element	: _____

No.	Description	Status	
		Availability	Detail
	PLANT & TOOLS		
1	Concrete Pumps		
2	Standby Concrete Pumps		
3	Cranes		
4	Truck Mixers		
5	Vibrators		
6	Trowlers		
7	Lighting		
8	Access Means		
9	Communications		
	QUALITY CONTROL		
1	Cubes		
2	Slump Apparatus		
3	Thermometer		
	STAFF & LABOUR		
1	Engineer		
2	Foreman		
3	Carpenter		
4	Steel Fixer		
5	Electrician		
6	Mechanic		
7	Vibrating Labour		
8	Trowling Labour		
9	Ordinary Labour		

SAMPLE FORM

SITE ENGINEER	CONTRACTOR MANAGER

CONSULTANT RECEIVED: _____ DATE: [/ /] TIME: _____

COMMENTS: _____

Signature of Resident Engineer: _____ Date: _____

FIGURE 4.74
Checklist for concrete casting.

Project Name

Consultant Name

QUALITY CONTROL OF CONCRETING

CONTRACTOR: DATE :
CONTRACT NO:

Site Engineer   :

Inspected Element   :

(N.B.: This form is to be prepared by the Site Engineer and submitted to the Consultant)

Starting Time
Maximum Recorded Temperature

Finishing Time
Total Concrete Quantity
Average Slump
Cubes IDs

Number of Vibrators Available through | S | U |
Process Compaction (S=Satisfactory, U=Unsatisfactory)

Preparation of Previous Construction Joints | Yes | No |
Stopped at Preplanned Locations | Yes | No |
Preparation of New Construction Joints | Yes | No |

Curing Starting Time

Measurements of Falsework / Formwork Deformation | S | U |
after casting (S=Satisfactory, U=Unsatisfactory).

R.E's Comments :

Resident Engineer :   Date :

FIGURE 4.75
Quality control of concreting.

Owner Name
Consultant Name

REPORT ON CONCRETE CASTING AT SITE (CC)

CONTRACTOR:

CONTRACT NO:

CC No.:_____

Date:_____

Time:_____

Location:	Zone:	Approved Check List Ref:	
		No.	Date :
Type of Concrete		(Required Strength)	
Starting Time			
Finishing Time			
Approximate Quantity		m3	No. of Trucks
Test Cubes / Cylinders	Total No.	Sr. Nos.	
	Cylinder Nos.	Slump	Average
SLUMP IN CENTIMETERS	SAMPLE FORM		
Engineer on Duty	Name :		Time:

Resident Engineer's Remarks :

Machinery: _____

Manpower: _____

Workmanship: _____

Rate of Casting: _____

Others:

Distribution: ☐ Engineer   ☐ Resident Engineer   ☐ Contractor

FIGURE 4.76
Report on concrete casting.

Project Name
Consultant Name

NOTICE FOR TESTING AT SITE LAB.

CONTRACTOR: **Date :** [/ /]
CONTRACT NO:

The following is the tentative schedule for testing activities in the Site Laboratory.

I.   **SOIL TESTING**

NO.	LOCATION	SAMPLE	TYPE OF TESTING

II.   **CONCRETE CUBES TESTING**

NO.	LOCATION	AGE OF CONCRETE	NO. OF CUBES

SAMPLE FORM

We request your to witness the testing at..........A.M./P.M.

SITE ENGINEER	CONTRACTOR MANAGER
_____	_____

CONSULTANT RECEIVED: _____ Date: [/ /]   Time:____

R.E's Comments : _____
Signature of Resident Engineer   Date:

FIGURE 4.77
Notice for testing at lab.

Project Name
Consultant Name

CONCRETE QUALITY CONTROL FORM

CONTRACTOR: Serial No.: _____

CONTRACT NO:

Building _____ Drawing Refer._____ Concrete Temp. _____

Particular Area_____ Type of Concrete _____ Slump (mm) _____

Date of Cube _____ Special Material _____ Drawing Refer. _____

Conc. Supplier _____ Truck # _____

➡ Date of Test ____/____/_____   Age of Cubes _____ Days

Cube No.	Width cm	Length cm	Height cm	Area cm2	Weight gram	Volume cm3	Density g/cm3	Failing Load kN	Compressive N/mm2 Strength	Type of Failure
1										
2										
3										

Required 7-Day Strength _____   Average Comp. Strength _____

Laboratory Technician   Consultant   Contractor

➡ Date of Test _____   Age of Cubes _____ Days

Cube No.	Width cm	Length cm	Height cm	Area cm2	Weight gram	Volume cm3	Density g/cm3	Failing Load kN	Compressive N/mm2 Strength	Type of Failure
1										
2										

Required 14-Day Strength _____   Average Comp. Strength _____

Laboratory Technician   Consultant   Contractor

➡ Date of Test _____   Age of Cubes _____

Cube No.	Width cm	Length cm	Height cm	Area cm2	Weight gram	Volume cm3	Density g/cm3	Failing Load kN	Compressive N/mm2 Strength	Type of Failure
1										
2										

Required 28-Day Strength _____   Average Comp. Strength _____

Laboratory Technician   Consultant   Contractor

Distribution: ☐ Owner   ☐ A/E   ☐ Contractor

FIGURE 4.78
Concrete quality control form.

Project Name
Project Name
CHECKLIST

CONTRACTOR : _____   CHECKLIST No. : [   ]

CONTRACT No. : _____   PREVIOUS C.L. No. : [   ]

TO   : **Resident Engineer**

CCS ACTIVITY NO : _____   SPECIFICATION DIVISION : _____   SECTION : _____

AREA :   [ ] Building Works   [ ] Electrical Works [ ] Mechanical Works

           [ ] HVAC Works   [ ] Finishes Works [ ]

Please inspect the following :-

Location : _____

Work : _____

_____

                              Sketch(es) attached (} No.

The work to be inspected has been coordinated with all related subcontractors.

Estimated Quantity of Work : _____   Date & Time Inspection Required : _____

Contractor Signature : _____   Date & Time _____

Received By : _____   Date & Time _____

C.C.: Owner Rep: _____   Date & Time _____

                  (All request must be submitted at least 24 hours prior to the required inspection).

Reply: The above is Approved/Not approved for the following :-

_____

_____

_____

_____

_____

Inspected by _____ Date & Time _____ Resident Engineer _____ Date & Time _____

Received by Contractor _____ Date & Time _____

C.C.: Owner Rep: _____ Date & Time _____

FIGURE 4.79
Checklist.

Project Name
Consultant Name

REMEDIAL NOTE (RN)

Contractor: _____   R.N. No.: _____

Contract No.: _____   DATE: _____

Your attention is drawn to the following works which have not been carried out in accordance with the Contract and are therefore not acceptable. Failure to carry out remedial works within a reasonable period of time may result either in additional work at your expense, or the Employer may elect to invoke Clause ---of the General Conditions of Contract.

LOCATION:

DEFECTS:

SAMPLE FORM

Signed: _____   Received by: _____
        Resident Engineer   Contractor/Date

Distribution: ☐ Owner   ☐ A/E   ☐ Contractor   ☐

FIGURE 4.80
Remedial note.

construction/installation process and avoid rejection of works by the supervision team (consultant). Figure 4.83 illustrates the PDCA cycle for execution/installation of works.

4.7.7.3 Contractor's Quality Control Plan

The contractor's quality control plan (CQCP) is the contractor's everyday tool to ensure meeting the performance standards specified in the contract documents. The efficient management of CQCP by the contractor's personnel

Project Name
Consultant Name

NON-CONFORMANCE REPORT
NO. [   ]

CONTRACTOR: DATE :

CONTRACT NO:

Location of
Non-Conformance :

Drawing / Specification :

SAMPLE FORM

Description of
Non-Conformance :

Resident Engineer : _____

Corrective / Preventive Action : (Proposed by Contractor)

Quality Control Engineer : _____

Contractor Manager: _____   Date: _____

Comments:

Resident Engineer: _____   Date: _____

FIGURE 4.81
Nonconformance report.

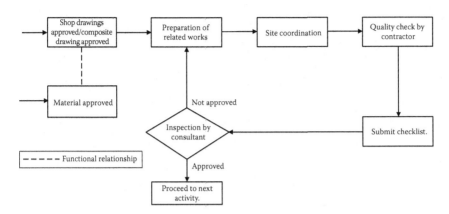

FIGURE 4.82
Sequence of execution of works.

has a great impact on both the performance of the contract and the owner's quality assurance surveillance of the contractor's performance.

According to Chung (1999), "A quality plan is a document setting out the specific quality activities and resources pertaining to a particular contract or project. Its contents are drawn from the company's quality system, the contract and related documents" (p. 45).

Chung further states that the quality plan is virtually a quality manual tailor-made for the project. The client, or the architect/engineer acting as his or her representative, may indicate in the contract what the quality plan must include and which items are subject to mutual agreement. For example, it is often specified that the inspection and test plans, which invariably form part of the quality plan, are to be approved by the architect/engineer before use.

The CQCP is the documentation of the contractor's process for delivering the level of construction quality required by the contract. It is a framework for the contractor's process for achieving quality construction. The CQCP does not endeavor to repeat or summarize contract requirements. It describes the process that the contractor will use to ensure compliance with the contract requirements. The quality plan is virtually a manual tailor-made for the project and is based on the contract requirements.

In the quality plan, the generic documented procedures are integrated with any necessary additional procedures peculiar to the project to attain specified quality objectives. Application of various quality tools, methods, and principles at different stages of construction projects is necessary to make the project qualitative, economical, and capable of meeting the owner's needs/specification requirements.

Based on contract requirements, the contractor prepares his or her QCP and submits it to the consultant for approval. This plan is followed by the contractor to maintain project quality.

TABLE 4.23

Reasons for Rejection of Executed Works

Serial Number	Description of Works	Probable Reasons for Rejection
Shoring		
1	Shoring	No adequate support for shoring. Vertical plumb level not proper. No proper facing. Not enough depth. Anchor test not approved.
2	Dewatering	Water level not under control.
Earth Works		
1	Excavation	Not as per specified level. Surface is not even. Excavated material not removed from the site.
2	Backfilling	Compaction is not proper. Backfilling thickness is not as specified. Soil is rubbish and loose. Soil test (strength) failed.
Concrete Sub Structure		
1	Blinding concrete	Thickness not as specified. Concrete strength not as specified.
2	Termite control	Uneven spray.
3	Reinforcement of steel	Reinforcement arrangement not as specified. Support not proper. Water stops are not provided. Construction joints are not provided.
4	Concrete casting	Concrete strength not as specified.
5	Shuttering for beams and columns	Shuttering dimensions not as specified. Shuttering is not strong enough. Shutters are not vertical. Shuttering height not proper.
Concrete Super Structure		
1	Reinforcement of steel for beams and columns	Reinforcement arrangement not as specified. Support not proper.
2	Shuttering for columns and beams	Shuttering dimensions not as specified. Shuttering is not strong enough. Shutters are not vertical. Shuttering height not proper.
3	Concrete casting of beams and columns.	Concrete strength not as specified.
4	Formwork for slab	Props spacing not correct. Props are not sturdy. Formwork surface is not clean.
5	Reinforcement for slab	Reinforcement arrangement not as specified. Reinforcement is not properly placed and secured. Spacers are not provided. Minimum concrete cover not provided. Construction joints not provided.
6	Concrete casting of slab	Concrete strength not as specified. Casting level in not proper.
7	Precast panels	Panels are not fixed properly. Load test on panel not performed before erection.

(Continued)

TABLE 4.23 (*Continued*)

Reasons for Rejection of Executed Works

Serial Number	Description of Works	Probable Reasons for Rejection
Masonry		
1	Block work	Block alignment is not proper. Joints are not aligned. Guidelines are missing. Anchor beads are not provided. Reinforcement mesh is not provided. Mortar is not as specified.
2	Concrete unit masonry	Block alignment is not proper. Joints are not aligned. Reinforcement mesh is not provided. Mortar is not as specified.
Partitioning		
1	Installation of frames	Stud spacing not correct. Fixation method is not as specified. Insulation is not provided.
2	Installation of panels	Alignment not proper. Joints are not proper. Panels are not painted.
Metal Work		
1	Structural steel work	Anchorage and fixing not proper. Method of erection not as specified. Fire protection is not applied. Finishing is different than specified.
2	Installation of cat ladders	Fixation is not proper. Alignment is not done. Finishing is not as specified.
3	Installation of balustrade	Fixation is not proper. Alignment is not done. Finishing is not as specified.
4	Installation of space frame	Fixation is not proper. Alignment is not done. Finishing is not as specified.
5	Installation of handrails and railings	Fixation is not proper. Alignment is not done. Finishing is not as specified.
Thermal and Moisture Protection		
1	Applying waterproofing membrane	Number of layers not as specified. There is no overlap between the rolls. Application not done properly. No. of coats as per specs not applied. There is no skirting. Leakage test failed.
2	Installation of insulation	Fixation is not proper. Insulation thickness is not as specified.
Doors and Windows		
1	Aluminum doors and windows	Dimensions are not as specified. Accessories and hardware are different than specified. Windows are not air/water tight and not sealed properly. Doors are not opening properly.
3	Glazing	Color and type is different than specified. Fixing method is not proper. Cracks are observed in some of the units.
4	Steel doors and windows	Alignment not proper. Hardware and accessories not as specified. Finishing is not proper.

(*Continued*)

TABLE 4.23 (*Continued*)

Reasons for Rejection of Executed Works

Serial Number	Description of Works	Probable Reasons for Rejection
6	Wooden frames	Dimensions are different. Finishing is not as specified.
7	Wooden doors	Alignment not proper. Finishing is not as specified. Fixation hardware is not as specified.
8	Curtain wall	Pattern is not as specified. Glazing is different than specified. Drainage system is not provided. Expansion joints not provided. Wind pressure and water test failed.
9	Hatch doors	Door is not aligned. Finishing is not matching as per architectural requirements. Fixing is not proper.
11	Louvers	Alignment and levels not proper. Accessories not as specified.
Internal Finishes		
1	Plastering	Cracks in the plaster. Voids are observed. Specified accessories not used. Hollow sound observed. Curing is not enough.
2	Painting	No. of layers not as specified. Color and texture is not proper.
3	Cladding-ceramic	Alignment, angles and joints are not proper. Grouting is not done properly. Color and texture pattern is not matching.
4	Cladding-marble	Fixation is not proper. Alignment, angles and joints are not proper. Lines are not matching.
5	Ceramic tiling	Alignment, angles and joints are not proper. Grouting is not done properly. Color and texture pattern is not matching
6	Stone flooring	Fixation is not proper. Alignment, angles and joints are not proper.
7	Acoustic ceiling	Suspension system is not as specified. Alignment, levels and joints are not proper. Services openings are not provided. Ceiling height not matching with approved level.
8	Demountable partitions	Fixation is not proper. Alignment is not done properly.
Furnishing		
1	Carpeting	Joints are not smooth. Carpet fixed without adhesive. Edges are not trim properly. Color and texture not matching.
2	Blinds	Stacking/rolling not proper. Fixing is not secured.
3	Furniture	Location is not as per approved layout. Dimensions are not correct. Finishing is not proper. Fixation is not secured properly. Accessories are not matching with the furniture.

(Continued)

TABLE 4.23 (*Continued*)

Reasons for Rejection of Executed Works

Serial Number	Description of Works	Probable Reasons for Rejection
External Finishes		
1	Painting	No. of layers not as specified. Color and texture is not proper.
2	Brickwork	Brick alignment is not proper. Joints are not aligned. Grouting is not proper.
3	Stone	Fixation is not proper. Alignment, angles and joints are not proper. Color and texture not matching.
4	Cladding-aluminum	Alignment is not proper. Cladding is not levelled and aligned. Fixation is not secured. Joints are without sealant.
5	Cladding-granite	Fixation is not proper. Alignment, angles and joints are not proper. Color and texture not matching.
6	Curtain wall	Accessories are not properly fixed. Levelling and alignment not done. Drainage system is not provided. Wind pressure and water tests failed.
7	Glazing	Glass type is different than specified and approved. Fixation is not secured. Levelling and alignment not done properly. Joints have gap. Frame finishing is not as per approved sample.
Equipment		
1	Installation of maintenance equipment	Brackets are not fixed at specified location.
2	Installation of kitchen equipment	Equipment are not fixed as per approved layout.
3	Installation of parking control	Barrier is not fixed at specified location. Control panel is not terminated.
Roof		
1	Parapet wall	Parapet level is not correct. Finishing is not as specified.
2	Thermal insulation	Type of insulation and thickness is different than specified.
3	Waterproofing	No. of layers are less than specified. Application is not smooth. There is no overlap between the sheets. Skirting is not provided. Water test failed.
4	Roof tiles	Thickness of tiles is less than specified. Tiles are not aligned and levelled. Joints are not proper.
5	Installation of drains	Drains are not located as per approved shop drawings. Fixation is not proper.
6	Installation of gutters	Type of gutters is different than specified. Gutters are not aligned and levelled.

(*Continued*)

TABLE 4.23 (*Continued*)

Reasons for Rejection of Executed Works

Serial Number	Description of Works	Probable Reasons for Rejection
Elevator System		
1	Installation of rails	Fixation is not proper.
2	Installation of door frames	Frame is not aligned properly.
3	Installation of cabin	Cabin is not levelled.
4	Cabin finishes	Cabin finishes are incomplete.
5	Installation of wire rope	Rope joints are weak.
6	Installation of drive machine	Drive is levelled properly. Equipment earthing is not done.
7	Installation of controller	Location of controller is not proper. Controller is not terminated. Equipment earthing is not provided. Identification label is not provided.
Mechanical Works		
A. Fire fighting System		
1	Installation of piping system	Pipes are dirty need cleaning and painting.
2	Installation of sprinklers	Sprinkler spacing not as per authorities approved shop drawing.
3	Installation of foam system	Foam agent concentration level not proper.
4	Installation of pumps	Eccentric reducer need to fixed at the suction side of the pump. Pumps are not levelled. Cable termination not complete. Equipment earthing not done.
5	Installation of hose reels/cabinet	Cabinet door not closing properly. Color of cabinet to be as per authorities requirement.
6	Installation of fire hydrant	Fire hydrant to be fixed properly at specified location.
7	External fire hose cabinet	External fire hose cabinet need hose rack.
B. Water Supply		
1	Installing of piping system	Pipe supports required to keep the pipe in straight position.
2	Installation of pumps	Pumps are not levelled. Cable termination not complete. Equipment earthing not done.
3	Filter units	Filter installed without bypass line and tab point for pressure gauge.
4	Toilet accessories	Towel rod not fixed.
5	Installation of insulation	Insulation has cuts. Insulation is not covered with canvas.
6	Water heaters	Water heater drain not installed.

(*Continued*)

TABLE 4.23 (*Continued*)

Reasons for Rejection of Executed Works

Serial Number	Description of Works	Probable Reasons for Rejection
7	Hand dryers	Installation height is more than specified. Termination is not proper.
8	Water tank	Float switch for pump not installed.
C. Drainage System		
1	Installation of pipes below grade	Drainage pipe to be tightened properly. Slope not proper.
2	Installation of pipes above grade	Slope not proper. More support are required.
3	Installation of manholes	Manhole level is not correct.
4	Installation of clean out, floor drains	Floor drains and clean out are not flush with the floor finish.
5	Installation of sump pumps	Height of float switch need adjustment.
6	Installation of gratings	Grating frame not flush with floor finish level.
D. Irrigation System		
1	Installation of piping system	Pipe crossing under road need sleeves.
2	Installation of pumps	Pumps installed without suction strainer.
3	Installation of controls	Controllers are not weatherproof. Installation height is not proper. Fixing method is not proper.
HVAC Works		
1	Installation of piping	Hanger supports are not properly fixed. Invert level is not correct. Piping is rusty, not clean and not painted.
2	Installation of ducting	Duct level not correct. Duct metal not clean. Suspension system not properly fixed. No proper sealant around duct joints.
3	Installation of insulation	Insulation has cuts. Straps not fixed at specified interval.
4	Installation of dampers, grills and diffusers	Material is squeezed. Color is faded. Material not matching with architectural requirements.
5	Installation of cladding	Cladding surface is not even. Cladding is not overlapped and sealed properly.
6	Installation of fans	Fans fixing is not secured.
7	Installation of Fan Coil Units (FCU)	FCU level not correct. FCU fixed without spring isolators. No slope for drain pipe.
8	Installation of Air Handling Units (AHU)	No flexible connector between chilled water piping and AHU. AHU body is damaged. Body paint is peeled off.
9	Installation of pumps	Pumps are not levelled. Cable termination not complete. Equipment earthing not done.

(*Continued*)

TABLE 4.23 (*Continued*)

Reasons for Rejection of Executed Works

Serial Number	Description of Works	Probable Reasons for Rejection
10	Installation of chillers	Rubber isolators not installed. Anchorage not fixed properly.
11	Installation of cooling towers	Vibration springs not installed. Damage in cooling tower surface. Spray nozzle damaged
12	Installation of thermostat and controls	Location not as specified. Thermostat body damaged.
13	Installation of starters	Location to be near the equipment. Installation height not as specified.
14	BMS	All the components not installed. Termination not complete. Wiring not dressed and bunched properly.
Electrical Works		
1	Conducting—raceways	Method of installation not proper. Conduit run is not parallel. Minimum clearance from other services less than specified. Supports are not secured, straps or clamp not provided. Spacing, spacing between embedded conduit and concrete aggregate is not as specified. Fastening of conduit with reinforcement steel is not secured. Location of sleeves not as per app roved shop drawings. Location of boxes is not as specified and is not coordinated.
2	Cable tray—trunking	Cable trays are not aligned properly. Cable tray run is not parallel and no proper bends. Supports are not as per approved shop drawings. Minimum spacing between any cable tray and other services not maintained. Supports are not secured.
3	Floor boxes	Location is not coordinated. Level of floor boxes is not proper
4	Wiring	No. of wires in the conduit are not as per approved shop drawing. Circuit wiring for switches not properly done. Termination method is not proper.
5	Cabling	Distance between two cables on cable tray is less than twice the diameter of larges cable diameter. Cable tie is not provided. Circuit identification is missing.
6	Installation of bus duct	Supports are not fixed properly. Connection between the ducts is not proper. Level of installation is not as specified. Spacing between parallel is less than specified. Supports are not secured.
7	Installation of wiring devices/accessories	Installation height is not as specified, coordination with architectural requirements not done to match with the finishes.
8	Installation of light fittings	Installation method is not correct. Proper size of hangers and support not provided. Location and levels are not matching with architectural requirements.

(Continued)

TABLE 4.23 (*Continued*)

Reasons for Rejection of Executed Works

Serial Number	Description of Works	Probable Reasons for Rejection
9	Grounding	Welding method for joining tapes is not thermoweld type. Connection methods between tape and clamp to be as specified. Provide clamp at specified distance.
10	Distribution switch boards	Installation height of boards not as specified. Termination of wires/cables not done properly. Shrouding for cable not provided. Wires and cables do not have continuity. Wires in distribution board are not properly dressed and bunched.
Fire Alarm System		
4	Installation of detectors, bells, pull stations, interface modules	Location is not as specified. Height of installation not matching with shop drawing and architectural requirements.
5	Installation of repeater panel	Installation height not as specified, method of installation not proper. Termination of cables is not done.
6	Installation of mimic panel	Installation height not as specified. Method of installation not proper. Termination of cables is not done.
7	Installation of fire alarm panel	Installation height not as specified. Method of installation not proper. Termination of cables is not done.
Telephone/Communication System		
1	Installation of racks	Method of installation not as per approved shop drawing. Termination of cables not done properly. Racks are not levelled. Cables are not labelled.
2	Installation of switches	Switches are not properly installed. Switched are not installed in sequence. Cabling is not complete. Identification label not provided for cables.
Public Address System		
5	Installation of speakers	Fixing of speakers is not secured. Height of installation is not as specified. Speakers installed in false ceiling do not have proper support.
6	Installation of racks	Method of installation not as per approved shop drawing. Termination of cables not done properly. Racks are not levelled. Cables are not labelled.
7	Installation of equipment	Equipment are not properly installed. Cabling is not proper.
Audio visual System		
5	Installation of speakers	Fixing of speakers is not secured. Height of installation is not as specified. Speakers installed in false ceiling do not have proper support.

(*Continued*)

TABLE 4.23 (*Continued*)

Reasons for Rejection of Executed Works

Serial Number	Description of Works	Probable Reasons for Rejection
6	Installation of monitors/screens	Method of fixing nor proper. Installation height is not as per approved shop drawings.
7	Installation of racks	Method of installation not as per approved shop drawing. Termination of cables not done properly. Racks are not levelled. Cables are not labelled.
8	Installation of equipment	Equipment are not properly installed. Cabling is not proper.
CCTV/Security System		
3	Installation of cameras	Fixing of cameras not proper. Installation height is not as specified. Termination of cameras not proper.
5	Installation of panels	Panels not properly fixed. Cabling not complete.
7	Installation monitors/screens	Method of fixing nor proper. Installation height is not as per approved shop drawings.
Access Control		
3	Installation of RFID proximity readers, finger print readers	Readers not installed properly.
4	Installation of magnetic locks, release buttons, door contacts	Magnetic locks not fixed properly. Release buttons not installed at specified height and location.
5	Installation of panel	Panels not properly fixed. Cabling not complete.
6	Installation of server	Cable termination not proper. Cables to be properly dressed. Identification labels not provided.
Systems Integration		
1	Installation of switches	Switches not stacked as per sequence. Identification labels not provided.
2	Installation of servers	Cable termination not proper. Cables to be properly dressed. Identification labels not provided.
External Works		
1	Site works	Compaction failure. Grading level not proper. Slope is not provided.
2	Asphalt work	Asphalt levels not proper. Asphalt material not as specified.
3	Pavement works	Pavement limits not as per approved layout. Curb stones not laid properly.
4	Piping works	Pipe not laid at specified depth. No protection on pipes. Pipe joints not done well. Leakage observed.
5	Electrical works	Cable not buries properly. Location of light poles not correct. Light poles are not vertically installed.
6	Manholes	Location of manholes not correct. Manhole level not correct.
7	Road marking	Marking color not as specified.

TABLE 4.24

Responsibility for Site Quality Control

| | | Linear Responsibility Chart | | | | | | |
| Serial Number | Description | Owner | Consultant | Contractor | | | | | |
		Owner/ Project Manager	Consultant/ Designer	Contractor Manager	Quality Incharge	Quality Engineers	Site Engineers	Safety Officer	Head Office
1	Specify quality standards	□	■						
2	Prepare quality control plan			□	■	□			□
3	Control distribution of plans and specifications			□	■	□			□
4	Submittals			■	□		■		
5	Prepare procurement documents			□			■		■
6	Prepare construction method procedures			□	□	□	■		
7	Inspect work in progress		■	■	□	□	■		
8	Accept work in progress		■						
9	Stop work in progress	■	□						
10	Inspect materials upon receipt		■	□	■		■		

(Continued)

TABLE 4.24 (*Continued*)
Responsibility for Site Quality Control

| | | | | Linear Responsibility Chart | | | | | |
| | | Owner | Consultant | Contractor | | | | | |
Serial Number	Description	Owner/ Project Manager	Consultant/ Designer	Contractor Manager	Quality Incharge	Quality Engineers	Site Engineers	Safety Officer	Head Office
11	Monitor and evaluate quality of works		■	□	■	■	■		
12	Maintain quality records			□	■				
13	Determine disposition of nonconforming items	□	□	■		□			
14	Investigate failures	□	■	■	■		■		
15	Site safety	□	□	□				■	
16	Testing and commissioning	□		□			■		
17	Acceptance of completed works	■	□	□					

■, Primary responsibility; □, Advise/assist.

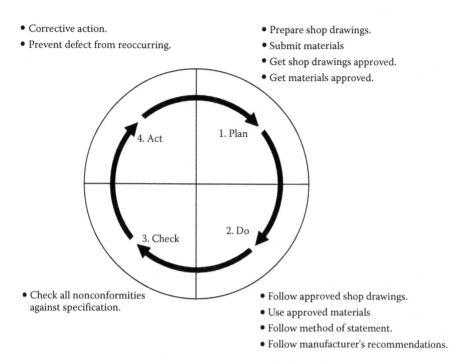

FIGURE 4.83
PDCA cycle (Deming wheel) for execution of works.

The CQCP is prepared based on the project-specific requirements as specified in the contract documents. The plan outlines the procedures to be followed during the construction period to attain the specified quality objectives of the project while fully complying with the contractual and regulatory requirements.

Appendix C illustrates an outline of a CQCP for a major building construction project. However, the contractor has to take into consideration the requirements listed under the contract documents, depending on the nature and complexity of the project.

The contract documents specify the details of the contents of the QCP to be prepared by the contractor for the construction project; the plan has to be submitted to the consultant for approval. The following is the outline for preparation of a QCP:

1. Purpose of the QCP

2. Project description

3. Site staff organization chart for quality control

4. Quality control staff and their responsibilities

5. Construction program and subprograms

6. Schedule for submission of subcontractors, manufacturer of materials, and shop drawings

7. QC procedure for all the main activities such as

   a. Procurement (direct bought out items)

   b. Off-site manufacturing, inspection, and testing

   c. Inspection of site activities (checklists)

   d. Inspection and testing procedure for systems

   e. Procedure for laboratory testing of material

   f. Inspection of material received at site

   g. Protection of works

8. Method statement for various installation activities

9. Project-specific procedures for SWIs and remedial notes

10. Quality control records

11. List of quality procedures applicable to project in reference to the company's quality manual and procedure

12. Periodical testing procedure for construction equipment and tools

13. Quality updating program

14. Quality auditing program

15. Testing

16. Commissioning

17. Handover

18. Health, safety, and environment

Specifications of work quality are an important feature of construction project design. Specifications of required quality and components represent part of the contract documents and are detailed under various sections of particular specifications. Generally, the contract documents include all the details as well as references to generally accepted quality standards published by international standards organizations. Proper specifications and contract documentations are extremely important as these are used by the contractor as a measure of quality compliance during the construction process.

A contract for construction commits the contractor to construct the facility and the owner to pay. Once the contract is signed, it commits all the parties to obligations and liabilities and is enforceable by law. A breach of contract by either party may make that party liable for payment of damages to the other.

There are standard sets of conditions of contract published by engineering institutes/societies and other bodies. Depending on the need for the construction project and the type of contract arrangements, an appropriate set of condition of contracts is selected. The contract document must include health and safety programs to be followed by the contractor during the construction process.

4.7.7.4 Inspection of Works

The inspection of construction work is performed throughout the execution of project. Inspection is an ongoing activity to physically check the installed works. Checklists are submitted by the contractor to the consultant to inspect the executed work/installations. If the work is not carried out as specified, then it is rejected and the contractor has to rework, or take remedial measures, to ensure compliance with the specifications. During construction, all the physical and mechanical activities are accomplished on the site. The contractor carries out the final inspection of the work to ensure full compliance with the contract documents.

4.7.8 Risk Management

Probability of occurrence of risk during the construction phase is very high compared to the design phases. During the construction phase, uncertainty comes from various sources as this phase has involvement of various participants. Risk in construction projects has already been discussed under Section 1.13.2. Since the duration of the construction phase is longer than that of earlier phases, the contractor has to also consider occurrence of financial, economical, commercial, political, and natural risks. The contractor has to develop risk management plan. Risk management plan identifies how risk associated with the project will be identified, analyzed, managed, and controlled. Risk management is an integral part of project management as the risk is likely to occur at any stage of the project. Therefore, the risk has to be continually monitored and response actions must be taken immediately. The risk management plan outlines how risk activities will be performed, recorded, and monitored throughout the life cycle of the project. It is intended to maximize the positive impact for the benefit of the project and decrease/minimize or eliminate the impact of events adverse to the project. The risk management must commence early in the project development stage (study stage) and proceed as the project evolves and more information about the project is available. The project plan should

- Define risk management strategy/approach
- Define project objectives, goals related to risk management
- Identify risk owner and team members
- Define risk decisions
- Detail about risk resources

- Include risk management process
  - Methods of risk identification
  - Methods of risk assessment
  - Level of risk
  - Response to risk
  - Management of risk
  - Control of risk
- Process of integrating risk management activities into project scope, schedule, cost, and quality
- Documenting and recording of risks
- Communication procedure for risk reporting
- Update of risk management plan

Figure 1.34, discussed earlier under Section 1.13, illustrates risk management cycle (process) and Figure 1.35, discussed earlier under Section 1.13, lists probability of occurrence of various categories of risks during construction project.

Table 4.25 lists the major risks during the construction phase and mitigation actions. The contractor has to monitor the occurrence of these risks and take necessary measures to mitigate the effects on the project.

4.7.9 Contract Management

Contract management during the construction phase is an organizational method, process, and procedure to manage all contract agreements involved between the owner, contractor, subcontractor, manufacturers, and suppliers. During the construction phase, contracts are managed mainly by the following parties who are directly involved for the execution of project:

- Consultant/construction (project) manager
- Contractor

Apart from these two parties, subcontractors and vendors also have their contract management system.

Contract management process starts once the contract is signed. The consultant is responsible to manage contract on behalf of the owner. The consultant monitors the scope, schedule, cost, and quality of the construction to ensure contract conditions are met. The contractor is responsible to ensure that all project works are executed within the agreed-upon time and cost in accordance with the contract conditions and specifications.

TABLE 4.25

Major Risks during Construction Phase and Mitigation Action

Serial Number	Description of Risk	Mitigation Action
1	Incompetent subcontractor	Contractor has to monitor the workmanship and work progress
2	Failure of team members not performing as expected	Select competent candidate. Provide training
3	Low-bid project cost	Contractor to try competitive material, improve method statement and higher production rate from its manpower
4	Delay in transfer of site	Contractor to adjust the construction schedule
5	Delay in mobilization	Adjust construction schedule accordingly
6	Scope/design changes	Resolve change order issues in order not to delay the project
7	Different site conditions to the information provided	Contractor to investigate site conditions prior to starting the relevant activity
8	Inadequate site investigation data	Contractor to investigate site conditions prior to starting the relevant activity
9	Conflict in contract documents	Amicably resolve the issue
10	Incomplete design	Raise request for information (RFI)
11	Incomplete scope of work	Raise RFI
12	Design changes	Follow contract documents for change order
13	Design mistakes	Raise RFI
14	Errors and omissions in contract documents	Raise RFI
15	Incomplete specifications	Raise RFI
16	Inappropriate construction method	Raise RFI and correct the method statement
17	Conflict with different trades	Coordinate with all trades while preparing coordination and composite drawings
18	Change in laws and regulations	Contractor to inform owner/consultant and raise RFI
19	Statutory/regulatory delay	Regular follow-up by the contractor, owner with the regulatory agency
20	Project schedule	Compress duration of activities
21	Inappropriate schedule/plan	Contractor to prepare schedule taking into consideration site conditions and all the required parameters
22	Delay in changer order negotiations	Request owner/supervisor/project manager to expedite the negotiations and resolve the issue
23	Resource availability (material)	Contractor to make extensive search

(Continued)

TABLE 4.25 (*Continued*)

Major Risks during Construction Phase and Mitigation Action

Serial Number	Description of Risk	Mitigation Action
24	Resource (labor) low productivity	Contractor to engage competent and skilled labors
25	Equipment/plant productivity	Contractor hire/purchase equipment to meet project productivity requirements
26	Insufficient skilled workforce	Contractor arrange workforce from alternate sources
27	Failure/delay of machinery and equipment	Contractor to plan procurement well in advance
28	Failure/delay of material delivery	Contractor to plan procurement well in advance
29	Delay in approval of submittals	Notify owner/project manager
30	Delays in payment	Contractor to have contingency plans
31	Quality of material	Locate suppliers having proven record of supplying quality product
32	Variation in construction material price	Contractor to negotiate with supplier/manufacturer for best price. Contractor to request for change order if applicable as per contract
33	Damage to equipment	Regularly maintain the equipment. Take immediate action to repair damaged equipment
34	Damage to stored material	Contractor to follow proper storage system
35	Structure collapse	Contractor to ensure that formwork and scaffolding is properly installed
36	Access to worksite	Access road to be planned in coordination with adjacent area and local authority
37	Leakage of hazardous material	Contractor to take necessary protection to avoid leakage. Store in safe area
38	Theft at site	Contractor to monitor access to site. Record entry/exit to the site. Provide fencing around project site
39	Fire at site	Contractor to install temporary firefighting system. Inflammable material to be stored in safe and secured place with necessary safety measures
40	Injuries	Contractor to keep First Aid provision at site. Take immediate action to provide medical aid
41	New technology	Owner/contractor to mutually agree for changes in the contract for better performance of project

For successful contract management, the contractor as well as consultant/ CM/PM has to consider the following points while executing the project:

1. Use of RFI to get clarification on some aspects of the project. There are two parts in the RFI:
    i. "Question" by the contractor
    ii. "Answer" by the owner (consultant)
2. Execution of project works using specified and approved materials, equipment, and systems
3. Developing project execution plan considering realistic duration for each activity
4. Execution of contracted works in a timely manner in accordance with agreed-upon schedule
5. Dealing variations to the specified product, method, work in accordance with related specification, contract clauses, and by providing substantiation and justifications that have resulted in proposing alternative or substitute material
6. Managing errors, omissions, and additions strictly in accordance with contract terms and avoiding any delays to the project
7. Conducting meetings to monitor progress and clarify prevailing project issues
8. Cooperating with all team members to fulfill their contractual obligations
9. Resolving disputes in an amicable way by adopting cooperative approach
10. Providing resources to ensure timely availability of competent workforce as per resource schedule
11. Taking action on all the transmittals within agreed-upon period
12. Timely reply to all correspondences, queries
13. Communicating issues and problems well in advance
14. Not to ignore problems/issues with the hope that they might go away
15. Arranging payment of progress payment as per contractual entitlement within stipulated time
16. Maintaining list of claims on monthly basis
17. Settling claims in an accordance with contract terms
18. Maintaining proper logs and records

4.7.10 Site Safety

The construction industry has been considered to be dangerous for a long time. The nature of work at site always presents some dangers and hazards.

There are a relatively high number of injuries and accidents at construction sites. Safety represents an important aspect of construction projects. Every project manager tries to ensure that a project is completed without major accidents on the site.

The construction site should be a safe place for those working there. Necessary measures are always required to ensure safety of all those working at construction sites. Effective risk control strategies are necessary to reduce and prevent accidents.

Contract documents normally stipulate that the contractor, upon signing the contract, has to submit a safety and accident prevention program. It emphasizes that all personnel have to put in efforts to prevent injuries and accidents. In the program, the contractor has to incorporate safety and health requirements of local authorities, manuals of accident prevention in construction, and all other local codes and regulations. A safety violation notice is issued if the contractor or any of his subcontractors are not complying with safety requirements. Figure 4.84 illustrates a safety violation report, on the basis of which action has to be taken by the contractor.

Penalties are also imposed on the contractor for noncompliance with the site safety program. The safety program shall embody the prevention of accidents, injury, occupational illness, and property damage. The contract specifies that a safety officer is engaged by the contractor to monitor safety measures. The safety officer is normally responsible for

1. Conducting safety meetings
2. Monitoring on-the-job safety
3. Inspecting the work and identifying hazardous areas
4. Initiating a safety awareness program
5. Ensuring availability of first aid and emergency medical services as per local codes and regulations
6. Ensuring that personnel are using protective equipment such as hard hat, safety shoes, protective clothing, life belt, and protective eye coverings
7. Ensuring that the temporary firefighting system is working
8. Ensuring that work areas are free from trash and hazardous material
9. Housekeeping

Construction sites have many hazards that can cause serious injuries and accidents. The contractor has to identify these areas and ensure that all site personnel and subcontractor employees working at the site are aware of unsafe acts, potential and actual hazards, the safety plan and procedures, and the immediate corrective action to be taken if something untoward occurs.

Project Name
Consultant Name

Contract No.:   SVN No.
Contractor: Date :
                                           Time:

SAFETY VIOLATION REPORT

SAFETY RELATED ITEMS

Sr.No.	Description	Sr.No.	Description
1	Access Facilities	13	Hygieninc
2	Barricade/Railing	14	Poor lighting
3	Construction Equipment	15	Protective Equipment
4	Crane	16	Lifting Gears
5	Earhtwork/Excavation	17	Poor lighting
6	Electrical	18	Protective Equipment
7	Fire Fighting/Protection	19	Safety Gears
8	First Aid	19	Scaffolding
9	Formwork	20	Site Fencing
10	Hand and Power Tools	21	Storage Facilities
11	Hazars/Imflamable Material	22	Vehicles
12	Hoist	23	Welding/Hot Work
12	Housekeeping	24	Others

VIOLATION DESCRIPTION   Action code: ☐

Item No.	Location	Description

ORIGINATOR: **RESIDENT ENGINEER:**

CONTRACTOR'S ACTION

Item No.	Location	Action	Date	Time

SAFETY OFFICER: **CONTRACTOR'S**
                                         PROJECT MANAGER:

Action Code: ☐ A   For immediate action /(   )hours   ☐ B   Within (   ) days

FIGURE 4.84
Safety violation notice.

The following are some of the common causes of injuries/accidents at site:

1. Unsafe access

2. Ladders

3. No barricades around excavated areas, trenches, openings, holes, and platforms

4. No barricades/railings on the stairs

5. Scaffolding
6. Lifting gear
7. Crane
8. Hoist
9. Welding
10. Hand and power tools
11. Poor lighting
12. Fire

The following are general safety guidelines the contractor may follow to avoid accidents/injuries:

1. Ensure safe access.
2. Ensure that ladders are in good condition and properly secured.
3. Provide barricades around openings, holes, and platforms.
4. Choose the right system of scaffolding for the job to be performed.
5. Check lifting gears for capacity, condition of wire rope, slings, hooks, eyebolts, shackles, proprietary lighting equipment, and spreader beam.
6. Ensure that the crane is
   a. On firm, level ground and outriggers are fully extended
   b. Operating with a safe working load
   c. Working with minimum load swing
   d. Operating with the right type of chain
   e. Working with the load kept clear of personnel
7. The hoist is certified by a third party, and the certificate is valid.
8. Wear all protective equipment and garments necessary to be safe on the job.
9. Wear and use eye protection coverings.
10. Use safety shoes.
11. Use hard hats.
12. Use safety belts.
13. Use respiratory masks whenever required.
14. Use the right size of hand tools.
15. Before using any plant or mechanized equipment, check that it is certified for safe operation.

16. Before using electric tools:
    a. Check that it is properly earthed.
    b. Ensure that cable, plugs, or connectors are in sound condition and properly wired up.
17. Use proper guards while using power tools such as circular saws, portable grinders, and bench grinders.
18. Use safe loading/unloading techniques.
19. All material stored at site should be stacked properly to ensure that it is stable and secured against sliding or collapse.
20. Work areas and means of access should be maintained in a safe and orderly condition.
21. Mark access and escape routes.
22. Keep passageways and access ways free from materials, supplies, and obstructions all the time.
23. Mark all the hazardous areas.
24. Prohibit storage of flammable and combustible material.
25. "No smoking" sign to be displayed.
26. Install sirens and alarm bells at site.
27. Ensure that the temporary firefighting system is working all the time.
28. All formwork, shoring, and bracing should be designed, fabricated, erected, supported, braced, and maintained so that it will safely support all vertical and lateral loads that might be applied, until such loads can be supported by the structure.
29. While placing concrete
    a. Make use of protective clothing and equipment
    b. Use appropriate gloves
    c. Use rubber boots
    d. Wear protective goggles
    e. Take necessary precautions while using concrete skips or concrete.
30. Display safety signs such as DANGER, CAUTION, and WARNING on all live electrical panels.
31. Post safety, warning signs, and notices of weekly project safety record.
32. Conduct "Safety Awareness" programs.

The contractor is responsible for ongoing maintenance of accident/incident records on the construction site and their notification to the consultant. Figure 4.85 shows an accident-reporting form.

Project Name		
ACCIDENT REPORT		

CONTRACT   : _____   CONTRACT NO. : _____

CONTRACTOR   : _____   REPORT #   : _____

SUBCONTRACTOR : _____   REPORT DATE   : _____

SAMPLE FORM

ACCIDENT DATE		
ACCIDENT TIME		
ACCIDENT LOCATION		
INJURED PERSON	I.D.#	
ADDRESS		AGE
ACCIDENT DETAILS :		
- CAUSE		
- PERSONAL INJURY		
- PROPERTY DAMAGE		
ACCIDENT REPORTED BY	I.D.#	
ACCIDENT REPORTED TO	I.D.#	
WITNESSED BY	I.D.#	
INVESTIGATED BY	I.D.#	
ACTION TAKEN :		
- MEDICAL AID		
- FOLLOW UP		
- LEGAL ACTION		

NOTE : IDENTIFY ENTRY PASS NUMBERS FOR ALL INDIVIDUALS INVOLVED.

_____
CONTRACTOR REP.SIGNATURE.

FIGURE 4.85
Accident report.

4.8 Testing, Commissioning, and Handover

Testing, commissioning, and handover constitute the last phase of a construction project's life cycle. This phase involves testing of electromechanical systems, commissioning of the project, obtaining authorities' approval, training user's personnel, handing over of technical manuals, documents, and as-built drawings to the owner/owner's representative. During this period, the project is transferred/handed over to the owner/end user for his or her use, and a substantial completion certificate is issued to the contractor.

4.8.1 Develop Testing and Commissioning Plan

The testing is carried out mainly in electromechanical works/systems and electrically operated equipment/systems, which are energized after connection of permanent power supply to the facility. These include the following:

1. Conveying system
2. Water supply, plumbing, and public health system
3. Fire suppression system
4. HVAC system
5. Integrated automation system (building automation system)
6. Electrical lighting and power system
7. Grounding (earthing) and lightning protection system
8. Fire alarm system
9. Telephone system
10. Information and communication system
11. Electronic security and access control system
12. Public address system
13. Emergency power supply system
14. Electrically operated equipment

The testing of these works/systems is essential to ensure that each individual work/system is fully functional and operates as specified. The tests are normally coordinated and scheduled with specialist contractors, local inspection authorities, third-party inspection authorities, and manufacturer's representatives. Sometimes, owner's representative may accompany the consultant to witness these tests.

Test procedures are submitted by the contractor along with the request for final inspection. Standard forms, charts, and checklists are used to record the testing results.

4.8.2 Identification of Project Team

The following project team members are involved during the testing, commissioning, and handover phase:

1. Owner
   - Owner's representative/project manager
2. Construction supervisor
   - Construction manager (agency)
   - Consultant (designer)
   - Specialist contractor
3. Contractor
   - Main contractor
   - Subcontractor
   - Testing and commissioning specialist
4. Regulatory authorities
5. End user
6. Third-party inspecting agency

4.8.3 Testing and Commissioning Quality Process

Commissioning is the orderly sequence of testing, adjusting, and balancing the system and bringing the systems and subsystems into operation and starts when the construction and installation of works are complete. Commissioning is normally carried out by the contractor or specialist in presence of consultant and owner/owner's representative and user's operation and maintenance personnel to ascertain proper functioning of the systems to the specified standards.

The commissioning of the construction project is a complex and intricate series of start-up operations and may extend over many months. Testing, adjusting, and operation of the systems and equipment are essential in order to make the project operational and useful for the owner/end user. It requires extensive planning and preparatory work, which commences long before the construction is complete. All the contracting parties are involved in the start-up and commissioning of the project as it is required for the project to be handed over to the owner/user. Figure 4.86 illustrates the testing and commissioning process.

4.8.4 Testing and Commissioning

Figure 4.87 illustrates a flowchart for inspection and testing plan.

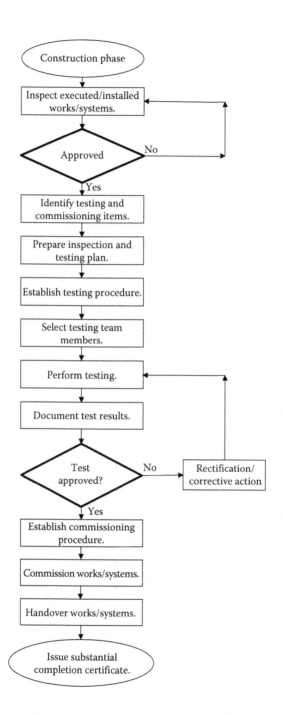

FIGURE 4.86

Logic flow process for testing, commissioning, and handover phase. (*Source*: Rumane, A.R. (2016), *Handbook of Construction Management: Scope, Schedule, and Cost Control*, CRC Press, Boca Raton, FL. Reprinted with permission from Taylor & Francis Group.)

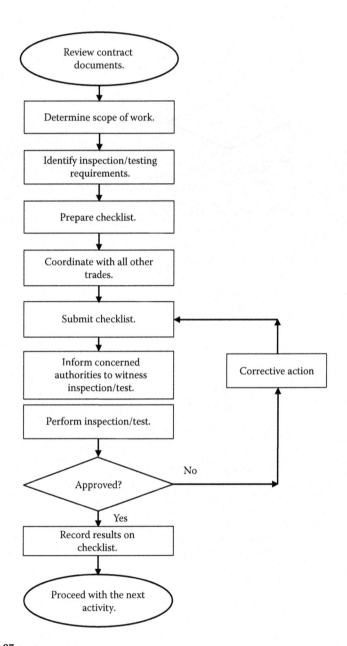

FIGURE 4.87
Development of inspection and test plan. (*Source*: Rumane, A.R. (2016), *Handbook of Construction Management: Scope, Schedule, and Cost Control*, CRC Press, Boca Raton, FL. Reprinted with permission from Taylor & Francis Group.)

4.8.4.1 Testing

All work is checked and inspected on a regular basis while the construction is in progress; however, there are certain inspections and tests to be carried out by the contractor in the presence of the owner/consultant. These are especially for electromechanical systems, conveying systems, and electrically operated equipment, which are energized after connection of permanent power supply. Testing of all these systems starts after completion of installation works. By this time, the facility is connected to a permanent electrical power supply, and all the equipment is energized. Figure 4.88 illustrates the procedure followed in Kuwait to get power supply. (This procedure may vary from country to country.)

The authors of *Civil Engineering Procedure* (Institution of Civil Engineers 1996) have described inspection and testing of works as follows:

> Inspection and Testing: During construction and on completion of parts of a project, inspection and testing are usually required in order to confirm compliance with the drawings and specification. Testing and inspection are generally required for static components of a project, while dynamic components such as machinery require testing and commissioning. The Project Manager or the Engineer, supported by his team, is usually responsible for inspection on and off-site and for testing of materials.
>
> Test Criteria and Schedules: The performance tests and criteria to be applied to any aspect of work should be specified in the contract for that work as far as possible, so as to enable the identification of the state at which an acceptable quality or degree of completion has been achieved. Depending on the type of project, samples and mock-up or factory inspections and acceptance test may be required. If they are, the responsibility for their cost should be defined.
>
> Schedules (lists) of the necessary inspections and tests should be agreed through collaboration between the Project Manager and all parties. (p. 105)

Testing is carried out mainly on electromechanical works/systems and electrically operated equipment, which are energized after connection of permanent power supply to the facility. These include the following:

1. HVAC system
2. Firefighting system
3. Water supply, plumbing, and public health system
4. Integrated automation system (building automation system)
5. Electrical lighting and power system
6. Grounding (earthing) and lightning protection system
7. Emergency power supply system
8. Fire alarm system

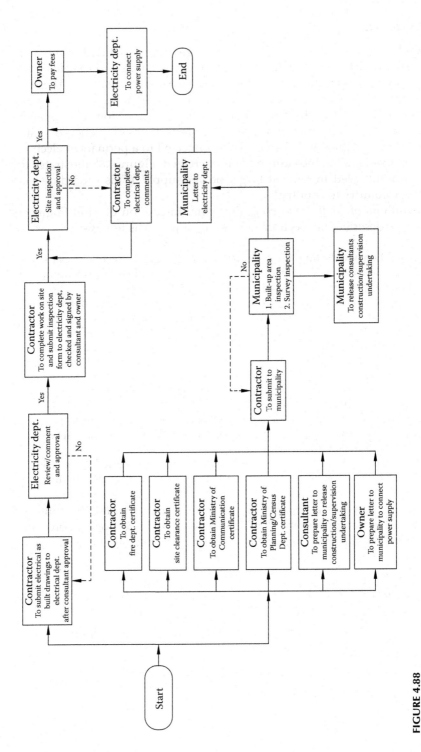

FIGURE 4.88
Electrical power supply connection procedure.

9. Telephone system
10. Communication system
11. Electronic security and access control system
12. Public address system
13. Conveying system
14. Electrically operated equipment

These tests are essential to ensure that each individual work/system is fully functional and operates as specified. These tests are normally coordinated and scheduled with specialist contractors, local inspection authorities, third-party inspection authorities, and manufacturers' representatives. Sometimes, the owner's representative may accompany the consultant to witness these tests.

Test procedures are submitted by the contractor along with the request for final inspection. Figure 4.89 is a request form for final inspection of electromechanical works.

4.8.4.2 Commissioning

Commissioning is the orderly sequence of testing, adjusting, and balancing the system and bringing the systems and subsystems into operation; it starts when the construction and installation of works are complete. Commissioning is normally carried out by the contractor or specialist in the presence of the consultant and owner/owner's representative and user's operation and maintenance personnel to verify proper functioning of the systems to the specified standards.

According to the authors of *Civil Engineering Procedure* (Institution of Civil Engineers 1996):

> Commissioning is the orderly process of testing, adjustment and bringing the operational units of the project into use. It is generally required where systems and equipment are to be brought into service following installation. Commissioning may be carried out by the promoters or contractor's staff by specialist personnel or by a mixed team. For complex industrial projects, a commissioning manager is usually appointed by the promoter to plan the commissioning preferably early in the project, establish budgets, lead a commissioning team, and procure the other necessary resources. For simple projects, commissioning is usually undertaken by the contractor, subject to the approval of procedures by the promoter or engineer. The commissioning of familiar or small process and industrial facilities is usually carried out by the operator with the advice and assistance of the suppliers of the equipment and systems. (p. 106)

The commissioning of the construction project is a complex and intricate series of start-up operations and may extend over many months. Testing,

---

Project Name
Project Name
ON SITE TESTING OF ELECTRO-MECHANICAL WORKS

CONTRACTOR   : _____   CHECK LIST No. : [   ]

SUBCONTRACTOR: _____   DATE [   ]

FOLLOWING WORKS ARE READY FOR INSPECTION ON:_____   TIME: _____

☐ ELECTRICAL   ☐ MECHANICAL
LOCATION:_____   DRAWING REF:_____

SPECIFICATION NO : _____   DIVISION :_____   SECTION : _____

SAMPLE FORM

Description of Work/System:
_____
_____
_____

TEST PROCEDURE: YES ☐   NO ☐

METHOD STATEMENT: YES ☐   NO ☐

CONTRACTOR SIGNATURE:_____   DATE:_____

A/E REMARKS:
_____
_____

CONSULTANT ENGINEER:_____   RESIDENT ENGINEER:_____
          DATE:_____   DATE: _____

Distribution   ☐ OWNER   ☐ R.E.   ☐ CONTRACTOR

FIGURE 4.89
Checklist for testing of electromechanical works.

adjustment, and operation of the systems and equipment are essential in order to make the project operational and useful for the owner/end user. It requires extensive planning and preparatory work, which commence long before the construction is complete. All the contracting parties are involved in the start-up and commissioning of the project as it is required for the project to be handed over to the owner/user.

4.8.5 As-Built Drawings

Most contracts require the contractor to maintain a set of record drawings. These drawings are marked to indicate the actual installation where the installation varies appreciably from installation shown in the original contract. Revisions and changes to the original contract drawings are almost certain for any construction project. All such revisions and changes are required to be shown on the record drawings. As-built drawings are prepared by incorporating all the modifications, revisions, and changes made during the construction. These drawings are used by the user/operator for reference purposes after taking over the project. It is a contractual requirement that the contractor hand over as-built drawings along with record drawings, record specifications, and record product data to the owner/user before handing over of the project and issuance of the substantial completion certificate. In certain projects, the contractor has to submit field records on excavation and underground utility services detailing their location and levels.

4.8.6 Technical Manuals and Documents

Technical manuals, design and performance specifications, test certificates, and warranties and guarantees of the installed equipment are required to be handed over to the owner as part of the contractual conditions.

According to the authors of *Civil Engineering Procedure* (Institution of Civil Engineers 1996):

> The commissioning and handover stage is the point for finalizing the project documentation. There are three categories of documents to be handed over by designers, equipment suppliers, and contractors:
>
> - Records of the equipment and services as installed
> - Commissioning installations, including safety rules
> - Operating and maintenance instructions, including safety rules
>
> Generally the first category will include design and performance specifications, test certificates, snagging lists, as-built drawings and warranties. In addition to these, documentation for the commissioning and operation will include permits, certificates of insurance, operation and maintenance manuals and handover certificates. (p. 109)

Systems and equipment manuals submitted by the contractor to the owner/end user generally consist of

- Source information
- Operating procedures
- Manufacturer's maintenance documentation

- Maintenance procedure
- Maintenance and service schedules
- Spare parts list and source information
- Maintenance service contract
- Warranties and guarantees

The procedure for submission of all these documents is specified in the contract document.

4.8.7 Training of User's Personnel

Normally, training of user's personnel is part of the contract terms. The owner's/user's commissioning, operating, and maintenance personnel are trained and briefed before commissioning starts in order to familiarize the personnel about the installation works and also to ensure that the project is put into operation rapidly, safely, and effectively without any interruption. The timing and details of training vary widely from project to project. Training must be completed well in advance of the requirement to make the operating teams fully competent to be deployed at the right time during commissioning. This needs to be planned from project inception, so that the roles and activities of the commissioning and operating staff are integrated into a coherent team to maximize their effectiveness.

4.8.8 Regulatory/Authorities' Approval

Necessary regulatory approvals from the respective concerned authorities are obtained so that the owner can occupy the facility and start using/operating it. In certain countries, all such approvals have to be obtained before electrical power supply is connected to the facility. It is also required that the building/facility be certified by the related fire department authority/agency that it is safe for occupancy.

4.8.9 Move-In Plan

No one wants to keep the facility unoccupied after completion of the construction project, and once it is declared fit for use/occupancy.

The move-in plan has been prepared based on the substantial completion of the facility or partial handover/taking over of the facility. With commercial facilities, the investor would like the tenant to occupy the building so that the investor starts getting revenues. With office buildings, the owner may be shifting from the existing offices to the newly built offices. The shifting of offices should be smooth and uninterrupted without affecting routine or day-to-day work carried out by the office staff. It is likely that certain furniture and equipment from the existing workplace will need to be transferred to the new offices, which requires close coordination to maintain uninter-rupted working conditions to achieve good results and successful transfer. Therefore, it is important that the move-in plan be prepared for occupation of the new facility taking into consideration all these factors.

4.8.10 Handover of Facility to Owner/End User

Once the contractor considers that the construction and installation of works have been completed as per the scope of contract, final tests have been per-formed, and all the necessary obligations have been fulfilled, the contractor submits a written request to the owner/consultant for handing over of the proj-ect and for issuance of a substantial completion certificate. This is done after testing and commissioning is carried out, and it is established that the project can be put into operation or the owner can take charge of the project. In most construction projects, there is a provision for partial handover of the project.

The owner/consultant inspects the work and informs the contractor of unfulfilled contract requirements. A punch list (snag list) is prepared by the consultant listing all the items still requiring completion or correction. The list is handed over to the contractor for rework/correction of the works men-tioned in the punch list. The contractor resubmits the inspection request after completing or correcting previously notified works. A final snag list is pre-pared if there are still some items that need corrective action/completion by the contractor; the remaining works are to be completed within the agreed period to the satisfaction of the owner/consultant.

The contractor starts handing over of all completed works/systems that are fully functional, and the owner agrees to take over the items handed over. A handing over certificate is prepared and signed by all the concerned par-ties. Figure 4.90 shows a handing over certificate.

Most contract documents include the list of spare parts, tools, and extra materials to be delivered to the owner/end user during the closeout stage of the project. The contractor has to properly label these spare parts and tools clearly indicating the manufacturer's name and model number if applicable. Figure 4.91 illustrates the spare parts handover form used by the contractor.

4.8.11 Substantial Completion

A substantial completion certificate is issued to the contractor once it is estab-lished that he or she has completed the work in accordance with the contract documents and to the satisfaction of the owner. The contractor has to submit all the required certificates and other documents to the owner before issu-ance of the certificate. Figure 4.92 illustrates the procedure for issuance of the substantial certificate normally used for building construction projects in Kuwait. Authorities' approval requirements mentioned in this figure may vary from country to country; however, other requirements need to be com-pleted by the contractor for issuance of a substantial completion certificate.

Project Name
Project Name
HANDING OVER CERTIFICATE

CONTRACTOR   : _____   CERTIFICATE No. : [   ]

SUBCONTRACTOR: _____   DATE [   ]

SPECIFICATION NO : _____   DIVISION : _____   SECTION : _____

DRAWING No.   BOQ REF: _____

AREA : [   ]   Building Works   [   ]   Electrical Works [   ] Mechanical Works

        [   ]   HVAC Works   [   ]   Finishes Works   [   ]

SAMPLE FORM

Description of Work/System:

The work/system mentioned above is completed by the contractor as specified and has been inspected and tested as per contract documents. The work/system is fully functional to the satisfation of owner/end user. The contractor hand over the said work/system to the owner/end user as on --------------. The guarentee/warranty of work/system shall start as of ------------ and shall be valid for a period of ------------------years(duration) from the date of issuance of substantial completion certificate. The contractor shall be liable contractually till the end of warranty/guarentee period.

SIGNED BY:

OWNER/END USER: _____   CONTRACTOR: _____

CONSULTANT: _____   SUBCONTRACTOR: _____

FIGURE 4.90
Handing over certificate.

Normally, a final walk-through inspection is carried out by a committee, which consists of the owner's representative, design and supervision personnel, and the contractor to decide the acceptance of works and whether the project is complete enough to be made operational. If there are any minor items remaining to be finished, then a list of such items is attached with the certificate of substantial completion for conditional acceptance of the project. Issuance of substantial completion certifies acceptance of the work. If the remaining work is

Project Name
Project Name
HANDING OVER OF SPARE PARTS

CONTRACTOR: _____ CERTIFICATE No. : [   ]

SUBCONTRACTOR:_____ DATE [   ]

SPECIFICATION NO : _____ DIVISION _____ SECTION : _____

DRAWING No. _____ BOQ Ref. _____

AREA : [   ]   Building Works   [   ]   Electrical Works [   ]   Mechanical Works

[   ]   HVAC Works   [   ]   Finishes Works   [   ]

Following Spare Parts have been handed over to the owner/end user

Description of Spare Parts

Sr.No.	Description	BOQ Reference	Spec. Ref.	Manufacturer	Specified Qty	Delivered Qty

(Attach additional sheet, if required)

SIGNED BY:

OWNER/END USER: _____   CONTRACTOR: _____

CONSULTANT: _____   SUBCONTRACTOR: _____

FIGURE 4.91
Handing over of spare parts.

of a minor nature, then the contractor has to submit a written commitment that he or she will complete the agreed-upon work within the stipulated period. A memorandum of understanding is signed between the owner and the contractor that the remaining works will be completed within the stipulated period.

Table 4.26 illustrates a list of activities that need to be considered for project closeout.

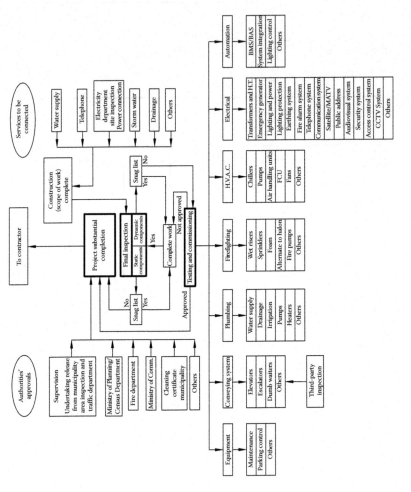

FIGURE 4.92
Project substantial completion procedure.

TABLE 4.26

Project Closeout Checklist

Serial Number	Description	Yes/No
Project Execution		
1	Contracted works completed	
2	SWIs completed	
3	Job site instructions completed	
4	Remedial notes completed	
5	Noncompliance reports completed	
6	All services connected	
7	All the contracted works inspected and approved	
8	Testing and commissioning carried out and approved	
9	Any snags	
10	Is project fully functional?	
11	All other deliverables completed	
12	Spare parts delivered	
13	Is waste material disposed?	
14	Whether safety measures for use of hazardous material established	
15	Whether the project is safe for use/occupation	
Project Documentation		
16	Authorities' approval obtained	
17	Record drawings submitted	
18	Record documents submitted	
19	As-built drawings submitted	
20	Technical manuals submitted	
21	Operation and maintenance manuals submitted	
22	Equipment/material warrantees/guarantees submitted	
23	Test results/test certificates submitted	
Training		
24	Training to owner/end user's personnel imparted	
Payments		
25	All payments to subcontractors/specialist suppliers released	
26	Bank guarantees received	
27	Final payment released to main contractor	
Handing over/Taking over		
28	Project handed over/taken over	
29	Operation/maintenance team taken over	
30	Excess project material handed over/taken over	
31	Facility manager in action	

The certificate of substantial completion is issued to the contractor, and the facility is taken over by the owner/end user. By this stage, the owner/end user has already taken possession of the facility, and operation and maintenance of the facility commence. The project is declared complete, and the construction project's life cycle is over.

5

Operation and Maintenance

5.1 Operation and Maintenance

Once the project is completed and fully functional, then it becomes part of the operation system. The operation and maintenance (O&M) of a completed project are associated with the quality of the constructed project.

The owner may employ the same contractor(s) who built the facility or engage another specialist operation and maintenance organization to manage the operation and maintenance. It is likely that the owner may have his or her own team to operate and maintain the facility. The following documents are required to operate and maintain the completed facility. These documents are normally submitted by the contractor during the testing, commissioning, and handover phase of the construction project:

1. As-built drawings
2. Operation and maintenance manuals
3. Operating procedures
4. Source information for availability of spare parts of the installed works

The cost of operation and maintenance should be considered in the feasibility studies of the project. Postconstruction costs have a great impact on the overall life cycle cost of the project.

5.2 Categories of Maintenance

Following are the different types of maintenance normally being carried out, depending on the nature of activity, system, or equipment, to avoid interference/interruption with essential installation operations, endanger life or property, or involving high cost for replacement of the item.

5.2.1 Preventive Maintenance

Preventive maintenance is the planned—with scheduled periodic inspection—adjustment, cleaning, lubrication, parts replacement of equipment and systems in operation. Its aim is prevention of breakdown and failure. Preventive maintenance consists of many activities required to be checked and performed to ensure that the operations are safe and can be performed without any danger to life or facility and will not warrant high cost or long lead time for replacement. Preventive maintenance is the cornerstone of any good maintenance program.

5.2.2 Scheduled Maintenance

Scheduled maintenance includes those maintenance tasks whose cycles exceed a specified period. This type of maintenance relates to finish items, such as painting, polishing of wood, parking and road markings, roof maintenance, and testing of fire alarm systems.

5.2.3 Breakdown Maintenance

Breakdown maintenance for systems or equipment is scheduled (planned) and unscheduled (unplanned). Unscheduled (unplanned) is unanticipated maintenance because of system or component failure. If the problem relates to an essential service or has any hazardous effect, then an emergency response is necessary. If the problem is not critical, then routine response is adequate.

5.2.4 Routine Maintenance

Routine maintenance is an action taken by the maintenance team to restore a system or piece of equipment to its original performance, capacity, efficiency, or capability.

5.2.5 Replacement of Obsolete Items

Replacement of obsolete items refers to the work undertaken by a maintenance team to bring a system or component into compliance with new codes of safety regulations or to replace an item that is unacceptable, inefficient, or for which spare parts are no longer available or have had technological changes. Early detection of problems may reduce repair and replacement costs, prevent malfunctioning, and minimize downtime as, for example, with communication systems, audiovisual systems, and equipment.

5.2.6 Predictive Testing and Inspection

Predictive testing and inspection refers to testing and inspection activities that involve the use of sophisticated means to identify maintenance requirements. For example, specialized tests are to be carried out:

- To locate wearing problems on bus bar bends of an electrical main's low tension panel
- To detect insulation cracks
- To locate thinning of pipe walls and cracks
- To identify vibration problems for equipment like chillers, diesel generator sets, pumps
- To locate heat buildup (rise in temperature) in electrical bus duct

5.3 O&M Program

Addressing operation and maintenance considerations at the start of the project contributes greatly to improved working environments, higher productivity, and reduced energy and resources costs. During the design phase of the project and up to handover of the facility, O&M personnel should be involved in identifying maintenance requirements for inclusion in the design. The goal of effective O&M is to achieve the intent of the original design team, that is, the systems and equipment delivery services to the user, enhancing a comfortable, healthy, and safe environment. O&M should also include long-term goals such as economy, energy conservation, and environmental protection. To create an effective O&M program, the following procedures should be followed:

- Ensure that up-to-date as-built drawings for all the systems are available
- Ensure that operational procedures and manuals for the installed equipment are available
- Prepare a master schedule for operation and preventive/predictive maintenance
- Implement preventive maintenance programs complete with maintenance schedules and records of all maintenance performed for all the equipment and systems installed in the project/facility
- Follow manufacturer's recommendations for maintenance of equipment and systems

- Ensure the maintenance personnel have full knowledge of the equipment and systems installed and for which he/she is responsible to operate and maintain
- Ensure that O&M personnel are provided training during the testing, commissioning, and handover phase
- Offer training and professional development opportunities for each of the O&M team members
- Implement a monitoring program that tracks and documents equipment performance to identify and diagnose potential problems
- Implement a monitoring program that tracks and documents systems performance to identify and diagnose potential problems
- Perform predictive testing and inspection for critical and important items
- Use preventive maintenance and standbys, etc., so that the failed components can be isolated and repaired without interrupting system performance, thus minimization of equipment failures

For an effective O&M program, the following specific items should be considered:

1. HVAC system and equipment
2. Indoor air quality system and equipment
3. Electrical systems and equipment
4. Water fixtures and system
5. Life safety (fire suppression systems)
6. Cleaning equipment and products
7. Landscape maintenance

A guideline to prepare a preventive maintenance program taking into consideration the most important equipment and systems of an office/commercial building is illustrated in Table 5.1.

TABLE 5.1

Preventive Maintenance Program

Sr. No.	Activities	Maintenance Frequency				
		Daily	Weekly	Monthly	Quarterly	Yearly
Chiller						
Cooling tower						
Air handling unit						
Fan coil unit						
Exhaust air fan						
Supply air fan						
Package unit						
DX unit						
Split unit						
VAVs						
Chilled water pump						
Condenser water pump						
Hot water pump						
Domestic water pump						
Booster pump						
Sump pump (drainage pump)						
Fire pumps						
Jockey pump						
Fire-fighting system						

(Continued)

TABLE 5.1 (*Continued*)

Preventive Maintenance Program

Sr. No.	Activities	Maintenance Frequency				
		Daily	Weekly	Monthly	Quarterly	Yearly
Foam system						
Building automation system						
Water boilers						
Sanitary and water fixtures						
Main low tension panel						
Main/submain switch boards						
Electrical distribution boards						
Bus duct						
Automatic transfer switch						
Diesel generator set						
Fire alarm system						
Communication system						
Security system						
Access control system						
Audiovisual system						
Public address system						
Dimmer system						
Uninterrupted power supply (UPS)						

(Continued)

TABLE 5.1 (*Continued*)

Preventive Maintenance Program

Sr. No.	Activities	Maintenance Frequency				
		Daily	Weekly	Monthly	Quarterly	Yearly
Cleaning equipment						
Kitchen equipment						
Elevator						
Escalator						
Landscape						

6

Facility Management

6.1 Facility Management

Facility management is the process by which an organization integrates all noncore specialist services to achieve strategic objectives. It enhances the performance of the organization. It is an integration of people, place, process, and technology.

Facility management is mainly applied to buildings/properties that are complex in nature, including more complex management and operation. Construction projects are becoming more sophisticated and more technical in nature; therefore, their performance has to be maintained and the level of services has to be improved to satisfy user needs.

Facility management is effectively managing physical assets of an organization to ensure their value is realized and their repair, maintenance, and replacement are planned in an economical manner. Facility management is a multidiscipline activity. It involves a number of disciplines and services. It encompasses multidisciplinary activities within the built environment and the management of their impact upon people and the workplace. It covers the development, coordination, and development of all the noncore specialist services of an organization, together with the buildings and their systems, environment, equipment, plant, health and safety, information technology, infrastructure, landscape, and all the services. Figure 6.1 illustrates the areas covered under facility management.

Facility management is gaining increasing recognition among large organizations as an important factor for business improvement. Effective facilities management, combining resources and activities, is vital to the success of any organization. At a corporate level, it contributes to the delivery of strategic and operational objectives. Facility management is a relatively new aspect in Asia and other developing countries. However, there has been an

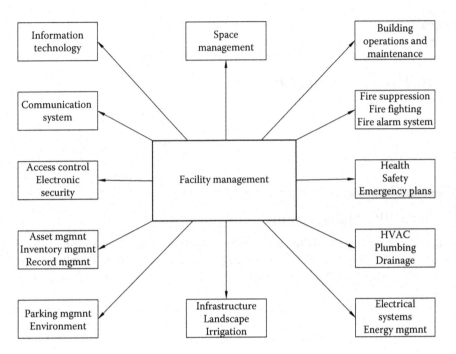

FIGURE 6.1
Areas under facility management.

enormous growth in facility management activities worldwide. Table 6.1 summarizes facilities management market characteristics between Europe and the Middle East.

Operation and maintenance expenses are normal. However, the cost can be minimized through an aggressive facility management program and the use of applicable diagnostic tools. The key to success of facility management is that the team members must be proactive and embark on a realistic, long-range facility management program. Timely and well-planned maintenance and repair will result in cost savings. The distinction between maintenance management and facilities management is illustrated in the conceptual graph of building capability versus time shown in Figure 6.2.

The facility management function is performed by either an in-house facility management team or an outsourced team.

In both cases, the team members must have innovative approach in day-to-day operations to reduce operating costs and adding value to the business. It is essential that facility management practitioners (in-house or outsourced team) are proactive while carrying out an innovative project, even in the day-to-day operations, and are aware of the trends and developments in

TABLE 6.1

FM Market General Characteristics—Europe versus Middle East

Europe	Middle East
• Matured & structured facility management (FM) Industry	• Immature & unstructured FM Industry
• Benchmark & historical data available	• No historical benchmarking data available
• Standard RFP's & structures Service Level Agreements are the norm	• Not generally available. Market characterized by poor quality trends & contract documentation
• Clients & Service Providers conversant with requirements including Performance Measurement Criteria	• Clients 7 Suppliers using "construction type" forms of contracts
• Best practice Quality Management System (QMS) & processes "imbedded" in service delivery models	• QMS not generally utilized or implemented
• Resources (internal & external) available & trained in FM systems & procedures	• Resources & skills available but not integrated into FM companies
• Experienced in use of systems technology as key management tools	• Lack of systems technology & understanding of application & benefits
• Experienced FM service providers available who can manage "bundled" services on single solution basis	• Lots of small players & single service providers • Some good FM companies exist, but limited number with international & local experience

Source: EMCOR Facilities Services Kuwait (www.emcorgroup.com). Reprinted with permission from EMCOR Facilities Services Kuwait.

the facility management and related industries. Following are the pros and cons of in-house facilities management.

Pros	Cons
More responsive to corporate culture and needs of organization	Lack of flexibility and availability of multidisciplinary skills
Sense of "belonging"/being part of client culture	Too often in-house staff "transferred" from other departments without right skills or training
Easier communication and change management decision and processes	Lack of exposure to best practice facility management processes and systems
Confidentiality issues/more secure	Specialist "in-house" skills not available

Facility managers are the professionals most responsible for integrating people with their physical environment. They come from a variety of professions. Facility management professionals are judged by their managerial capabilities rather than their technical competence. Facilities professionals need to be trained and equipped with tremendous amount of knowledge and management ability to cope with and solve a multitude of complex problems and challenges. Table 6.2 illustrates typical requirements of a facility management professional.

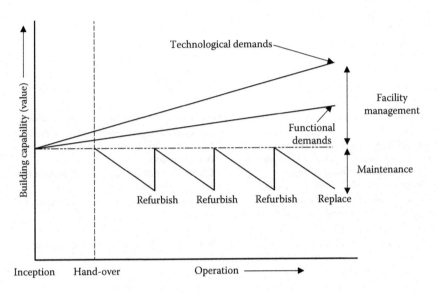

FIGURE 6.2

Distinction between maintenance management and facility management. (*Source*: Keith Alexander (1996), *Facilities Management, Theory and Practice*. Reprinted with permission from Taylor & Francis Group.)

TABLE 6.2

Typical Requirements of Facility Management Professional

Sr. No.	Requirements
1	Intuitive and managerial skills
2	Ability to improve business performance
3	Ability to satisfy customers
4	Capability of seeking new methods of doing work in order to bring about marked and sustainable improvements in performance of business
5	Ability to develop new or improved facilities, services, technologies that add to the value
6	Ability to perform things happen with level of technical understanding
7	Ability to accept full accountability for all actions they perform
8	Should be flexible having visionary skills
9	Willing to accept changes in business and social practices
10	Leadership skills
11	Communication skills
12	Technical and analytical skills
13	Knowledge of industry standards and practices as well as regulatory mandate
14	Purchasing and contracting skills
15	Personal and interpersonal relationship skills

6.2 Facility Management Outsourcing

Facility management must achieve an organization's objectives and promote an organization's growth. The facility management companies must be able to pool resources to achieve the objectives. While selecting a facility management firm, it is necessary to study the competency of the firm in the following areas:

- Operation and maintenance
- Planning and project management
- Technology
- Quality assessment and innovation
- Management and leadership
- Communication
- Environmental factors

Due to globalization and the competitive market, business outsourcing has become a common factor. The following are the benefits from outsourcing facility management:

- Reduce/control costs
- Access to leading-edge technologies
- Focus on core business
- Free resources for other purposes
- Availability of resources not available internally
- Improved, proactive service
- Meet and exceed organization strategic plans

Before deciding to outsource, organizations must decide the following:

- What are the core competencies?
- What are current costs and quality of services provided in-house?
- Would it be better to reengineer processes internally to achieve goals?
- What services may be better delivered by a facility management service provider?
- What are the goals to be achieved by outsourcing?
- Is a service readily available from suppliers who are better structured to efficiently deliver service at the right price?
- How will the contract (Service Level Agreement) be structured?
- What "in-house" capability/management will be retained/provided to manage service relationships?

If the organization decides to outsource facility management, it has to

- Appoint a dedicated project "champion" from within client organization to define and manage the process
- Develop a proper Request for Proposal (RFP) and evaluation criteria
- Engage in a structured and detailed process with a "preferred bidder" to develop
  - "Single source" solutions
  - Properly structured Service Level Agreement (SLA)
- Ensure dedicated and ongoing management of facility management services and SLA

6.3 Computer-Aided Facility Management

The facilities management profession has come of age. Its practitioners require skill and knowledge to manage the increasingly broad range of services being included in the facilities management areas. One of the key tools of facilities management included in these different services is an integrated CAFM system. In order to achieve better results, many of the organizational functions are integrated with the help of CAFM. There are special software such as Maximo, Concept 500, Datastream7i, Frontline/Shire systems available to manage these functions. Figure 6.3 illustrates elements of CAFM.

CAFM enables the facility manager and his or her team to track the status of work, assets, and records in order to manage and communicate an organization's day-to-day operations. The system can provide reports to use in allocating the organization's resources, preparing key performance indicators (KPIs) metrics to use in evaluating the effectiveness of current operations, and for making organizational decisions. The system can be used for space planning, procurement and stores management, assets management budget preparation, statistical reporting, resource allocation and management, security management, waste management, and energy management. CAFM provides integrated facility management.

6.4 Benefits of Facility Management

The following are the benefits of facility management:

- Deliver strategic and operational objectives of the organization
- Increase profits of the organization

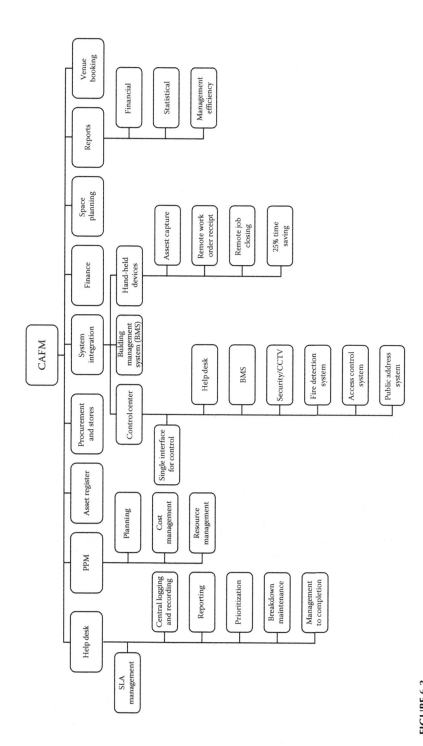

FIGURE 6.3
Elements of computer-aided facility management.

- Obtain client satisfaction
- Enhance and project an organization's identity and image
- Deliver business continuity and workforce protection in an era of heightened security threats
- Help the integration process associated with change, postmergers, or acquisitions
- Enable new working styles and processes—vital in this technology-driven age
- Deliver effective management of an organization's assets
- Manage energy efficiently
- Bring about a safe and efficient working environment

TABLE 6.3

Quality Control Requirements in Facility Management

Sr. No.	Work Area	Acceptable Quality Level	Quality Assurance Method
1	Access control	Not more than 1 defect per month	100% inspection
2	Asset management	No defect	100% inspection
3	Building operations and maintenance	Not more than 5% defect rate	Random sample
4	Communication system	Not more than 2% defect rate	Random sample
5	Drainage	No defect	100% inspection
6	Electrical system	Not more than 5% defect rate	Random sample
7	Electronic security	No defect	100% inspection
8	Emergency plans	Not more than 1 defect per month	100% inspection
9	Environment	As per regulatory requirements	As per regulatory requirements
10	Fire alarm system	Not more than 5% defect rate	Random sample
11	Fire-fighting plumbing	Not more than 5% defect rate	Random sample
12	Health and safety	As per regulatory requirements	As per regulatory requirements
13	HVAC	Not more than 1 defect per month	Random sample Customer complaint
14	Information technology	No defect	100% inspection
15	Infrastructure	Not more than 1 defect per month	Random sample
16	Inventory management	No defect	100% inspection
17	Irrigation	Not more than 1 defect per month	100% inspection
18	Landscape	No defect	100% inspection
19	Parking management	Not more than 1 defect per month	Random sample
20	Plumbing	Not more than 1 defect per month	100% inspection
21	Record management	No defect	100% inspection
22	Space management	Not more than 1 defect per month	100% inspection

6.5 Quality Requirements of Facility Management

The requirements for facility management are set by the owner of the facility. A facility management organization is required to establish a quality control system to ensure that facilities management work is performed satisfactorily and there is no interruption or breakdown of the systems. Table 6.3 illustrates an example of quality control requirements for operation and management of the facility. The actual requirements may vary from facility to facility and reflect the importance of various systems and services.

7

Assessment of Quality

7.1 Introduction

In today's global competitive environment, with the increase in customer demand for higher-performing products, organizations are facing many challenges. They are finding that their survival in the competitive market is increasingly in doubt. To achieve a competitive advantage, effective quality improvement is critical for an organization's growth. This can be achieved through development of long-range strategies for quality. Assessment of existing quality systems is helping the development of long-term organizational strategies for quality.

7.2 Assessment Categories

Assessment of quality comprises

1. Cost of poor quality
2. Organization's standing in the marketplace
3. Quality culture in the organization
4. Operation of quality management system

7.2.1 Cost of Poor Quality

This assessment is mainly to find out

1. Customer dissatisfaction resulting due to product failure after sales
2. Annual monetary loss of product and processes that are not achieving their objectives

In order to know customer satisfaction, an independent product survey can be carried out to determine the reasons and improvement measures to be adapted. Such assessment can be done through a questionnaire related to quality of work (workmanship) and functioning, and getting the customer's opinion/reaction. Table 7.1 lists customer satisfaction questionnaires that can be used to get related information about completed building construction projects.

Management can use the same for quality improvement. Necessary improvement if required for proper functioning of the installed system should be carried out to maintain customer satisfaction. Normally, the contractor is required to maintain the completed project for a certain period under the maintenance contract, and therefore necessary modifications can be carried out during this period. Such defects/modifications have to be recorded by the organization as feedback information and necessary measures can be taken to avoid occurrence of similar things in other projects. The cost involved to carry out such repairs/modifications is borne by the contractor, which helps improve customer satisfaction.

Poor quality results in higher maintenance and warranty charges. In construction projects, these costs are mainly due to the following:

1. Rejection of work during construction: This results in rework by the contractor to meet the specification requirements. If any extra time and cost is required to redo the work, the contractor is not compensated for the same.

2. Rejection of shop drawings: Repeated rejection of shop drawings results in the contractor spending extra time and cost, and a delay in start of relevant activities.

Preventive measures are required for overcoming this problem. Necessary data can be collected to analyze the reasons for rejections. With a cause-and-effect diagram, the reasons for rejection can be analyzed and necessary improvements can be implemented.

7.2.2 Organization's Standing in the Marketplace

A company's standing in the marketplace fetches business opportunities for the organization. The organization needs to know where it stands on quality in the marketplace. This can be done through market study. Numerous questions related to the product should be considered for such studies, which can be carried out by the marketing department of the company. The input from the study should be considered for quality improvement. Various quality tools can be used to analyze the situation and the system can be improved, taking into consideration the analysis results.

TABLE 7.1

Customer Satisfaction Questionnaire

Sir/Madam

We are conducting a survey to improve our performance. We would appreciate your response to the following questions. Your valuable information will help us to satisfy your requirements in a better way. (The information provided by you shall be kept secret)

Customer Satisfaction Questionnaire

Name of Organization:

Name of Completed Project:

Q1   Workmanship of the following installed items

Sr. No.	Work Description	Very Good	Good	Average	Poor	No Comment
1	Structural work					
2	Precast work					
3	Internal finishes					
4	External finishes					
5	Plumbing					
6	Drainage					
7	Fire fighting					
8	HVAC					
9	Elevator					
10	Escalator					
11	Electrical work					
12	Landscape work					
13	Pavements					
14	Streets/roads					
15	Water proofing					

(Continued)

TABLE 7.1 (*Continued*)

Customer Satisfaction Questionnaire

Q2	How are the following systems functioning?					
Sr. No.	Systems	Very Good	Good	Average	Poor	No Comment
1	HVAC					
2	Water system					
3	Lighting					
4	Power					
5	Fire alarm system					
6	Telephone system					
6	Public address system					
7	Security system					
8	Kitchen equipment					
9	Window cleaning system					
10	Irrigation system					
11	Elevators					
12	Escalators					

Q3	How was the behavior of our site staff?			
Sr. No.	Position	Co-operative	Polite	Unpolite
1	Managers			
2	Engineers			
3	Foreman			
4	Labors			
5	Office staff			

(*Continued*)

TABLE 7.1 (*Continued*)

Customer Satisfaction Questionnaire

Q4   **How was site safety**

        Good ☐   Average ☐   Poor ☐

Q5   **Whether project was completed as per schedule**

        Yes ☐   No ☐

Q5A   **If your reply to Q5 is No, then please let us know:**

        Have you incurred any revenue loss/any other damages due to noncompletion in time?

        Yes ☐   No ☐

Q6   **How was the response to your problems at site?**

        Immediate ☐   Within one week ☐   Within two weeks ☐   No response

Q7   **Whether our variation claims were as per contract**

        Yes ☐   No ☐

Q8   **Are you satisfied with our performance?**

        Satisfied ☐   Not satisfied ☐   Need improvement ☐

Q9   **Are you satisfied with our performance?**

        Satisfied ☐   Not satisfied ☐   Need improvement ☐

Q10   **How was the response to your problems from our Head Office**

        Good ☐   Poor ☐   Need improvement ☐

Q11   **How do you compare our performance with other competitors?**

        Better than others ☐   Same as others ☐   Less than others ☐

Q12   **Whether you will consider inviting us for your new tenders**

        Yes ☐   Can't say ☐

Q13   **Any suggestion(s)?**

7.2.3 Quality Cultures in the Organization

The concept of quality culture has changed during the past three decades. The quality approach is centered on organization-wide participation aimed at long-term success through customer satisfaction. It is focused on participative management and strong operational accountability at the individual contribution level.

Employees' opinion and understanding of the company's products and implementation of a quality system help in developing a long-term strategy based on company-wide assessments of quality. This can be achieved through considerate focused group discussion or using questionnaires to survey employees' opinions or by doing both.

The following is a list of examples of questions that can be used by building construction organizations asking their employees to respond:

1. Do you know what is good quality for this project?
2. Are you aware of the company's emphasis on quality?
3. Are you familiar with the project quality?
4. Do you know the inspection procedures for this project?
5. Do you know how you can provide high-quality workmanship?
6. How you will do the remedial work?
7. How can you improve project quality?

7.2.4 Operation of Quality Management System

This element of assessment is related to the evaluation of present quality management system activities. Assessment of present quality activities can be evaluated from two perspectives:

1. Assessment that focuses on customer satisfaction results but includes an evaluation of the present quality system
2. Assessment that focuses on evaluation of the present quality system with little emphasis on customer satisfaction results

In either case, the assessment can be performed by the organization itself or by an external body. The assessment performed by the organization is known as "self-assessment," whereas an assessment performed by an external body is referred to as a quality audit. A quality audit is formal or methodical, examining, reviewing, and investigating the existing system to determine whether agreed-upon requirements are being met.

Audits are mainly classified as follows:

First party—audits their own organization (internal audit)
Second party—customer audits the supplier
Third party—audits performed by independent audit organization

Third-party audits may result in independent certification of a product, process, or system, such as ISO 9000 quality management system certification. Third-party certification enhances an organization's image in the business circle. The assessment provides feedback to management on the adequacy of implementation and effectiveness of the quality system.

7.3 Self-Assessment

Self-assessment is a comprehensive, systematic, and regular review of an organization's activities, the results of which are referenced against a specific model. The goal of this assessment is to identify the following:

- What your organization is doing well?
- What it is not doing well?
- What it is not doing at all?
- Where and how can it make measurable improvements?

As a means of improving the management of organizations companies, use models related to quality awards:

1. Malcolm Baldrige National Quality Award (USA)
2. European Quality Award (Europe)
3. Deming Prize (Japan)

The Baldrige Award is presented annually by the President of the United States to organizations that demonstrate quality and performance excellence in manufacturing, service, small business, education, health care, and nonprofit organizations that apply and are then judged to be outstanding in seven categories. These categories are

Category No. 1—Leadership
Category No. 2—Strategic planning

Category No. 3—Customer and market focus

Category No. 4—Measurement, analysis, and knowledge management

Category No. 5—Workforce focus

Category No. 6—Process management

Category No. 7—Results

The Baldrige National Quality Program promotes an understanding of requirements for performance excellence and fosters sharing information about successful performance strategies and the benefits derived from using these strategies. Many organizations use Baldrige criteria to conduct self-assessments without submitting an application for the award and use the result to identify improvement opportunities within the organization for performance excellence.

Appendix A: Design Review Checklists

A.1 Design Checklist—Architectural Work

Serial No.	Item	Yes	No	Action Required
1	Check property limits			
2	Check building dimensions as per approval from authorities			
3	Check if modules match with structural layout			
4	Check for energy conservation measures			
5	Check for constructability			
6	Check user requirements about zoning area			
7	Check no. of rooms and room size as per occupancy and purpose			
8	Check relationship between all the services requirements and project user items (rooms, etc.)			
9	Check horizontal and vertical transportation (stairs, elevators, escalators) and escape routes as per architectural norms			
10	Check no. of floors and area of each floor			
11	Check level of each floor			
12	Check axis, grid			
13	Check elevations, sections, and details			
14	Check system ventilation			
15	Check floor plans			
16	Check roof plan			
17	Check wall openings			
18	Check suitable location and spaces for equipment			
19	Check finishes of all areas with respect to cost range with the finishing material			
20	Check finishes schedule			
21	Check schedule of doors and windows			
22	Check access for maintenance of equipment/ services requirement			
23	Check for suitability of appropriate roofing system			
24	Check whether room/area identification is provided			
25	Check fire protection requirements			
26	Check safety requirements			
27	Check requirements for special needs people			

A.2 Design Checklist—Structural Work

A.2.1 Foundation Layout

Serial No.	Item	Yes	No	Action Required
1	Check that latest architectural drawing is used			
2	Check if axis and grids match with architectural layout			
3	Check location of columns axis			
4	Check if depth of excavation matches with required depth for no. of basements			
5	Check for thickness of blinding concrete			
6	Check whether waterproofing and moisture protection material is suitable for subsurface conditions			
7	Check if foundation design is consistent with geotechnical report/soil report			
8	Check footing details for foundation			
9	Check whether seismic requirements for foundation are considered			
10	Check footing design			
11	Check minimum concrete cover considered			
12	Check if total load is calculated and considered			
13	Check whether coefficient of sliding on foundation soil is considered			
14	Check if levels are shown			
15	Check if expansion joints are shown			
16	Check if reinforcement details are shown			
17	Check spacing arrangement for reinforcement			
18	Check if foundation for pits, tanks is considered			
19	Check for type of concrete and steel			
20	Check for earthquake, seismic control requirements			

A.2.2 Basement Design

a) Basement Retaining Wall				
Serial No.	Item	Yes	No	Action Required
1	Check that latest architectural drawing is used			
2	Check whether axis and grids match with architectural layout			
3	Check property limit/boundary limit			
4	Check whether columns' axes match with architectural drawings			
5	Check thickness of retaining wall against all type of forces/loads			
6	Check waterproofing, dampproofing requirements			
7	Check insulation requirements			
8	Check if seismic load is considered			
9	Check required openings for services, sleeves			
10	Check reinforcement for retaining wall			
11	Check for expansion joints			
12	Check for wall footing			
13	Check levels are shown			
14	Check if fair-faced wall is needed			

b) Basement Columns				
Serial No.	Item	Yes	No	Action Required
1	Check that latest architectural drawing is used			
2	Check whether axis and grids match with architectural layout			
3	Check load factor is considered			
4	Check whether columns' axes match with architectural drawings			
5	Check whether column orientation matches with architectural drawings			
6	Check axial load and bending moment as per ACI			
7	Check reinforcement area			
8	Check minimum concrete cover			
9	Check for type of concrete and steel			
10	Check column dimensions			
11	Check for deflection			
12	Check for temperature effect			
13	Check tie size and spacing			
14	Check levels are shown			

c) Basement Beams

Serial No.	Item	Yes	No	Action Required
1	Check that latest architectural drawing is used			
2	Check if axis and grids match with architectural layout			
3	Check floor framing plan			
4	Check location of columns, walls, stairs			
5	Check minimum depth/height of beam			
6	Check applied moment			
7	Check steel ratio			
8	Check design for longitudinal bars			
9	Check deflection			
10	Check stirrup design			
11	Check dimensions of beam			
12	Check if levels are shown			

d) Basement Slab

Serial No.	Item	Yes	No	Action Required
1	Check that latest architectural drawing is used			
2	Check whether axis and grids match with architectural layout			
3	Check total load and load factor			
4	Check if span length matches with architectural requirements			
5	Check type of slab as per design requirements			
6	Check if moment and other forces are considered			
7	Check steel area			
8	Check minimum concrete cover area			
9	Check if openings are considered for services requirement			
10	Check for deflection			
11	Check if slope is required			
12	Check provision for floor drains for drainage, sanitary system			
13	Check if levels are shown			
14	Check expansion joints are shown			
15	Check authorities' requirements for fire protection			

e) Basement Stairs

Serial No.	Item	Yes	No	Action Required
1	Check that latest architectural drawing is used			
2	Check if axis and grids match with architectural layout			
3	Check location of stairs matches with architectural, and as per authorities requirements			
4	Check if width of stairs matches with architectural requirements			
5	Check riser height			
6	Check if landing platform is provided as per code			
7	Check reinforcement details for stairs			
8	Check if balustrade is required as per architectural requirements			

A.2.3 Superstructure

a) Superstructure (Ground Floor, Typical Floors)

Serial No.	Item	Yes	No	Action Required
1	Check as per Table 4.3.1.3 B. b), c), d), and e) for relevant items			

A.2.4 RoofSlab

Serial No.	Item	Yes	No	Action Required
1	Check as per Table 4.3.1.3 B. d) for relevant items			
2	Check for skylight, if any			
3	Check for parapet wall			
4	Check if equipment loads are considered			
5	Check for openings for services			
6	Check openings for hatches, lightwells, stairwells			
6	Check foundation for equipment			
7	Check for levels			
8	Check if any slope is required			

A.2.5 General Requirements

Serial No.	Item	Yes	No	Action Required
1	Check elevator shaft dimensions			
2	Check elevator shaft reinforcement			
3	Check opening for elevator door and its level			
4	Check openings in the slab for elevator works			
5	Check pit and supports for escalator (if required)			

A.3 Design Checklist—Fire Protection System and Plumbing Works

A.3.1 Fire Protection System

a) Sprinkler Layout				
Serial No.	Item	Yes	No	Action Required
1	Check that latest architectural drawing is used			
2	Check if legends match with layout			
3	Check whether sprinkler is route properly shown			
4	Check levels are shown			
5	Check if sprinkler system matches with authorities requirements			
6	Check if sprinkler location is coordinated with false ceiling plan and architectural requirements			
7	Check if sprinkler locations coordinated with other trades			
7	Check if required shafts are shown			
8	Check if piping layout is coordinated with other trades			
9	Check if piping is connected to the related equipment			

b) Pump Selection				
Serial No.	Item	Yes	No	Action Required
1	Check if fire pump size is as per NFPA/KFD requirements			
2	Check if normal and standby pumps are selected and shown			
3	Check if jockey pump is selected and shown			
4	Check if pumps are shown in the pump room			
5	Check if pumps are connected to piping system			
6	Check if pumps are provided with power and are coordinated with electrical trade for power supply requirement (both normal and emergency)			
7	Check if starter/control panel is provided			
8	Check if interface requirements are shown			
9	Check technical data sheet for selected fire pump			
10	Check if valves and other accessories are shown			
11	Check pump detail plans are shown			

A.3.2 Special Systems

Serial No.	Item	Yes	No	Action Required
1	Check for FM 200 system			
2	Check for foam system			
3	Check fire hydrant layout			
4	Check hose reel layout			

A.3.3 Plumbing System

a) Water Supply Layout				
Serial No.	Item	Yes	No	Action Required
1	Check that latest architectural drawing is used			
2	Check if legends match with layout			
3	Check water supply is meeting the usage requirements			
4	Check if cold water supply route is properly shown			
5	Check if levels are shown			
6	Check if pipe sizes are suitable for proper flow of water			
7	Check if hot water supply route is properly shown			
8	Check if water storage capacity is adequate to meet the required demand			
9	Check if required shafts are shown			
10	Check if piping layout is coordinated with other trades			
11	Check if piping is connected to the related equipment			
12	Check for water pressure requirements			
13	Check if all the fittings and accessories are considered			

b) Selection of Pumps and Plant Layout				
Serial No.	Item	Yes	No	Action Required
1	Check that latest architectural drawing is used			
2	Check if legends match with layout			
3	Check all plant room equipment are included			
4	Check calculation for selection of pumps			
5	Check if pipe connections are shown			
6	Check piping accessories are included and shown			
7	Check if necessary foundations for equipment are shown			
8	Check power supply for pumps is available			
9	Check if interface of equipment with BMS is provided			

A.3.4 Drainage System

a) Sanitary Drainage Layout

Serial No.	Item	Yes	No	Action Required
1	Check that latest architectural drawing is used			
2	Check if legends match with layout			
3	Check calculations for foul drainage demand and frequency factor for the following items: a) Wash basin b) Showers c) Urinals d) Water closets e) Kitchen sinks Other equipment such as dish washers, washing machines			
4	Check if inspection manhole/chamber is provided in the vicinity of the facility			
5	Check if vent for toilet is provided			
6	Check if pipe size and slope is adequate to ensure self-cleaning			
7	Check if floor drain is provided in every toilet and kitchen			
8	Check if oil/grease interceptors are provided (if applicable)			
9	Check sump pump size calculations			
10	Check sump pit dimensions calculations			
11	Check for relevant public authority approval for connection of drainage system to area network			

b) Storm Drainage Layout

Serial No.	Item	Yes	No	Action Required
1	Check that latest architectural drawing is used			
2	Check if legends match with layout			
3	Check if rainfall intensity matches with the recommendation by the authority			
4	Check if the no. of roof drains are enough to meet the storm water flow			
5	Check if storm water pipe size meets the requirements			
6	Check size of gullies and gutters			
7	Check necessary foundations for equipment shown			
8	Check for no. of manholes			
9	Check if connection to area network is shown			

A.4 Design Checklist—HVAC Work

A.4.1 HVAC Duct Layout

Serial No.	Item	Yes	No	Action Required
1	Check that latest architectural drawing is used			
2	Check if legends match with layout			
3	Check if ducting route is properly shown			
4	Check if levels are shown			
5	Check if air inlet/outlets are coordinated with false ceiling plan and architectural requirements			
6	Check if air inlet/outlets locations are coordinated with other services			
7	Check if required shafts are shown			
8	Check if duct layout is coordinated with other trades			
9	Check if duct layout is coordinated with false ceiling plans			
10	Check if all the duct accessories are shown			
11	Check if air inlet/outlets are connected to the duct			
12	Check if duct is connected to the related equipment			

A.4.2 HVAC Piping Layout

Serial No.	Item	Yes	No	Action Required
1	Check that latest architectural drawing is used			
2	Check if legends match with layout			
3	Check if piping route is properly shown			
4	Check if levels are shown			
5	Check if valve and other piping accessories are shown			
6	Check if required shafts are shown			
7	Check if piping layout is coordinated with other trades			
8	Check if piping layout is coordinated with false ceiling plans			

A.4.3 HVAC Equipment Selection

Serial No.	Item	Yes	No	Action Required
a) Check selection of following (applicable) equipment				
1	Chiller			
2	Cooling tower			
3	Chilled water/condensing water pumps			
4	Air handling units (AHU)/heat recovery unit (HRU)			
5	Fan coil units (FCU)			
6	DX mini split units			
7	Fans			
8	Heat exchanger			
9	Duct accessories a) VAV b) Sound attenuator			
10	Pipe accessories			
b) Check equipment schedule for (applicable) equipment as per requirements under a)				

A.4.4 HVAC Control Schematic

Serial No.	Item	Yes	No	Action Required
1	Check control schematic of chiller with pumps			
2	Check control schematic of cooling tower with pumps			
3	Check control schematic for chemical treatment for chilled water/condensing water pumps			
4	Check control schematic of air handling units (AHU)/heat recovery unit(HRU) with related accessories			
5	Check control schematic for VAV with accessories			
6	Check control schematic of smoke exhaust/fresh air fans system			
7	Check control schematic of heat exchanger			
8	Check control schematic of duct heater with accessories			
9	Check control schematic of BMS with accessories			

A.4.5 HVAC Building Management System

Serial No.	Item	Yes	No	Action Required
1	Check if all the devices/sensors/ detectors are connected in the loop			
2	Check if devices/sensors/detectors are shown			
3	Check if the location of DDC Panels is shown			
4	Check for interface with other services			
	Check if the following equipment/systems are included in the BMS point schedule: a) HVAC equipments b) Sump pumps c) Diesel generator d) Main low-tension panels e) Distribution boards, if required f) Other			

A.4.6 HVAC Sections and Plants

Serial No.	Item	Yes	No	Action Required
1	Check whether all related sections and details for equipment, duct, and piping are shown			

A.4.7 HVAC-Related Electrical Works

Serial No.	Item	Yes	No	Action Required
1	Check HVAC-related electrical works and applicable interface with other systems (BMS)			

A.5 Design Checklist—Electrical System

a) Lighting Layout				
Serial No.	Item	Yes	No	Action Required
1	Check that latest architectural drawing is used			
2	Check if legends match with layout			
3	Check if summary of luminaire includes all the fixtures shown on the drawing			
4	Check whether all the areas are provided with lighting			
5	Verify if illumination calculation/isolux diagram requirements match with fixtures shown on the lighting layout			
6	Check if specified fittings match with reflected ceiling plan/architectural requirements			
7	Check if circuiting references for all fixtures (both normal and emergency) are shown			
8	Check if switch arrangements and location of switches are shown			
9	Check if push buttons, dimmer, and LCP (if any) are located			
10	Check if exit lights are shown			
11	Check if emergency lights/E-symbol are shown			
12	Check if lights are provided in elevator shaft			
13	Check whether wire sizes for circuits are correct			
14	Verify voltage drop for circuit wiring			
15	Check if consistency is maintained to indicate type of light fixtures for all the circuits			
16	Check if all DBs are located and demarcated			
17	Check all dimmer panels, if any, are located			
18	Check if lighting control panel, if any, is located			
19	Check if the circuiting references match with DB schedules			
20	Check that the no. of fittings in a circuit does not exceed the MEW specified limit			
21	Check DB schedule for contactor, switches, push button, and interface with BMS (if any)			
22	Check for DB schedule load balancing			

b) Power				
Serial No.	Item	Yes	No	Action Required
1	Check that latest architectural drawing is used			
2	Check if legends match with layout			
3	Check if outlets are provided for general in rooms/general areas and locations are coordinated			
4	Check if outlets are provided in kitchen			
5	Check if outlets/isolators are provided in kitchen for equipment/appliances as per kitchen equipment layout			
6	Check if ring circuiting is as per codes, regulations			
7	Check for location of socket outlets/isolators for PH equipment and if the ratings are suitable for connected load			
8	Check for location of socket outlets/isolators for firefighting equipment and if the ratings are suitable for connected load			
9	Check for location of socket outlets/isolators for HVAC equipment and if the ratings are suitable for connected load			
10	Check for location of socket outlets/isolators for low-voltage equipment and if the ratings are suitable for connected load			
11	Check for location of isolators for conveying system equipment and if the ratings are suitable for connected load			
12	Check if circuit references are correct as per DB schedule			
13	Check if wire/cable sizes are correct as per voltage drop calculation			
14	Check if socket outlet is provided in elevator pit			
15	Check if all EDB/DBs are located and demarcated			
16	Verify voltage drop for circuit wiring			
17	Check if circuit references are correct as per SMSB schedule/schematics			
18	Check if all ESMSB/SMSBs are located and demarcated			
19	Check if all EMSB/MSBs are located and demarcated			
20	Check if breaker sizes of EDB/DBs are correct			
21	Check if breaker sizes of ESMSB/SMSB are correct			

(Continued)

b) Power				
Serial No.	Item	Yes	No	Action Required
22	Check if breaker sizes of EMSB/MSB are correct			
23	Check if emergency power is proved for all equipment to be operated under fire mode/emergency			
24	Check if emergency power is proved for all security equipment			
25	Check if cable tray sizes are as per codes, regulations			
26	Check if cable tray routes are shown			
27	Check if risers are shown			
28	Check if switch board schematic matches with load schedule			
29	Check if the breakers' rating shown on schematic is suitable for connected load			
30	Check if the bus bar rating of schematic is 1.25 times of incomer breaker size			
31	Check if ELR is shown for spare breaker			
32	Check if AV Alarm is shown for fire alarm panel circuit			
33	Check if AV alarm is shown for sanitary and storm water pump circuit			
34	Check if AV alarm is shown for smoke fan and fresh air supply fan circuits			
35	Verify voltage drop calculation for outgoing feeders			
36	Check if equipment/isolators are identified for all outgoing circuits of schematics			
37	Check if earthing is shown in the schematics for all equipment and isolators			
38	Check schematic for MCC panel, type of starters, size and no. of wires, cable size and verify if the breaker sizes are suitable for connected load			
39	Check if the size of downstream breakers of LT panel schematic are suitable for connected load			
40	Check for shunt trip, if any			
41	Check LT panel schematic with load schedule			
42	Check if ACB size, bus coupler (if any), and bus bar rating of LT panel schematic are as per codes, regulations			
43	Check if all metering is shown on the LT schematic			

(Continued)

b) Power				
Serial No.	**Item**	**Yes**	**No**	**Action Required**
44	Check if ATS (if any) is shown and connected to the LT panel/MSB			
45	Check D.G. set (if required) and its panel is located			
46	Check no. of transformer and ratings			
47	Check substation layout			
48	Check earthing/grounding system			
49	Check lightning protection system			

A.6 Design Checklist—Landscape

Serial No.	**Item**	**Yes**	**No**	**Action Required**
1	Check property limits			
2	Check landscape settings and levels			
3	Check water percolation in landscape area			
3	Check grass area and edges			
4	Check location/demarcations of plants, shrubs, trees			
5	Check irrigation system, water distribution to plants, water reservoirs, and pumping system			
6	Check location of special features			
7	Check location of sidewalk, pedestrian paths			
8	Check pavement, curb stones details			
9	Check lighting bollard, light poles and signage			
10	Check utility services both under- and aboveground			
11	Check material storage and workshop facility for landscape maintenance			

A.7 Design Checklist—Infrastructure

Serial No.	Item	Yes	No	Action Required
1	Check grading details			
2	Check asphalt/paving area			
3	Check asphalt/paving details			
4	Check pavement, curb stone area and details			
5	Check location of light poles			
6	Check utility services routes			
7	Check trenches and tunnels for services			
8	Check location of manholes and handholes			
9	Check traffic signs and markings			
10	Check security system requirements			

Appendix B: Major Activities during the Construction Process in Building Construction Project

Sr. No.	Description
Shoring	
1	Shoring
2	Dewatering
Earth Works	
1	Excavation
2	Backfilling
Concrete Substructure	
1	Blinding concrete
2	Termite control
2	Reinforcement for raft
3	Casting of raft foundation
4	Reinforcement for grade beams
5	Shuttering for grade beams
6	Concrete casting of grade beams
7	Reinforcement for footings
8	Shuttering for footings
9	Concrete casting of footings
10	Reinforcement for wall and columns
11	Shuttering for wall and columns
12	Concrete casting of retaining wall
Concrete Superstructure	
1	Reinforcement for columns
2	Shuttering for columns
3	Concrete casting of columns
4	Formwork for slab
5	Reinforcement for slab
6	Shuttering for beams
7	Reinforcement for beams
8	Concrete casting of slab
9	Shuttering for stairs
10	Reinforcement for stairs
11	Concrete casting of stairs
12	Shuttering for wall and columns

(Continued)

Sr. No.	Description
13	Reinforcement for wall and columns
14	Concrete casting of wall and columns
15	Precast panels
Masonry	
1	Block work
2	Concrete unit masonry
Partitioning	
1	Installation of frames
2	Installation of panels
Metal Work	
1	Structural steel work
2	Installation of cat ladders
3	Installation of balustrade
4	Installation of space frame
5	Installation of handrails and railings
Thermal and Moisture Protection	
1	Applying waterproofing membrane
2	Installation of insulation
Doors and Windows	
1	Aluminum windows
2	Aluminum doors
3	Glazing
4	Steel doors
5	Steel windows
6	Wooden frames
7	Wooden doors
8	Curtain wall
9	Hatch doors
10	Revolving door
11	Louvers
Internal Finishes	
1	Plastering
2	Painting
3	Cladding—ceramic
4	Cladding—gypsum
5	Cladding—marble
6	Ceramic tiling
7	Stone flooring
8	Acoustic ceiling
9	Gypsum board ceiling
10	Glass reinforce gypsum ceiling

(Continued)

Sr. No.	Description
11	Metal tile ceiling
12	Demountable partitions
Furnishing	
1	Carpeting
2	Blinds
3	Furniture
External Finishes	
1	Painting
2	Brickwork
3	Stone
4	Cladding—aluminum
5	Cladding—granite
6	Curtain wall
7	Glazing
Equipment	
1	Installation of maintenance equipment
2	Testing of maintenance equipment
3	Installation of kitchen equipment
4	Testing of kitchen equipment
5	Installation of parking control
6	Testing of parking control system
Roof	
1	Parapet wall
2	Thermal insulation
3	Waterproofing
4	Roof tiles
5	Installation of drains
6	Installation of gutters
Elevator System	
1	Installation of raceways
2	Installation of rails
3	Installation of door frames
4	Installation of cabin
5	Cabin finishes
6	Installation of wire rope
7	Installation of drive machine
8	Installation of controller
9	Testing of elevator
10	Third-party inspection
Mechanical Works	
A. Fire Suppression System	
1	Installation of piping
2	Installation of valves

(*Continued*)

Sr. No.	Description
3	Installation of flow switches
4	Installation of sprinklers
5	Installation of foam system
6	Installation of pumps
7	Installation of hose reels/cabinet
9	Electrical works/termination of pumps
10	Installation of fire hydrant
11	External fire hose cabinet
12	Testing of firefighting system
B. Water Supply	
1	Installing of piping
2	Installation of valves
3	Installation of pumps
4	Electrical works/termination of pumps
5	Filter units
6	Toilet accessories
	(a) Wash basin
	(b) Water closet
	(c) Urinals
	(d) Valves
	(e) Mixers
	(f) Kitchen sinks
7	Installation of insulation
8	Water heaters
9	Hand dryers
10	Water tank
11	Electrical works/termination of pumps, heaters
12	Controls—level indicator
13	Flushing of system
14	Testing of water supply system
C. Drainage System	
1	Installation of pipes below grade
2	Installation of pipes above grade
3	Installation of manholes
4	Installation of clean out
5	Installation of floor drain
6	Installation of sump pumps
7	Electrical works/termination of pumps
8	Installation of gratings
9	Testing of sump pumps

(Continued)

Sr. No.	Description
D. Irrigation System	
1	Installation of pipes
2	Installation of valves
3	Installation of pumps
4	Installation of controls
5	Installation/connection with weather station
6	Installation of central control system
7	Testing of irrigation system
HVAC Works	
1	Installation of piping
	(a) Chilled water
	(b) Condensing water
	(c) Hydrostatic test
2	Installation of ducting
	(a) Air-conditioning duct
	(b) Exhaust duct
3	Installation of insulation
	(a) Pipe
	(b) Duct
4	Installation of dampers
5	Installation of grills and diffusers
6	Installation of cladding
7	Installation of fans
	(a) Exhaust
	(b) Fresh air
	(c) Ventilation
	(d) Pressurization
8	Installation of fan coil units (FCU)
9	Installation of air handling units (AHU)
10	Installation of pumps
11	Installation of chillers
12	Installation of cooling towers
13	Installation of thermostat and controls
14	Installation of starters
15	Electrical works/termination of pumps
16	Electrical works/termination of AHU, FCU
17	Electrical connection/termination of chillers, cooling towers
18	HVAC water treatment
19	Start-up of pump
20	Flushing of piping system
21	Start-up of chillers
22	Start-up of cooling towers

(Continued)

Sr. No.	Description
23	Start-up of AHUs
24	Start-up of fan coil unit
25	Building management system (BMS)
	(a) Conduiting
	(b) Wiring–cabling
	(c) Components
	(d) Termination
26	Testing of HVAC system
27	Testing of BMS
28	Testing adjusting and balancing
Electrical Works	
1	Conduiting–raceways
2	Cable tray–trunking
3	Floor boxes
4	Wire pulling
5	Cable pulling
6	Installation of bus duct
7	Meggering of wires and cables
8	Installation of wiring devices/accessories
9	Installation of light fittings
10	Installation of earthing tape
11	Earth pits
12	Distribution boards (DB) installation
13	DBs termination
14	Installation of lighting control panel
15	Testing of lighting control panel
16	Installation of isolators
17	Installation of submain switch boards (SMSB)
18	Installation of MSBs
19	Installation of motor control center (MCC) panels
20	Termination of SMSBs
21	Termination of MSBs
22	Testing of MCC panels
23	Installation of automatic transfer switch (ATS)
24	Termination of ATS
25	Testing of DBs
26	Testing of SMSBs
27	Testing of MSB
28	Installation of main low-tension boards (MLTBs)
29	Installation of generator
30	Testing of generator
31	Testing of ATS

(Continued)

Sr. No.	Description
32	Testing of MLTBs
33	Installation of central battery system
34	Testing of central battery system
Fire Alarm System	
1	Conduiting–raceways
2	Installation of cable trunking
3	Wiring–cabling
4	Installation of detectors, bells, pull stations, interface modules
5	Installation of repeater panel
6	Installation of mimic panel
7	Installation of fire alarm panel
8	Interface with other equipment
9	Testing of fire alarm system
Telephone/Communication System	
1	Conduiting–raceways
2	Installation of cable trunking–cable tray
3	Wiring–cabling
4	Termination of cables
5	Installation of wiring devices
6	Installation of racks
7	Termination of cable at the racks
8	Testing of structured cabling
9	Installation of switches
10	Installation of telephone sets
11	Testing of system
Public Address System	
1	Conduiting–raceways
2	Cable tray–trunking
3	Wiring–cabling
5	Installation of speakers
6	Installation of racks
7	Installation of equipment
8	Testing of public address system
Audio Visual System	
1	Conduiting–raceways
2	Cable tray–trunking
3	Wiring–cabling
5	Installation of speakers
6	Installation of monitors/screens
7	Installation of racks

(*Continued*)

Sr. No.	Description
8	Installation of equipment
9	Testing of audio visual system
CCTV/Security System	
1	Conduiting–raceways
2	Wiring–cabling
3	Installation of cameras
4	Termination of cameras
5	Installation of panels
6	Termination of panels
7	Installation monitors/screens
8	Testing of security system
Access Control	
1	Conduiting–raceways
2	Wiring–cabling
3	Installation of RFID proximity readers, fingerprint readers
4	Installation of magnetic locks, release buttons, door contacts
5	Installation of panel
6	Installation of server
7	Testing of system
Systems Integration	
1	Installation of switches
2	Installation of servers
3	Systems integration programming
Landscape	
1	Site clearance
2	Excavation
3	Soil preparation
4	Plantation
5	Pavement and sidewalk
6	Special features (if any)
7	Irrigation system
8	Lighting poles/bollards
External Works (Infrastructure)	
1	Site clearance
2	Cutting and filling works
3	Grading
4	Asphalt works
5	Light poles
6	Water supply piping works
7	Firefighting works

(Continued)

Sr. No.	Description
8	Storm water piping works
9	Drainage works
10	Curb stones and pavement works
11	Hard landscape works
12	Road marking works
13	Traffic signs and signals

Appendix C: Content of Contractor's Quality Control Plan

The contractor's quality control plan is prepared based on project-specific requirements as specified in the contract documents.

The plan outlines the procedures to be followed during the construction period to attain the specified quality objectives of the project fully complying with the contractual and regulatory requirements.

The following is an outline of such a plan, based on contract documents.

C.1 Introduction

The contractor's quality control plan is developed to meet contractor's quality control (CQP) requirements of (project name) as specified under clause (—) and section (—) of contract documents.

The plan provides the mechanism to achieve the specified quality by identifying the procedures, control, instructions, and tests required during the construction process to meet the owner's objectives. This QCP does not endeavor to repeat or summarize contract requirements. It describes the process M/S ABC (contractor name) will use to assure compliance with those requirements.

C.2 Description of Project

(Owner name) has contracted M/S ABC to construct (name of facility) located at (site plan). The facility consists of approximately (—) m² gross building area and approximately (—) m² of two basements for car parking and other services.

The facility is to be used as office premises to accommodate (—) personnel. The architecture of the building is of very high quality with a spacious atrium between two towers interconnected with a bridge inside the building, and a well-designed internal landscape area with plants and sky garden to provide a pleasant view inside the building. The building has glazed walls and curtain walling for the atrium area.

Vertical transportation is through panoramic elevators and a designated elevator for VIPs. Additionally, there is a goods elevator and firefighters' elevator to meet an emergency situation.

The structure of the building is of reinforced concrete, and the atrium is of structured steel. External cladding includes a spandrel curtain wall for the block structure and special curtain walling for the atrium. Interior finishes are painted plastered walls and marble and stone finish. Internal partitions have flexibility to adjust the office area. The entire building has a raised floor system, and space under the raised floor is used for distribution of electromechanical services.

The HVAC system is comprised of water-cooled chillers, and electric ducts heaters are provided for heating during winter. The building management system takes full control of HVAC. Building is provided with all the safety measures against fire. A fire protection system, smoke exhaust fans, and fire alarm system with voice evacuation system are provided to meet an emergency situation. The fire protection system also includes automatic sprinklers, hose reels, extinguishers, and a foam system.

The building is equipped with all the amenities required for a public health system. The plumbing system is consistent with the requirement of the building and includes cold and hot water, drainage, rain water collection, and an irrigation system. Water fountains are provided at various locations for drinking purposes.

Electrical systems consist of energy-saving lighting with centralized controls. Electrical distribution has all required safety features. A diesel generator system and uninterrupted power supply system is provided to meet the emergency due to power failure from the main electricity provider. The building to be constructed shall be equipped with the latest technological systems. It shall have an IP-based communication system, and all the low-voltage systems shall be fully integrated. Authorized persons shall have access from anywhere, either from within the building or outside.

The building has fully integrated low-voltage systems.

Apart from training and conference facilities, a fully functional auditorium having capacity for (—) people and conferencing is available in the building. It has a sophisticated conferencing system with a rear projection screen.

The contract documents consist of total (—) contract drawings. These are:

Architectural	(—) nos.
Architectural interior	(—) nos.
Structural	(—) nos.
Firefighting	(—) nos.
Public health	(—) nos.
HVAC	(—) nos.
Electrical	(—) nos.
Smart building	(—) nos.
Traffic signage	(—) nos.
Landscape	(—) nos.

C.3 Quality Control Organization

The quality control (QC) organization is independent of those persons actually performing the work. They shall be responsible for implementing the quality plan for the entire contract/project-related activities by scheduling the inspection, testing, sampling, and preparation of mock-up and to assure that works are performed as per approved shop drawings and contract documents.

An organization chart showing the line of authority and functional relationship is shown in Figure C.1.

The quality control incharge shall be responsible to ensure implementation of project quality. He or she will be supported by quality control engineers as follows:

1. Quality control engineer (civil works)
2. Quality control engineer (concrete works)
3. Quality control engineer (mechanical works)
4. Quality control engineer (electrical works)
5. Foreman (concrete works)

These engineers will be responsible to implement three phases of the quality system at site. The quality control incharge will coordinate with the company's head office for all the support and necessary actions. Respective quality control engineers shall be responsible to implement the quality program in their respective fields.

C.4 Qualification of QC Staff

All the QC personnel will have adequate experience in their respective fields. The following is the qualification of each of the QC staff. Additional staff shall be provided if required.

Quality control incharge: He or she shall be a qualified civil engineer having minimum 10 years of experience as a quality control engineer in major construction projects.

Quality control engineer (civil works): Graduate civil engineer with minimum 3 years' experience as a quality engineer on similar projects.

Quality control engineer (concrete works): Graduate civil engineer with minimum 3 years' experience as a quality engineer on similar projects.

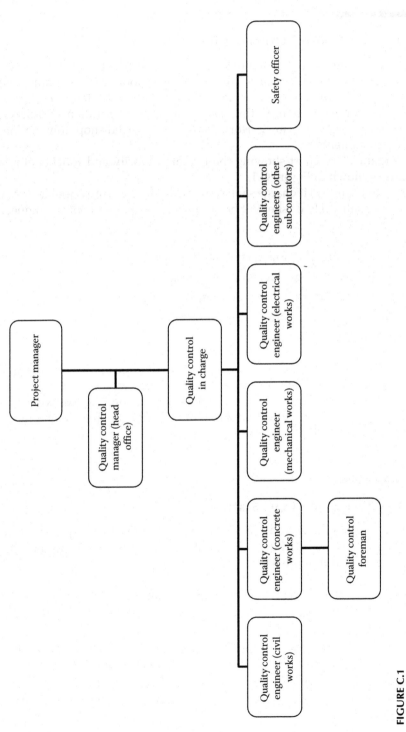

FIGURE C.1
Site quality control organization.

Quality control engineer (mechanical works): Graduate mechanical engineer with minimum 3 years' experience as a quality engineer on similar projects.

Quality control engineer (electrical works): Graduate electrical engineer with minimum 3 years' experience as a quality engineer on similar projects.

Foreman (concrete works): Diploma in civil engineering with minimum 5 years' experience as a quality supervisor on similar projects.

C.5 Responsibilities of QC Personnel

Project manager: Overall project responsibilities.

Quality control incharge: He/she shall be responsible for the following:
- Preparation of QC plan
- Responsible for overall QC/QA responsibilities
- Responsible for monitoring and evaluation CQP
- Implementation of QC plan
- Responsible for implementing QC procedure
- Maintain QC records and documents
- Inspection of works
- Responsible for off-site inspection
- Inspection of incoming material
- Responsible for subcontractor's QC plan
- Coordinating with safety officer to implement safety plan
- Calibration of measuring instruments
- Monitoring equipment certification

Quality control engineers: Each trade shall have a responsible quality control engineer to ensure that works are carried out in accordance with the contract documents. The following quality control engineers shall be available at site. A foreman will assist the quality control engineer to maintain the quality of concrete works.

1. Quality control engineer (civil works)
2. Quality control engineer (concrete works)
3. Quality control engineer (mechanical works)
4. Quality control engineer (electrical works)
5. Quality control engineer (other subcontractors)
6. Foreman (concrete works)

These personnel will be responsible for the following:

- Overall quality of the respective trade
- Preparation of method statement
- Preparation of mock-up
- Monitor and inspect quality of site works
- Inspection of incoming material
- Coordinate with the quality incharge for preparation of CQP
- QC records and documents

C.6 Procedure for Submittals

M/S ABC shall be responsible for timely submission of subcontractors, shop drawings, and materials to be used in the project.

C.6.1 Submittals of Subcontractor(s)

Prior to submission of the subcontractor's name to the owner/consultant, M/S ABC shall review the capabilities of the subcontractor to perform the contracted works properly per the specified quality and within the time allowed. M/S ABC shall also consider the contractor's performance in the previously executed project. Companies implementing a quality management system plan and having ISO certification shall be given preference while selecting the subcontractor(s) to work on this project.

C.6.2 Submittals of Shop Drawings

Before the start of any activity, M/S ABC shall submit shop drawings based on contract drawings and other related documents. The number of shop drawings shall depend on the requisite details to be prepared against each contract drawing. The shop drawings shall be prepared on the agreed-upon format and shall have coordination certification from all the trades. All the shop drawings shall be submitted through site transmittal for shop drawings. Shop drawing approval transmittals shall be numbered serially for each section of the particular specification. M/S ABC shall be responsible for shop drawings related to subcontracted works.

C.6.3 Submittals of Materials

M/S ABC shall submit materials, products, and systems for owner's/consultant's approval from the recommended/nominated manufacturers specified in the contract documents. If material is proposed from a manufacturer who is not nominated, then all the documents to justify that the proposed product is approved and equal to the specified shall be submitted along with the site transmitted for materials approval.

C.6.4 Modification Request

If during construction, the contractor observes that some work as mentioned in the contract needs modification, then such a modification request shall be submitted to the owner/consultant for their review and approval. Financial obligations shall be clearly specified in the request form.

C.6.5 Construction Program

M/S ABC shall prepare and submit a construction program based on the contracted time schedule and submit the same to the consultant for approval. Progress of works shall be monitored on a regular basis and the construction program shall be updated as and when required to overcome the delay (if any). The following reports shall be submitted to the consultant as per the following schedule:

1. Daily report on a daily basis
2. Weekly report on a weekly basis
3. Monthly report on a monthly basis
4. Progress photographs on a monthly basis

C.7 Quality Control Procedure

C.7.1 Procurement

Prior to placing a purchase order or supply contract with the supplier/vendor, ABC will ensure that the material has been approved and it complies with the requirements of the contract. A procurement log shall be maintained for all the items for which action has been taken. The procurement log shall be submitted to the owner/consultant every 2 weeks for review and information.

C.7.2 Inspection of Site Activities (Checklists)

All the works at site shall be performed as per approved shop drawings by using approved material. At least three phases of control shall be conducted by the quality control incharge for each activity prior to submission of the checklist to the consultant.

These will be

- Preparatory phase
- Start-up phase
- Execution phase

Preparatory phase: The concerned engineer shall review the applicable specifications, references and standards, approved shop drawings and materials, and other submittals (method of statements, etc.) to ensure compliance with the contract.

Start-up phase: The concerned engineer shall discuss with the foreman responsible to perform the work and shall establish the standards of workmanship.

Execution phase: The work shall be executed by continuous inspection during this phase.

All the execution/installation works shall be carried out as per approved shop drawings utilizing approved material. All the works will be checked for quality before submitting the checklist to the consultant to inspect the works.

The following main categories of site works shall be performed by conducting the QC as per the phases mentioned earlier.

C.7.2.1 Definable Feature of Work

Procedure for scheduling, reviewing, certifying, and managing QC for a definable feature of work shall be agreed upon during the coordination meetings. The definable features of work for concrete shall include formwork, reinforcement, and embedded items, design mix, placement of concrete, concrete finishes, etc.

Likewise, the detail list for other trades shall be prepared during the coordination meetings. This will consist of embedded conducting works, embedded sleeves, ducts, underground piping, etc.

C.7.2.2 Earthworks and Site Works

This category shall consist of the following subgroups:

1. Excavation and backfilling
2. Compaction
3. Dewatering

A checklist shall be submitted to the consultant at every stage of work, that is, after backfilling and compaction. Samples shall be taken to make a compaction test. If the compaction test fails the required specifications, then remedial action shall be taken to obtain specified results.

C.7.2.3 Concrete

M/S (approved subcontractor to supply ready-mix concrete) will provide ready-mix concrete for entire concrete structured works of the project. M/S (ready-mix subcontractor) will have their own QC system to maintain the mix design.

Reinforcement steel shall be inspected upon receipt of material at site for proper size and type, and the factory test certificates received with the supply. Small pieces of sample shall be taken from each lot and size for laboratory testing. The testing shall be performed by an approved testing laboratory.

Formwork and the reinforcement steel process shall be performed under qualified civil engineer(s) and foreman (foremen). They will be continuously inspecting the work for each operation and check list shall be submitted to the consultant for their approval.

Placement of concrete shall be supervised by the civil engineer along with the foreman and other team members. The respective consultant will witness concrete casting and take concrete samples during casting for laboratory testing and crushing tests. Slump tests shall be performed on the concrete to verify the strength. Curing shall be supervised by the respective foreman.

In case the test results do not comply with the specification limit, the results shall be discussed with the consulting engineer. Failing tests will be followed by appropriate corrective (reworking) efforts, and retesting and remedial measures.

M/S ABC shall maintain all the records, laboratory tests, and results regarding the concrete works and submit the same to the consultants for their information.

C.7.2.4 Masonry

The concrete block bricks used on the project shall be procured from the approved source. Material for mortar aggregate and other accessories such as inserts, reinforcement material, and wire mesh shall be submitted to the consultant for approval.

Prior to the start of masonry work, all the related and coordinated drawings shall be reviewed. Marking shall be made for masonry layout. Necessary mock-ups shall be submitted for each type of unit masonry work. A checklist shall be submitted for visual and other type of inspection.

C.7.2.5 Metal Works

Metal fabrication and installation work shall be carried out by certified welders, as per manufacturer's recommendations and approved shop drawings.

Samples of fittings, brackets, fasteners, anchors, and different types of metal members including plates, bars, pipes, tubes, and any other type of material to be used in the project shall be submitted to the consultants prior to start of fabrication works at site. Finishing material such as primer and paint shall be submitted for approval to the consultant/owner.

In case the fabrication is carried out at the subcontractor(s)'s workshop, then ABC shall take full responsibility to control the quality.

A checklist shall be submitted at different stages of work to ensure compliance with the specifications and to avoid any rework at the end of completion of fabrication and installation.

C.7.2.6 Wood Plastics and Composite

Prior to the start of any woodwork, the material shall be physically and visually inspected for the timber quality to confirm its compliance with the type of the approved material and also to ensure that the wood is free from decay and insects, and that necessary treatment has already been done on the wood to be used for the project. A product certificate signed by the woodwork manufacturer certifying that the products comply with specified requirements shall be submitted along with material submittal. All types of fasteners and other hardware to be used shall be submitted for approval as per applicable standards, and checklists shall be submitted at various stages of work prior to applying the paint or any other finishes.

C.7.2.7 Doors and Windows

This category shall consist of the following subcategories:

1. Wooden doors and windows
2. Steel doors
3. Aluminum windows
4. Glazing

All doors and windows shall be fabricated from a specified type and material and as per approved shop drawings.

Fire-rated types of doors shall comply with relevant standards and local regulations.

Samples of materials used in fabrication of doors and windows shall be submitted to the consultant/owner for approval.

Wooden doors and windows shall be fabricated at an approved subcontractor for carpentry works and aluminum doors and windows shall be fabricated at an approved subcontractor for aluminum works. Finishes shall be as approved by the owner/consultant. Samples of ironmongery coating material and finishing material shall be submitted for approval. Special care shall be taken to maintain the acoustic nature of doors and their sound transmission properties.

Glass and glazing material shall be submitted for approval and shall take care of heat transfer properties to maintain the inside temperature of the building. The glass used in the project shall comply with specified strength, safety, and impact performance requirements.

Inspection at different stages shall be arranged by field inspection at the factory.

The entire fabrication shall be carried out under supervision of ABC to control the quality of finished products.

Doors and windows fabrication shall be coordinated for security requirements. Mock-ups shall be prepared before start of installation work.

C.7.2.8 Finishes

This category shall consist of the following subgroups:

1. Acoustic ceiling
2. Specialty ceiling
3. Masonry flooring
4. Tiling
5. Carpet
6. Wall cladding
7. Wall partitioning
8. Paints and coating

All the finishes shall be as specified and approved. Execution of finishing work shall be coordinated with all other trades. Samples and mock-ups shall be submitted for approval before start of any finishing work. All the finishing work shall be performed by skilled workmen. Specialist subcontractor(s) shall be submitted for approval to carryout finishing works. QC shall be under direct supervision of ABC.

C.7.2.9 Landscape

Landscape work shall be executed by a specialist-approved subcontractor as per the approved shop drawings under direct supervision of ABC. Samples of each and every type of plant, tree, etc., shall be submitted for approval. Soil preparation shall be done per specified standards.

C.7.2.10 Furnishing

All the furnishing shall be from the specified recommended manufacturer. Special care shall be taken to protect the furnishings. Furnishing material sample(s) shall be submitted for approval. Fabrication of furniture shall be at the approved subcontractor's factory. ABC shall be responsible for controlling the quality.

C.7.2.11 Equipment

This category shall consist of the following subgroups:

1. Maintenance equipment such as window and facade cleaning
2. Stage equipment
3. Parking control equipment
4. Kitchen equipment

All the equipment shall be from specified manufacturers. If the items specified are commercial items—that is, the materials are manufactured and sold to the public as against the materials made to specifications—necessary precaution shall be taken to verify that materials shall be installed per manufacturer's recommended procedure under the supervision of a qualified engineer.

Items to be manufactured per contract documents shall be procured from the manufacturer producing products complying with recognized quality standards. The product and manufacturer's data shall be submitted for approval with all the technical details. Installation of the equipment shall be carried out by skilled workmen under supervision of the manufacturer's authorized representative and under the control of ABC.

C.7.2.12 Conveying System

This category consists of the following subgroups:

1. Passenger elevators
2. Goods elevators
3. Kitchen elevators
4. Escalators

Prior to submission of shop drawings, full technical details along with the catalogs and compliance statement shall be submitted for approval. The elevators shall be from one of the recommended manufacturers specified in the contract documents.

Fabrication of conveying systems shall comply with relevant codes and standards as applicable. Regulatory approval shall be obtained from local authorities for installation of the conveying systems to assure their compliance with local codes and regulations.

Finishing material for the cab, landing door, car control station, and car position indicator shall be submitted for approval.

Coordinated factory shop drawings shall be submitted for approval. Size of shaft and door openings shall be coordinated during structural work. Necessary tests shall be performed by the manufacturer's authorized personnel.

Factory fabrication works shall be performed as per the manufacturer's QC plan, and installation at site shall be carried out by personnel authorized by the manufacturer.

Third-party inspection shall be arranged as per specification requirements.

C.7.2.13 Mechanical Works

This category shall consist of the following subgroups:

1. Fire suppression
2. Water supply, plumbing, and public health
3. HVAC

Mechanical works shall be executed by approved mechanical subcontractor. All the works shall be carried out as per approved shop drawings using approved material.

A mechanical engineer from the subcontractor shall be responsible for controlling the quality of work. A quality control engineer (mechanical works) shall coordinate all quality-related activities on behalf of ABC. Fire suppression or firefighting equipment shall be carried out as per approved drawings by the relevant authority. Materials shall be from the manufacturers approved by the relevant authority having jurisdiction over such materials. Fire pumps and firefighting materials shall comply with regulatory requirements.

Piping for water supply, plumbing, and public health shall be installed with approved material. Leakage and pressure tests shall be performed as specified.

Checklists shall be submitted after installation of piping, accessories, and fixing of equipment. Drainage systems shall be executed as per approved shop drawings.

HVAC works shall be supervised by a qualified mechanical engineer. Ductwork and duct accessories shall be fabricated at the subcontractor's workshop and shall be installed at site as per approved shop drawings. A mechanical engineer from subcontractor shall be responsible to control the quality at the workshop. A quality control engineer (mechanical works) shall coordinate all quality-related activities on behalf of ABC. Duct material shall be submitted for approval prior to start of fabrication works. Chilled/hot water piping shall be from the approved manufacturer. Installation of piping shall be carried out by certified pipe fitters.

Selection of HVAC pumps shall be done with the help of performance curves and technical data from the pump manufacturer. Chillers shall be from the specified manufacturer.

Installation of chillers shall be carried out as per manufacturer's recommendation.

C.7.2.14 Automation System

Automation work shall be carried out under supervision of a specialist engineer. The system shall be from the specified manufacturer.

Shop drawings and schematic diagrams shall be submitted for approval configuring approved components/items/equipment.

C.7.2.15 Electrical Works

This category shall consist of the following subgroups:

1. Electrical lighting and power
2. Fire alarm system
3. Communication system (telephone)
4. Public address system
5. Access control and security system
6. Audiovisual system
7. MATV system
8. Emergency generating system
9. Alternate energy system

All types of electrical works shall be executed by an approved electrical subcontractor. All the materials/products to be used shall be from the specified manufacturer and shall be installed by skilled workmen under the supervision of a qualified engineer.

An electrical engineer from the subcontractor shall be responsible to control the quality at the workshop. A quality control engineer (electrical works) shall coordinate all quality-related activities on behalf of ABC. Work shall be carried out as per approved shop drawings. Quality of the works shall be controlled at different stages such as after installation of conducting/raceways, pulling of wires/cables, and installation of accessories/equipment/panels. Works shall comply with the specifications and recognized codes and standards.

Specified tests shall be performed at every stage. Coordination will be done with other trades to assure that required power supply for their equipment will be available at designated locations.

Checklists shall be submitted after completion of activities at each stage.

Work not accepted or rejected shall be repaired/reworked or replaced and checklist shall be resubmitted.

Guarantees and warrantees for installed products/items shall be submitted as per contract requirements during handover of the project.

C.7.3 Inspection and Testing Procedure for Systems

The QC procedure for systems (low voltage/low current) and their equipment/components shall be performed in the following stages.

The proposed manufacturer's complete data giving full details of the capability of the proposed manufacturer to substantiate compliance with the specifications and to prove that the proposed manufacturer is following quality management system and producing the product to recognized quality standards as specified shall be submitted. Components/equipment data shall be submitted for all the specified items of the system.

A schematic diagram configuring all the items for the system shall be submitted along with a detailed technical specification and confirmation from the manufacturer that it complies with the contract documents and is equal or better than the specified products.

Shop drawing(s) fully coordinating with other trades, showing the system location, and the raceways used for installation of the system shall be submitted.

All the work at site shall be submitted for inspection at different stages of execution of work.

Cable testing shall be performed prior to installation of equipment. Checklists shall be submitted to the consultant to witness these tests.

Installation of equipment shall be done as per manufacturer's recommendations and approved installation methods.

Checklists along with testing procedures shall be submitted prior to inspect final inspection and testing.

C.7.4 Off-Site Manufacturing, Inspection, and Testing

QC procedures for off-site manufacturing/assembling shall be as per contract documents. Factory control inspection for items fabricated or assembled shall be carried out to comply with all the specified requirements and standards under the responsibility of QC personnel at the fabrication/assembly plant. Specified tests shall be carried out as per the manufacturer's quality procedures and specification requirements. Test reports for items fabricated/assembled off-site shall be submitted along with the delivery of material at site. Witnessing/observation of inspection and tests by the consultant/client, if it is required as per specifications, shall be arranged.

Procedure for such tests and inspection shall be submitted in advance of factory/plant visit.

Quality control for off-the-shelf items shall be standard quality control practices followed by the manufacturer.

C.7.5 Procedure for Laboratory Testing of Materials

M/S ABC shall submit name(s) of testing laboratories for approval for performing tests specified in the contract documents.

C.7.6 Inspection of Material Received at Site

All the material received at site shall be submitted for inspection by the consultant. A material on-site inspection request (MIR) shall be submitted to the consultant. Prior to dispatch of material from the manufacturer's premises, all the tests shall be carried out, and test certificates shall be submitted along with the MIR. Apart from this, all the relevant documents such as the packing list, delivery note, certificate of compliance to recognized standards, country of origin, and any other documents shall be submitted along with the MIR.

Material received at the site shall be properly stored at designated locations and shall maintain original packing until its installation/use.

Inventory records shall be maintained for all the material stored at the site. In case of materials to be directly installed at the site upon its receipt, the inspection shall be carried out before its use or installation on the project.

In case of concrete ready-mix, samples will be tested as per contract documents before pouring the concrete for castings.

Specified tests shall be performed after delivery of material at site.

C.7.7 Protection of Works

All the works shall be protected at site as per specified methods in the contract documents.

C.7.8 Material Storage and Handling

M/S ABC shall arrange for a storage facility on the project site to keep the material in safe condition. Material at site shall be fully protected from damages and properly stored. Necessary care shall be taken to store paints and liquid types of materials by maintaining appropriate temperature to prevent damage and deterioration of such products. Materials, products, and equipment shall be properly packed and protected to prevent damage during transportation and handling.

C.8 Method Statement for Various Installation Activities

Method statement shall be submitted to the consultant for their approval as per the contract documents. The method statement shall describe the steps involved for execution/installation of work by ensuring safety at each stage. It shall have the following information:

1. Scope of work: Brief description of work/activity
2. Documentation: Relevant technical documents to undertake this work/activity
3. Personnel involved
4. Safety arrangement
5. Equipment and plant required
6. Personal protective equipment
7. Permits/authorities' approval to work
8. Possible hazards
9. Description of the work/activity: Detail method of sequence of each operation/key steps to complete the work/activity

Figures C.2 through C.9 are illustrative examples of the method of sequence for major activities, for different trades, performed during the construction process.

C.9 Project-Specific Procedures for Site Work Installations Remedial Notice

QC for such installations shall be as specified in the contract documents or as mentioned in the instructions.

Remedial notices received shall be actioned immediately and replied to inspect the works having corrected/performed as per contract documents. Similarly nonconformance report shall be actioned and replied.

All repaired and reworked items shall be inspected by the concerned engineer and quality control engineer before submitting a "Request for Inspection" or "Checklist."

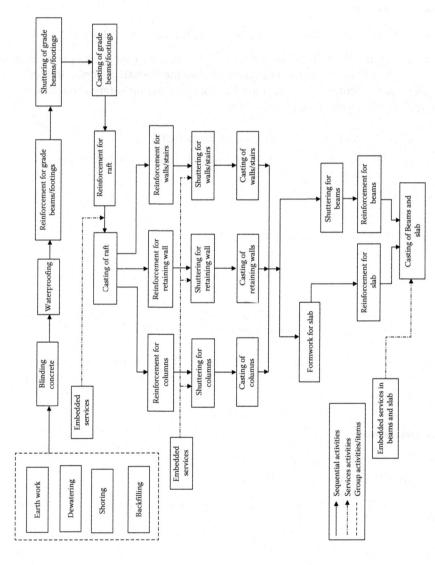

FIGURE C.2
Method of sequence for concrete structure work.

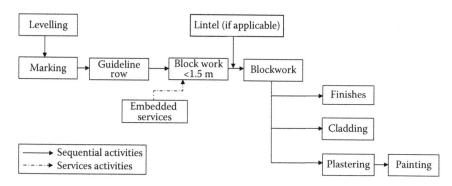

FIGURE C.3
Method of sequence for block masonry work.

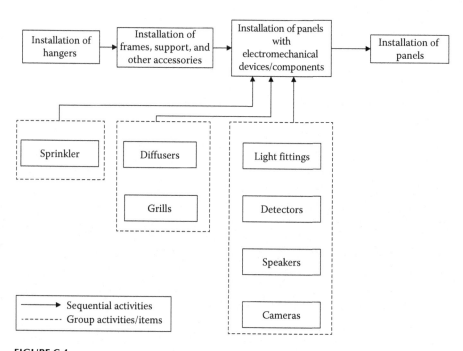

FIGURE C.4
Method of sequence for false ceiling work.

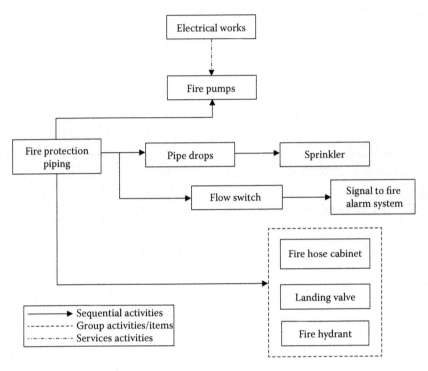

FIGURE C.5
Method of sequence for mechanical work (fire protection).

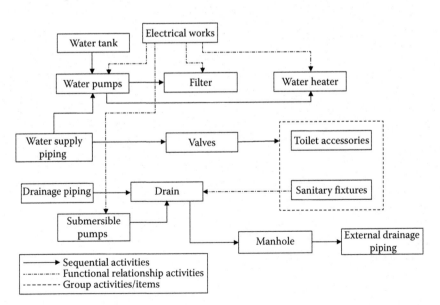

FIGURE C.6
Method of sequence for mechanical work (public health).

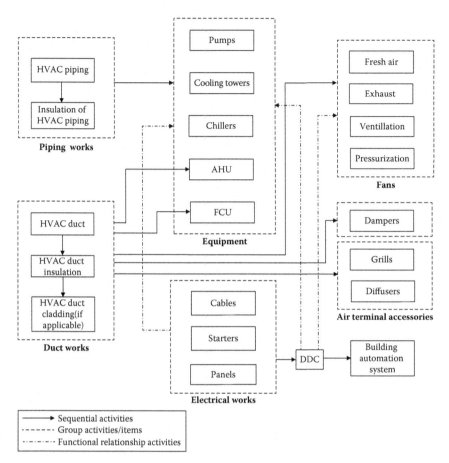

FIGURE C.7
Method of sequence for HVAC work.

Appropriate preventive actions shall be taken to avoid repetition of nonconformance work. The following steps shall be taken to deal with any problem requiring preventive action:

- Detect
- Identify the potential causes
- Analyze the potential causes
- Eliminate the potential causes
- Ensure the effectiveness of preventive action

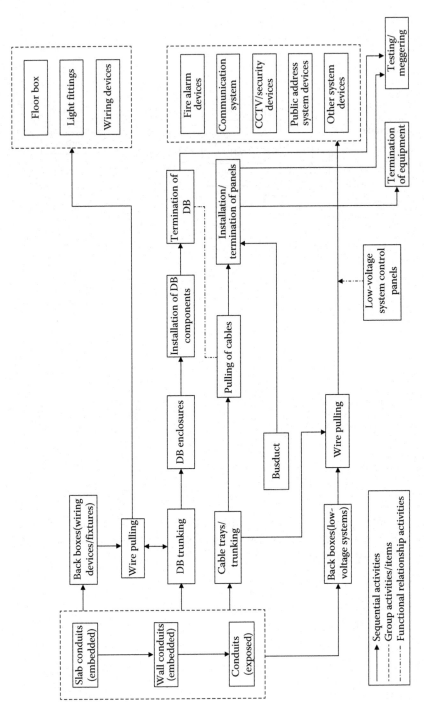

FIGURE C.8
Method of sequence for electrical work.

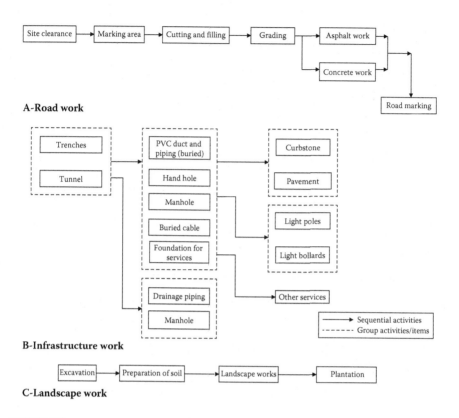

A-Road work

B-Infrastructure work

C-Landscape work

FIGURE C.9
Method of sequence for external works.

C.10 Risk Management

The contractor shall maintain risk register throughout the contract period and shall resolve/mitigate the risk with appropriate response(s).

C.11 Quality Control Records

QC records shall be maintained by the quality control incharge and shall be accessible to authorized personnel for review and information. The records shall include:

- Contract drawings and revisions
- Contract specifications and revisions
- Approved shop drawings

- Record drawings
- Material approval record
- Checklists
- Test reports
- Material inspection reports
- Minutes of QC meetings
- Site works instructions
- Remedial notes
- Other records specified in the contract documents or as requested by the owner/consultant

C.12 Company's Quality Manual and Procedures

The following documents from the company's manual shall be part of contractor's QC plan:

1. Documents for specific projects
2. Subcontractor evaluation
3. Supplier evaluation
4. Client supplied items
5. Receiving inspection and testing
6. Site inspection and testing
7. Final inspection, testing, and commissioning
8. Material storage and handling
9. Risk management system
10. Control of quality records
11. Internal audit

C.13 Periodical Testing of Construction Equipment

Hoists, cranes, and other lifting equipment used at site shall be tested periodically by third-party inspection authority and test certificates shall be submitted to the consultant. Measuring instruments shall have calibration tests performed by the approved testing agencies/government agencies at regular intervals, and a calibration certificate shall be submitted to consultants. Other equipment shall be tested/calibrated as per contract requirements and the schedule for such tests.

C.14 Quality Updating Program

The QC program shall be reviewed every 6 months and necessary updates shall be done, if required. All such information shall be given to the consultants and a record shall be made for such revisions.

C.15 Quality Auditing Program

Internal auditing shall be performed by the company's internal auditor to ensure that specified quality procedures are followed by the site quality personnel.

C.16 Testing, Commissioning, and Handover

The test shall be carried out as per contract documents. Necessary test procedures shall be submitted prior to start of testing of installed systems/equipment.

Engineers from the respective trades will be responsible for arranging orderly performance of testing and commissioning of the systems/equipment.

C.17 Health, Safety, and the Environment

The HSE officer shall be responsible for implementing health and safety measures as per OHSAS 18000 and the project's environmental requirements as per local regulatory authority. ABC shall prepare and submit a health and safety management and accident prevention program. The safety program shall embody the prevention of accidents, injuries, occupational illnesses, and property damage. All personnel at the site shall be provided with safety gear. Regular meetings and awareness training shall be conducted at the site on a regular basis. ABC shall ensure that first aid and emergency medical facilities are available at the site all the time. ABC shall comply with all safety measures specified in the contract documents. ABC shall take all necessary measures to comply with regulatory requirements and environmental considerations.

Bibliography

Alexander, K. (1996). *Facilities Management: Theory and Practice*. London, U.K.: E & FN SPON Press.

American Society of Civil Engineers. (2000). *Quality in the Constructed Project: A Guide for Owners, Designers, and Constructors*, Second edition. Reston, VA: American Society of Civil Engineers.

American Society for Quality. (2001). Quality 101, Exhibit 17, Evolution of Quality, ASQ. www.asq.org.

American Society for Quality. (2001). Quality 101, Quality during World War II, ASQ. www.asq.org.

American Society for Quality. (2007). Quality basics. http://www.asq.org.

American Society for Quality. (2017). Total Quality Management in Quality Glossary, ASQ. www.asq.org.

Ashford, J.L. (1989). *The Management of Quality in Construction*. London, U.K.: E & FN SPON.

Balanchard, B.S. and Fabrycky, W.J. (1998). *Systems Engineering and Analysis*, Third edition. Upper Saddle River, NJ: Prentice-Hall.

Bennet, F.L. (2003). *The Management of Construction: A Project Life Cycle Approach*. Oxford, U.K.: Butterworth-Heinemann.

Chase, R., Aquilano, N., and Jacobs, F. (2001). *Operations Management for Competitive Advantage*, Ninth edition. New York: Irwin, McGraw-Hill.

Chung, H.W. (1999). *Understanding Quality Assurance in Construction: A Practical Guide to ISO 9000 for Contractors*. London, U.K.: E & FN SPON.

Construction Industry Institute. (February 1990a). *The Quality Performance Management System*. CII Publication 10-3. Austin, TX: University of Texas.

Construction Industry Institute. (April 1990b). *Total Quality Management: The Competitive Edge*. CII Publication 10-4. Austin, TX: University of Texas.

Construction Industry Institute. (November 1990c). *Potential for Construction Industry Improvement*, Vol. 1. CII Source Document 61. Austin, TX: University of Texas.

Construction Industry Institute. (June 1992a). *Guidance for Implementing Total Quality Management in Engineering and Construction Industry*. CII Source Document 74. Austin, TX: University of Texas.

Construction Industry Institute. (October 1992b). *Quality Performance Measurement of EPC Process*. CII Source Document 79. Austin, TX: University of Texas.

Construction Industry Institute. (1994). *Quality Performance Management System: Measurement*. CII Product No. EM-4A. Austin, TX: University of Texas.

Corrie, R.K. (1991). *Engineering Management Project Evaluation*. London, U.K.: Thomas Telford Ltd.

Crosby, P.B. (1979). *Quality Is Free: The Art of Making Quality Certain*. New York: McGraw-Hill.

Dell'Isola, A.J. (1997). *Value Engineering: Practical Applications for Design, Construction, Maintenance and Operations*. Kingston, MA: RSMeans.

Deming Institute. (2000). Out of the crisis. www.deming.org.

Feigenbaum, A.V. (1991). *Total Quality Control*, Third edition. New York: McGraw-Hill.

Gryna, F.M. (2001). *Quality Planning and Analysis*, Fourth edition. New York: Irwin, McGraw-Hill.

Institution of Civil Engineers. (1996). *Civil Engineering Procedure*. London, U.K.: Thomas Telford Ltd.

International Council on Systems Engineering. http://www.incose.org.

International Organisation for Standardisation, from ISO. http://www.iso.org.

Juran, J.M. and Godfrey, A.B. (1999). *Juran's Quality Handbook*, Fifth edition. New York: McGraw-Hill.

Kerzner, H. (2001). *Project Management: A Systems Approach to Planning, Scheduling and Controlling*, Seventh edition. New York: John Wiley & Sons.

Kuehn, A.A. and Day, R.L. (1954). Strategy of product quality. *Harvard Business Review*, Nov.–Dec., p. 831.

MasterFormat®. (2016), The Construction Specifications Institute and Construction Specifications Canada. http://www.csinet.org.

Oakland, J.S. (2003). *TQM*. Oxford, U.K.: Butterworth-Heinemann.

Oberlender, G.D. (2000). *Project Management for Engineering and Construction*. Singapore: McGraw-Hill.

Oswald, T.H., Burati, Jr., J.L., and Construction Industry Institute. (January 1989). *Cost of Quality Deviation in Design and Construction*. Publication 10-1. Austin, TX: University of Texas.

Project Management Institute. (2000). *A Guide to the Project Management Body of Knowledge (PMBOK)*. Newtown Square, PA: PMI.

Pyzdek, T. (1999). *Quality Engineering Handbook*. Boca Raton, FL: CRC Press.

Ritz, G.J. (1990). *Total Engineering Project Management*. New York: McGraw-Hill.

Rumane, A.R. (2013). *Quality Tools for Managing Construction Projects*. Hoboken, NJ: CRC Press (A Taylor & Francis Group Company).

Rumane, A.R. (2016). *Handbook of Construction Management: Scope, Schedule, and Cost*. Boca Raton, FL: CRC Press (A Taylor & Francis Group Company).

Sebestyen, G. (1998). *Construction-Craft to Industry*. London, U.K.: Spon Press SAVE International (The Value Society), http://www.value-eng.org.

Shtub, A., Bard, J.F., and Globerson, S. (1994). *Project Management: Engineering, Technology and Implementation*. Upper Saddle River, NJ: Prentice Hall.

Sullivan, W.G., Wicks, E.M., and Luxhoj, J.T. (2003). *Engineering Economy*, Twelfth edition. Upper Saddle River, NJ: Prentice Hall.

Tague, N.R. (2005). *The Quality Toolbox*. Milwaukee, WI: American Society for Quality Press.

Thorpe, B., Sumner, P., and Duncan, J. (1996). *Quality Assurance in Construction*. Surrey, U.K.: Gower Publishing Ltd.

Turner, J.R. (2003). *Contracting for Project Management*. Surrey, U.K.: Gower Publishing Ltd.

Westcott, R. (2014). *The Certified Manager of Quality/Organizational Excellence Handbook*, Fourth edition. Quality Press.

Index

A

Acceptable quality, 220, 429
Access control, 275, 410, 508
Accidents, 151, 420–421
Accreditation, 146–147
Action learning, 90–91
Agency construction management
    (Agency CM), 232–233
Agency construction managers (ACM),
    173
American National Standards Institute
    (ANSI), 123
American Society for Heating,
    Refrigerating, and Air-
    Conditioning Engineers
    (ASHRAE), 123, 125
American Society for Quality (ASQ),
    6–7, 11, 20, 52, 123
American Society for Testing and
    Materials (ASTM), 123, 125
American Society of Civil Engineers
    (ASCE), 8–9
American Society of Mechanical
    Engineers (SME), 123
Architect and engineer (A&E)
    type, 171
Ashford, J.L., 28–29
Assessment of quality
    cost of poor quality, 459–463
    marketplace, 460
    operation, 464–465
    quality culture, 464
    self-assessment, 465–466
AutoCAD, 59

B

Bidding and tendering
    contract award process, 312
    contract documents, 304
    contract forms and flexible,
        farsighted, ex post governance,
        307–308
    contractor selection, 312
    contract risks, 305
    contract schema, ex ante
        incentivization, 306–307
    contract selection methodology, 306
    identification of bidder, 309
    identification of project team, 309
    negotiated contract systems, 305
    organizing tender documents, 309
    risk identification/management, 311
    tender documents management,
        309–311
    tendering stages, 308
Bill of quantity (BOQ), 101–102
Black Belt Six Sigma teams, 91
Blitz teams, 92–93
Breakdown maintenance, 442
Breakthrough teams, 92
British Standards Institution (BSI), 123
Building construction project, 174,
    485–493
Building information modeling (BIM),
    62
Build–own–operate–transfer (BOOT),
    230–231

C

Cash flow, 295
Cause-and-effect analysis, 41
Certified value specialist (CVS), 110
Classical brainstorming, 110–111
Computer-aided design (CAD)
    application software packages, 60
    AutoCAD, 59
    ISO/IEC 9126, 60–61
    manual drafting, 59
    operating system software, 59
    quality factors, 61
Computer-aided facility management
    (CAFM), 454–455
Concept 500, 454
Conceptual design phase
    designer quality management, 253

development
  design consideration, 250
  house of quality, office building
    project, 245, 249
  implementation considerations,
    250–251
  project objectives, 251
  requirements, 251–252
  sustainability, 250
feasibility, 240–241
finalizing, 254
identification of alternatives, 242–243
identification of need, 237, 239
identification of project team
  contractor, 243
  contribution of various
    participants, 245, 247–248
  designer/consultant, 243–244
  owner, 243
  requirements of project team
    members, 244
  responsibilities of different parties,
    244, 246
  responsibilities of project team
    members, 244–245
  TOR, 244–245
initial planning, 237
logic flow, 237–238
project cost estimation, 252–253
resource estimation, 253
risk identification/management,
  253–254
time schedule estimation, 252
Construction
  activities, 313, 315–316, 335
  construction/project manager,
    313–314
  contract management, 416, 419
  identification of project team, 317, 319
  mobilization
    activities, 315, 317
    job site instruction form, 317, 320
    on-site facilities and services, 315
    site administration and
      communication matrix,
      317–319
  monitoring and control
    activities status collection and
      recording, 340

budget control, 346
cash and time control, 353, 357
CII Source Document 61, 345
contents of progress report, 340,
  342
equipment status log, 350–351
functions, 345
manpower comparison histogram,
  350, 352
milestones, 345
monthly progress report, 340,
  343–344
objectives, 340, 345
procurement log, 347, 350
progress reports and meetings,
  353, 357–361
project documents, 346
project payment/progress curve
  (S-curve), 350, 353–354
quality control, 346
schedule, 346
shop drawings and materials logs,
  347, 349
subcontractors submittal and
  approval log, 346–348
time control status, 353, 355
updating, network diagram, 345
variation orders, 361–370
Oberlender's description, 313
project planning and scheduling
  activities, 322–323
  analytical and graphical
    techniques, 325
  bar chart, 325
  benefits, 321
  construction program summary,
    331–334
  CPM, 325–326, 331
  definition, 321, 323
  electrical works, logic flow
    diagram, 326, 330
  electromechanical activities, 326
  finish-to-finish relationship, 326
  finish-to-start relationship, 326
  firefighting works, logic flow
    diagram, 326–327
  HVAC works, logic flow diagram,
    326, 329
  key principles, 323, 325

logical process, 319
minimum completion time, 331
network diagrams, 325
PERT, 325
plumbing works, logic flow
    diagram, 326, 328
slack time, 331
start-to-finish relationship, 326
start-to-start relationship, 326
steps, 323–324
WBS, 326
resident engineer, 314
resources/procurement management
    equipment schedule, 338–339
    equipment set, 338
    human resource, categories, 336
    manpower chart, 338
    material approval and
        procurement procedure,
        340–341
    minimum core staff needed, 336
    procurement log, 335
    procurement strategies, 340
    staff approval request form,
        336–337
    workmanship, 335
risk management
    major risks, 416–418
    plan, 415–416
    probability of occurrence, 415
site safety, 419
    accident-reporting form, 423–424
    causes of injuries/accidents,
        421–422
    penalties, 420
    risk control strategies, 420
    safety and accident prevention
        program, 420
    safety guidelines, 422–423
    safety officer responsibilities, 420
    safety violation notice, 420–421
supervision consultant's
    responsibilities, 313–314
supervision team, 314–315
Construction documents
    cost/budget estimation, 301
    development, 301
    finalizing, 304
    identification of project team, 300–301

identification of requirements, 300
project schedule, 301
quality management
    control quality, 302–303
    planning, 301–302
    quality assurance, 302
resource estimation, 303
risk management, 303
Construction industry institute (CII), 159
Construction management-at-risk
    (CM-at-risk), 173, 232–234
Construction project life cycle
    activities/components, 213–215
    benefits, 208
    classification, 205
    cost reduction, 205
    phases, 208–210
    process, 206–207
    WBS, 206, 208, 210–212, 215
Construction projects
    activities, 485–493
    A&E type, 171
    applications, 177
    architecture and construction,
        170–171
    building construction, 174
    CM-at-risk, 173
    construction manager, 173
    corrective action, 200–201
    definition, 169–170
    DOE, 189–190
    environment and culture, 172–173
    groups, 171–172
    heavy engineering construction, 174
    industrial construction, 174
    Industrial Revolution, 171
    Kaizan management
        activities, 199
        just-in-time production system,
            199
        organization politics management,
            199
        suggestion system, 199
        total productive maintenance, 199
        TQM, 198–199
    Lean tools
        concurrent engineering, 191–192
        5S, 192, 194–197
        value stream mapping, 192–193

manufacturing, 174, 176–177
Middle Ages, 171
mistake proofing, 200
monitoring and controlling process, 190–191
New Stone Age, 170
organization, 171–172
PDCA cycle, 197
preventive action, 202
QFD
    components, 178
    correlation matrix, 179
    house of, 178–179
    smart building system, 179–180
R&D projects, 171
residential construction, 174
significant changes, 171
Six Sigma
    benefits, 181
    customer needs and specifications, 184
    design phase, 186–189
    process analysis, 184–186
    project charter, 182
    project goals and customer deliverables, 181
    project plan, 182
    team selection, 181–183
systems engineering
    application areas, 205–206
    definition, 203–204
    human-made systems, 203
    INCOSE, 204–205
    life cycle, 205–215
    product competitiveness, 202
twentieth century, 171
types, 174–175
Construction project trilogy, 9, 81
Contractor's construction schedule (CCS)
    analyze phase, 102–104
    define phase, 100
    design phase, 103
    measure phase, 100–102
    products, services/processes, 99
    verify phase, 105
Contractor's quality control plan (CQCP), 168, 399
    contract documentations, 414–415

contractor's process documentation, 401
description, 495–496
execution/installation, 511–517
HSE, 519
inspection and testing, 509
preparation outline, 413–414
project-specific requirements, 413
quality control
    auditing, 519
    automation system, 508
    company's manual, 518
    concrete, 503
    conveying system, 506–507
    definable feature of work, 502
    doors and windows, 504–505
    earthworks and site works, 502–503
    electrical works, 508–509
    engineers, 497–498
    equipment, 506
    finishes, 505
    furnishing, 505–506
    landscape, 505
    masonry, 503
    material storage and handling, 510–511
    mechanical works, 507–508
    metal works, 503–504
    MIR, 510
    off-site manufacturing, 509–510
    personnel, 499–500
    phases of control, 501–502
    procurement, 501
    program updates, 519
    protection, 510
    quality control staff, 497, 499
    records, 517–518
    testing laboratories, 510
    wood plastics and composite, 504
risk management, 517
site work installations, 511, 515
specifications, 414
submittals, 500–501
testing agencies/government agencies, 518
Critical path method (CPM) technique, 100–101, 258
CVS, *see* Certified value specialist

D

Datastream7i, 454
Deming, W.E., 28–29
Deming's 14 principles for
    transformation, 34, 37
Deming wheel, *see* Plan-do-check-act
    (PDCA) cycle
Design development
  budget
    bill of material, 293, 295
    cash flow, 295
    identification of need, 293
    project S-curve, 295–296
  design of work
    accuracy, 264
    architectural design, 266–267
    audiovisual system, 274–275
    bridges, 277–278
    comprehensive design, 263
    concrete structure, 267–269
    drainage system, 270
    electrical system, 272–273
    elevator works, 269
    external works (infrastructure and
      road), 277
    fire alarm system, 273
    fire protection system, 269–270
    functions and responsibilities of
      design professional, 265
    highways, 279
    HVAC works, 271–272
    landscape, 276–277
    loose furnishings/furniture,
      279–288
    plumbing works, 270
    public address system, 274
    security system access control,
      275–276
    security system/CCTV, 275
    telephone/communication
      system, 274
    WBS, 264
    work packages, 264
  finalizing, 300
  identification of project team, 263–264
  identification of requirements, 261,
    263
  major activities, 261–262

  preliminary work program, 293–294
  preparation of contract documents
    and specifications
    MasterFormat® 2016 division
      numbers and titles, 287,
      289–290
    master format specifications, 287
    shop drawing and materials
      submittals, 290–293
  project plan, 293–294
  quality management
    control quality, 297–299
    planning, 295
    quality assurance, 297
  regulatory/authorities' approval, 287
  resource estimation, 297
  risk identification/management,
    298–300
Design for Six Sigma (DFSS), 94
Design of experiments (DOE), 189–190
Design review checklists
  architectural work, 467
  drainage system, 475
  electrical system, 479–482
  fire protection system, 472–473
  HVAC
    building management system, 478
    control, 477
    duct layout, 476
    electrical works, 478
    equipment selection, 477
    piping layout, 476
    plants, 478
  infrastructure, 483
  landscape, 482
  plumbing system, 474
  special systems, 473
  structural work
    basement design, 469–471
    foundation layout, 468
    requirements, 472
    roofslab, 471
    superstructure, 471
Detailed design/detailed engineering,
    see Design development
DFSS, *see* Design for Six Sigma
DMADDD tool, 94–95
DMADV tool, 94–95
DMAIC tool, 94–99

Documentation
    commissioning and handover stage, 433
    construction projects, 223–224
    contract, 414–415
    contractor's process, 401
    execution/installation, 511
    quality system, 140–146
    shipping, 69
DOE, *see* Design of experiments

E

Electrical system, 257, 272–273, 444, 479–482
Engineering, procurement, and construction (EPC), 231
Environmental management system, 150
European Committee for Electrotechnical Standardization (CENELEC), 123
European Committee for Standardization (CEN), 123
External works, 323, 335, 346

F

Facility management (FM)
    areas, 449–450
    benefits, 454, 456
    building capability *vs.* time, 450, 452
    CAFM, 454–455
    characteristics, 450–451
    corporate level, 449
    operation and maintenance, 450
    outsourcing, 453–454
    quality control, 456–457
    requirements, 451–452
    team members, 450–451
Feigenbaum, A.V., 27
Fire protection system, 472–473
Fishbone diagram, 41–42
Frontline/Shire systems, 454

G

GHQ's Civil Communication Section (CSS), 4
Green Belt Six Sigma teams, 92
Gryna, F.M., 28

H

Health, safety, and the environment (HSE), 519
Heating, ventilation, and air-conditioning (HVAC)
    design review checklists
        building management system, 478
        control, 477
        duct layout, 476
        electrical works, 478
        equipment selection, 477
        piping layout, 476
        plants, 478
    quality in construction projects, 271–272
Heavy engineering construction, 174
House of quality, 178–179

I

Industrial construction, 174
Institute of Electrical and Electronic Engineers (IEEE), 123, 125–126
Integrated project delivery
    contractual relationship between parties, 233–234
    generic life cycle phases, 233
    logic flow diagram
        design-bid-build system, 234–235
        design-build system, 234, 236
Integrated quality management system (IQMS), 166–168
International Council on Systems Engineering (INCOSE), 204–205
International Electrotechnical Commission (IEC), 123, 126
International Organization for Standardization (ISO), 7
    certification
        advantages, 150
        audits, 147
        certification schedule, 147, 149
        customer expectations, 147
        globalization and competitive market, 147
        process, 147–148
        website, 146

ISO 9000:2000, 128–130
ISO 9001:1994, 127
ISO 9001:2008, 128, 131–139
ISO 9001:2015, 128, 131–139
ISO 14000, 127, 150
ISO 14001, 131
ISO 22000:2005, 131
ISO 27001:2005, 131
ISO 28000:2005, 131
ISO 9000 family, 127, 131, 140–146
nongovernmental organization, 126
Ishikawa's Seven Tools of Quality
    Control, 40
ISO 9000
    customer's requirements, 127
    document control procedure, 140
    ISO 9001:2000, 140
    pyramid, 131, 140
    quality system, 140–146
ISO 14000, 127, 150
ISO certification
    advantages, 150
    audits, 147
    certification schedule, 147, 149
    customer expectations, 147
    globalization and competitive
        market, 147
    process, 147–148
    website, 146

J

Juran's quality trilogy, 42–44
Juran's triple role concept, 221
Juran's triple role-functional
    relationship, 224

K

Kerzner, Harold, 5
Key performance indicators (KPIs), 454

L

Landscaping, 115, 251, 323
Lean tools
    concurrent engineering, 191–192
    5S, 192, 194–197
    value stream mapping, 192–193

Log E-1, 347–349
Log E-2, 347, 350

M

Maintenance
    breakdown, 442
    preventive, 442
    routine, 442
    scheduled, 442
    total productive maintenance, 199
Masonry works, 503
Material on-site inspection request
    (MIR), 510
Materials approval
    checklist for concrete casting, 390, 393
    checklist for form work, 390–391
    checklist for general works, 390, 398
    concrete quality control form, 390,
        397
    concrete work, 390
    contractor's QCP, 386
    elements, 386
    material inspection report, 386, 389
    nonconformance report, 390, 400
    notice for daily concrete casting, 390,
        392
    notice for testing at lab, 390, 396
    PDCA cycle, 392, 399, 413
    quality control of concreting, 390, 394
    rejection of executed works, reasons
        for, 392, 402–410
    remedial note, 390, 399
    report on concrete casting, 390, 395
    request for substitution, 386–388
    sequence of execution of work, 390,
        401
    site quality control, responsibility for,
        392, 411–412
    site transmittal form, 382, 385
    specification comparison statement,
        382, 386
Maximo, 454
MIR, *see* Material on-site inspection
    request
Monitoring and control
    activities status collection and
        recording, 340
    budget control, 346

cash and time control, 353, 357
CII Source Document 61, 345
contents of progress report, 340, 342
equipment status log, 350–351
functions, 345
manpower comparison histogram,
     350, 352
milestones, 345
monthly progress report, 340,
     343–344
objectives, 340, 345
procurement log, 347, 350
progress reports and meetings, 353,
     357–361
project documents, 346
project payment/progress curve
     (S-curve), 350, 353–354
quality control, 346
schedule, 346
shop drawings and materials logs,
     347, 349
subcontractors submittal and
     approval log, 346–348
time control status, 353, 355
updating, network diagram, 345
variation orders, 361–370

N

National Fire Protection Association
     (NFPA), 123–124, 126, 269
Nominal group technique (NGT),
     111–112
Nonconformance report, 318,
     390, 400
Nonconformance request (NCR), 344

O

Occupational Health and Safety
     Assessment Series (OHSAS)
     18000, 151
Operation and maintenance (O&M)
     breakdown maintenance, 442
     construction project, 441
     postconstruction costs, 441
     predictive testing and inspection, 443
     preventive maintenance, 442
     preventive program, 443–447
     replacement of obsolete items, 442
     routine maintenance, 442
     scheduled maintenance, 442

P

Personnel, 215, 256, 425, 497, 499–500
Plan–do–check–act (PDCA) cycle, 31–33,
     197, 392, 399, 413
Plan–do–study–act (PDSA) cycle, 31–33
Plumbing system, 474
Poka-Yoke system, 46
Pollo, Marcus Vitruvius, 170–171
Preliminary design phase
     architect's design concept, 254
     budget, 258
     contract terms and conditions,
          257–258
     CPM schedule, 258
     deliverables, 255
     development, 256–257
     finalizing, 261
     identification of project team, 256
     identification of requirements, 256
     preparation, 255
     quality management, 258–259
     regulatory approval, 257
     resource estimation, 259
     risk identification/management,
          259–260
     VE study, 260
Preventive maintenance, 442
Progress reports and meetings
     daily checklist status report, 353, 361
     daily progress report, 353, 357–359
     progress meetings, 353, 360–361
     work-in-progress report, 353, 360
Pyzdek, Thomas, 5

Q

QC, *see* Quality control
QFD, *see* Quality function deployment
QIS, *see* Quality information system
QMS, *see* Quality management system
QPMS, *see* Quality performance
          management system
Quality
     assurance, 5, 20–21

control process
   applications, 14
   control charts, diagrams,
      checklists, 14
   feedback loop, 12–13
   industrial terminology, 12
   inspection-based approach, 11
   stability, 13
   tools, 15–19
customer relationship
   construction projects, 63–65
   customer satisfaction, 62–63
definition, 6–9
engineering technology, 22–24
history
   Industrial Revolution, 2–3
   issues, 1
   Japanese companies, 4
   manufacturing process, 3
   principles, 5
   quality control, 2
   quality eras, 5–6
   sample of surviving writings, 1–2
   Scandinavian shipbuilders, 1
   scientific management, 3
   SQC, 3–4
   TQM, 6
inspection, 10–11
management
   Ashford description, 28–29
   evolution, 25
   Feigenbaum description, 27–28
   Japanese products, 28
   principles, 29
   total-quality-control work, 27
   TQS, 25–26
QIS
   BIM, 62
   CAD, 59–61
   process, 57–58
   questions and response, 57–58
quality gurus
   Crosby's philosophy, 30–31
   Edward's philosophy, 30–37
   features, 47–48
   Feigenbaum's philosophy, 34, 39
   Ishikawa's approach, 34, 36–38,
      40–42
   Juran's philosophy, 42–44

Oakland's philosophy, 45–46
   organizational management,
      29–30
   Shingo's philosophy, 45–46
   Taguchi's philosophy, 46–47
risk management
   construction projects, 80–87
   identification, 73–75
   management cycle, 72–73
   monitor and control, 80
   plan risk response, 77, 79
   prioritization, 77
   qualitative analysis, 74, 76
   quantitative analysis, 76–78
   reducing risk, 80
Six Sigma
   CCS, 99–105
   DMADDD tool, 94–95
   DMADV tool, 94–95
   DMAIC tool, 94–99
   engaged teams, 91–93
   Ford Global 8D Tool, 93–94
   leadership principles, 90–91
   organizations, 88
   roadmap, 88–89
   standard deviation, 87
   universal measurement scale,
      87–88
supply chain management, 64, 66
   construction project, 70–72
   performance improvement, 70
   product handling, 69
   product quality, 68–69
   product specifications, 67–68
   purchasing process, 67–69
   stakeholders, 66–67
   supplier selection, 67–68
TQM
   changing views, 49–51
   periodical changes, 49
   principles, 51–57
TRIZ
   application, 107
   low-level problems, 107
   methodology, 105–106
   QFD matrix, 107
   quality improvement solution, 105
value, 108
value engineering

approach, 110–112
benefits, 117, 119–120
goal of, 109
manufacturer/client/end user,
    108
objective of, 109
stages, 115–118
team members, 109–110
timings, 112–115
Quality assurance, 20–21
Quality audits
    assessment, 152–153, 465
    contractor, 154
    design phase, 153–154
    external assessment, 154–155
    first party, 152
    vs. inspection, 155–156
    measurement, 152–153
    process, 152–153
    risks, 156
    second party, 152
    third party, 152
    types, 151
Quality circles, 38, 40
Quality control (QC)
    applications, 14
    auditing, 519
    automation system, 508
    company's manual, 518
    concrete, 503
    control charts, diagrams, checklists,
        14
    conveying system, 506–507
    definable feature of work, 502
    doors and windows, 504–505
    earthworks and site works, 502–503
    electrical works, 508–509
    engineers, 497–498
    equipment, 506
    feedback loop, 12–13
    finishes, 505
    furnishing, 505–506
    industrial terminology, 12
    inspection-based approach, 11
    landscape, 505
    masonry, 503
    material storage and handling,
        510–511
    mechanical works, 507–508

metal works, 503–504
MIR, 510
off-site manufacturing, 509–510
personnel, 499–500
phases of control, 501–502
procurement, 501
program updates, 519
protection, 510
quality control staff, 497, 499
records, 517–518
stability, 13
testing laboratories, 510
tools, 15–19
wood plastics and composite, 504
Quality costs
    annual monetary loss, 156–157
    appraisal costs, 157–158
    CII, 159
    components, 159
    construction, 162–165
    external failure costs, 157–158
    internal failure costs, 157–158
    poor quality, 159–162
    prevention costs, 157–158
Quality function deployment (QFD)
    components, 178
    correlation matrix, 107, 179
    house of, 178–179
    smart building system, 179–180
Quality gurus
    Crosby's philosophy, 30–31
    definitions, 7–8
    Edward's philosophy
        Deming's 14 principles for
            transformation, 34, 37
        PDCA cycle, 30–33
        seven-point action plan,
            34, 38
        SPC, 32, 34–36
    features, 47–48
    Feigenbaum's philosophy, 34, 39
    Ishikawa's approach, 34, 36–38,
        40–42
    Juran's philosophy, 42–44
    Oakland's philosophy, 45–46
    organizational management,
        29–30
    Shingo's philosophy, 45–46
    Taguchi's philosophy, 46–47

Quality information system (QIS)
  BIM, 62
  CAD
    application software packages, 60
    AutoCAD, 59
    ISO/IEC 9126, 60–61
    manual drafting, 59
    operating system software, 59
    quality factors, 61
  process, 57–58
  questions and response, 57–58
Quality management system (QMS)
  IQMS, 166–168
  ISO
    certification, 146–150
    ISO 9000, 127, 131, 140–146
    ISO 9000:2000, 128–130
    ISO 9001:1994, 127
    ISO 9001:2008, 128, 131–139
    ISO 9001:2015, 128, 131–139
    ISO 14000, 127, 150
    ISO 14001, 131
    ISO 22000:2005, 131
    ISO 27001:2005, 131
    ISO 28000:2005, 131
    nongovernmental organization,
      126
  OHSAS 18000, 151
  QPMS, 165–166
  quality audits
    assessment/measurement,
      152–153
    contractor, 154
    design phase, 153–154
    external assessment, 154–155
    first party, 152
    vs. inspection, 155–156
    process, 152–153
    risks, 156
    second party, 152
    third party, 152
    types, 151
  quality costs
    annual monetary loss, 156–157
    appraisal costs, 157–158
    CII, 159
    components, 159
    construction, 162–165
    external failure costs, 157–158

    internal failure costs, 157–158
    poor quality, 159–162
    prevention costs, 157–158
  standards
    Chung definition, 123
    ISO, 122
    organizations, 123–126
    Pyzdek definition, 121–122
    setting, 122–123
Quality of construction projects
  allowable defective items, 220
  bidding and tendering
    contract award process, 312
    contract documents, 304
    contract forms and flexible,
      farsighted, ex post governance,
      307–308
    contractor selection, 312
    contract risks, 305
    contract schema, ex ante
      incentivization, 306–307
    contract selection methodology,
      306
    identification of bidder, 309
    identification of project
      team, 309
    negotiated contract systems, 305
    organizing tender documents, 309
    risk identification/management,
      311
    tender documents management,
      309–311
    tendering stages, 308
  Chung's description, 219, 370
  CII Source Document 79, 220
  conceptual design phase
    development, 245, 249–252
    feasibility, 240–241
    finalizing, 254
    identification of alternatives,
      242–243
    identification of need, 237, 239
    identification of project team,
      243–248
    initial planning, 237
    logic flow, 237–238
    project cost estimation, 252–253
    quality management, 253
    resource estimation, 253

risk identification/management,
    253–254
time schedule estimation, 252
construction documents
    cost/budget estimation, 301
    development, 301
    finalizing, 304
    identification of project team,
        300–301
    identification of requirements, 300
    project schedule, 301
    quality management, 301–303
    resource estimation, 303
    risk management, 303
CQCP, 399
    contract documentations, 414–415
    contractor's process
        documentation, 401
    preparation outline, 413–414
    project-specific requirements, 413
    specifications, 414
Crosby's quality definition, 222
design development
    budget, 293, 295–296
    design of work, 263–288
    finalizing, 300
    identification of project team,
        263–264
    identification of requirements,
        261, 263
    major activities, 261–262
    preliminary work program, 293–294
    preparation of contract documents
        and specifications, 287, 289–293
    project plan, 293–294
    quality management, 295–299
    regulatory/authorities' approval,
        287
    resource estimation, 297
    risk identification/management,
        298–300
divisional values, 217–218
electromechanical services, 217
handover
    final snag list, 435
    handing over certificate, 435–436
    punch list, 435
    of spare parts, 435, 437
    substantial completion, 435–440

inspection of works, 415
Juran's triple role concept, 221
Juran's triple role-functional
    relationship, 224
key quality assurance activities,
    224–226
materials approval
    checklist for concrete casting, 390,
        393
    checklist for form work, 390–391
    checklist for general works, 390,
        398
    concrete quality control form, 390,
        397
    concrete work, 390
    contractor's QCP, 386
    elements, 386
    material inspection report, 386,
        389
    nonconformance report, 390, 400
    notice for daily concrete casting,
        390, 392
    notice for testing at lab, 390, 396
    PDCA cycle, 392, 399, 413
    quality control of concreting, 390,
        394
    rejection of executed works,
        reasons for, 392, 402–410
    remedial note, 390, 399
    report on concrete casting, 390, 395
    request for substitution, 386–388
    sequence of execution of work,
        390, 401
    site quality control, responsibility
        for, 392, 411–412
    site transmittal form, 382, 385
    specification comparison
        statement, 382, 386
minimum standards, 219
Oberlender's observation, 222
on-site inspection, 370
organizational framework, 221
phases, 227
philosophy of quality, 222–223
preliminary design
    architect's design concept, 254
    budget, 258
    contract terms and conditions,
        257–258

CPM schedule, 258
deliverables, 255
development, 256–257
finalizing, 261
identification of project team, 256
identification of requirements, 256
preparation, 255
quality management, 258–259
regulatory approval, 257
resource estimation, 259
risk identification/management,
259–260
VE study, 260
primary reasons, 221–222
project delivery system categories,
228
Agency CM, 232–233
BOOT, 230–231
CM-at-risk, 232–234
design–bid–build contract, 229
design–build contract, 230
integrated project delivery,
233–236
project manager contract, 231–232
Turnkey contract, 231
quality assurance and quality control
responsibilities, 373
quality plan, 223–224
quality principles, 227
shop drawings approval
architectural works, 374–376
builder's workshop drawings, 382
composite/coordination shop
drawings, 382, 384
electrical works, 380
elevator works, 378
external works (infrastructure),
381–383
HVAC electrical works, 380
HVAC works, 379
landscape works, 381
low-voltage systems, 381
mechanical works, 378–379
structural works, 376–378
testing and commissioning
as-built drawings, 433
commissioning, 431–432
electrically operated equipment/
systems, 425, 429

electrical power supply
connection procedure, 429–430
electromechanical works/systems,
425, 429
identification of project team, 426
inspection and testing plan
flowchart, 426, 428
move-in plan, 434–435
quality process, 426–427
regulatory/authorities' approval,
434
technical manuals and documents,
433–434
testing checklist, 431–432
testing procedure, 425, 429, 431
user's personnel training, 434
testing facilities, 370
TQM, 221
zero defects goal, 220
Quality performance management
system (QPMS), 165–166
Quality plan, 370

R

Replacement of obsolete items, 442
Request for information (RFI) form,
362–363
Request for Proposal (RFP), 454
Research and development (R&D)
projects, 171
Resident engineer, 314–315
Residential construction, 174
Resource loading, 350
Risk management
construction
major risks, 416–418
plan, 415–416
probability of occurrence, 415
construction projects
categories, 81–83
construction-related industries/
manufacturers, 80–81
potential risks, 81, 84–87
project owners, 80
risk factors, 81
CQCP, 517
identification, 73–75
management cycle, 72–73

monitor and control, 80
plan risk response, 77, 79
prioritization, 77
qualitative analysis, 74, 76
quantitative analysis, 76–78
reducing risk, 80
Routine maintenance, 442

S

Scandinavian shipbuilders, 1
Scheduled maintenance, 442
Schematic design, *see* Preliminary design
    phase
Self-assessment, 464–466
Service Level Agreement (SLA), 454
Single-Minute Exchange of Die (SMDE)
    system, 46
Site work instruction (SWI) form, 362,
    365
Six Sigma
    CCS, 99–105
    construction projects
        benefits, 181
        customer needs and specifications,
            184
        design phase, 186–189
        process analysis, 184–186
        project charter, 182
        project goals and customer
            deliverables, 181
        project plan, 182
        team selection, 181–183
    DMADDD tool, 94–95
    DMADV tool, 94–95
    DMAIC tool, 94–99
    engaged teams, 91–93
    Ford Global 8D Tool, 93–94
    leadership principles, 90–91
    organizations, 88
    roadmap, 88–89
    standard deviation, 87
    universal measurement scale, 87–88
Slack time, 331
Standard deviation, 87
Standards
    Chung definition, 123
    ISO, 122
    organizations, 123–126

Pyzdek definition, 121–122
    setting, 122–123
Statistical process control (SPC), 8, 32,
    34–36
Statistical quality control (SQC), 3–4
Substantial completion
    certificate, 435–436, 440
    final walk-through inspection, 436
    memorandum, 437
    project closeout checklist, 437, 439
    project substantial completion
        procedure, 435, 438
Subsystem, 204, 326, 426, 431
Supply chain management, 64
    construction project, 70–72
    performance improvement, 70
    product handling, 69
    product quality, 68–69
    product specifications, 67–68
    purchasing process, 67–69
    stakeholders, 66–67
    supplier selection, 67–68
Systems engineering
    application areas, 205–206
    definition, 203–204
    human-made systems, 203
    INCOSE, 204–205
    life cycle
        activities/components, 213–215
        benefits, 208
        classification, 205
        cost reduction, 205
        phases, 208–210
        process, 206–207
        WBS, 206, 208, 210–212, 215
    product competitiveness, 202

T

Terms of reference (TOR), 244–245
Three-dimensional (3D) solid modeling,
    59
Total quality management (TQM), 220
    changing views, 49–51
    construction projects, 198–199
    history, 6
    periodical changes, 49
    principles, 51–57
    QPMS, 165–166

Total quality system (TQS), 25–26
Traditional contracting system, 257
TRIZ
    application, 107
    low-level problems, 107
    methodology, 105–106
    QFD matrix, 107
    quality improvement solution, 105

V

Value, 108
Value engineering (VE)
    approach, 110–112
    benefits, 117, 119–120
    goal of, 109
    manufacturer/client/end user, 108
    objective of, 109
    stages, 115–118
    study, 260
    team members, 109–110
    timings, 112–115
Variation orders, 370–371
    attachment, 370, 372
    contract changes, 361–362
    cost and time implementation, 362
    processing RFI, flow diagram, 362, 364
    processing variation order, flow
        diagram, 362, 369

    proposal, 362, 368
    request for modification,
        362, 367
    request for variation, 362, 366
    RFI form, 362–363
    SWI form, 362, 365
Voice of the customer, 178–180

W

Work breakdown structure (WBS)
    benefits, 208
    bidding and tendering, 212
    conceptual design, 211
    construction documents, 212
    construction schedule, 101
    design development, 211–212
    design of work, 264
    hierarchical decomposition, 210
    Kaizan management, 199
    large-scale problem, 206
    project planning and scheduling, 326
    schematic design, 211
    testing, commissioning, and
        handover, 212, 215

Z

Zero Defect policy, 69, 163

Printed in the United States
by Baker & Taylor Publisher Services